Writing about Business and Industry

Writing about Business and Industry

Edited by
Beverly E. Schneller

New York   Oxford
Oxford University Press
1995

Oxford University Press

Oxford New York Toronto
Delhi Bombay Calcutta Madras Karachi
Kuala Lumpur Singapore Hong Kong Tokyo
Nairobi Dar es Salaam Cape Town
Melbourne Auckland Madrid

and associated companies in
Berlin Ibadan

Copyright © 1995 by Oxford University Press, Inc.

Published by Oxford University Press, Inc.,
200 Madison Avenue, New York, New York 10016

Oxford is a registered trademark of Oxford University Press

All rights reserved. No part of this publication may be reproduced,
stored in a retrieval system, or transmitted, in any form or by any means,
electronic, mechanical, photocopying, recording, or otherwise,
without the prior permission of Oxford University Press.

Library of Congress Cataloging-in-Publication Data
Writing about business and industry / edited by Beverly E. Schneller.
    p. cm.   ISBN 0-19-507378-9.
    ISBN 0-19-507379-7 (pbk.)
    1. Business writing.   2. Technical writing.
    3. Business literature.   I. Schneller, Beverly E.
        HF5718.3.W74   1995
        808'.06665—dc20   94-20890

Since this page cannot legibly accommodate all the copyright
notices, the following page constitutes an extension of the
copyright page.

The editor wishes to thank Philip Leininger of Oxford University Press for
his encouragement and assistance throughout this project.

1 3 5 7 9   8 6 4 2
Printed in the United States of America
on acid-free paper

Selections from works of the following publications and authors are reprinted by the permission of their respective publishers and representatives:

An Inquiry into the Nature and Causes of the Wealth of Nations by Adam Smith reprinted from *Great Books of the Western World*. © 1952, 1990 Encyclopaedia Brittanica, Inc.

The Building of Renaissance Florence: An Economic and Social History by Richard Goldthwaite. The Johns Hopkins University Press, Baltimore/London, pp. 29–51. Reprinted by permission of the publisher.

The Cambridge Economic History of India, Vol. 2, Dharma Kumar, ed. Copyright 1989 by Cambridge University Press. Reprinted by permission of Cambridge University Press and Dharma Kumar.

Compleat English Tradesman by Daniel Defoe. (Alan Sutton Publishing Limited, Gloucestershire, 1987.)

Conditions of the Working Class in England by Chaloner [Friedrich Engels]. Reprinted by permission of Basil Blackwell Publishers.

The Entrepreneurs: Explorations within the American Business Tradition. by Robert Sobel, copyright 1974 by Weybright & Talley Publishers.

Freedom and Co-ordination, by Mary Parker Follett, Chapter IV, Pages 47–60; edited by L. Urwick. Published by Management Publications Trust, Ltd., London, 1949. Lecture delivered at Department of Business Administration, London School of Economics and Political Science, February 1933. Permissions held by the editor, The University of London, and Sir Isaac Pitman & Sons, Ltd.

Gestures by Roger Axtell, Copyright © 1991 by Roger Axtell. Reprinted by permission of John Wiley & Sons, Inc.

Global Factory by Joseph Grunwald et al. Copyright 1985. Used by permission of The Brookings Institution.

The Journals of Lewis and Clark, edited by Bernard DeVoto. Copyright 1953 by Bernard DeVoto. Copyright © renewed 1981 by Avis Devoto. Reprinted by permission of Houghton Mifflin Company. All rights reserved.

The Man Who Discovered Quality by Andrea Gabor. Copyright © 1990 by Andrea Gabor. Reprinted by Permission of Times Books, a division of Random House, Inc.

My Life and Work by Henry Ford and Samuel Crowther. Copyright 1922 by Doubleday, a division of Bantam Doubleday Dell Publishing Group, Inc. Used by permission of Doubleday, a division of Bantam Doubleday Dell Publishing Group, Inc.

My Years With General Motors by Alfred P. Sloan, Jr. Copyright © 1963 by Alfred P. Sloan, Jr. Reprinted by permission of Harold Matson Company, Inc.

"The New Product Development Map" by Steven C. Wheelwright and W. Earl Sasser, Jr., reprinted by permission of *Harvard Business Review*, May/June 1989. Copyright © 1989 by the President and Fellows of Harvard College; all rights reserved.

Organizations and Organization Theory by Jeffrey Pfeffer. Copyright 1982 by Pitman Publishing Inc, Boston.

"The Parable of the Sadhu" by Bowen H. McCoy, reprinted by permission of *Harvard Business Review*, September/October 1983. Copyright © 1983 by the President and Fellows of Harvard College; all rights reserved.

Pensées by Blaise Pascal transl. by A. J. Krailsheimer, Penguin Books, Ltd., 1965. Published by Penguin Books Limited.

"Pouring Ideas into Tin Cans" by Arthur Pound originally published in the May 1935 issue of *The Atlantic Monthly*.

The Protestant Ethic and the Spirit of Capitalism by Max Weber, translated by Talcott Parsons. Copyright © 1958 Charles Scribner's Sons. Reprinted with the permission of Macmillan College Publishing Company.

Renewing American Industry: Organizing for Efficiency and Innovation by Paul R. Lawrence and Davis Dyer. Copyright © 1983 by The Free Press. Reprinted with the permission of The Free Press, a Division of Macmillan, Inc.

The Road to Wigan Pier by George Orwell. Reprinted by permission of Harcourt Brace & Company.

Small Business in American Life, Stuart Bruchey, ed. Copyright 1980 by Columbia University Press, New York. Reprinted with the permission of the publisher.

"Strategic Intent" by Gary Hamel and C. K. Prahalad, reprinted by permission of *Harvard Business Review*, May/June 1989. Copyright © 1989 by the President and Fellows of Harvard College; all rights reserved.

Women's Fabian Tracts, Sally Alexander, ed. London, New York: Routledge, 1988. Reprinted by permission of Routledge.

Workers: World of Labor by Eric Hobsbawm, copyright 1978. George Weidenfeld & Nicolson, London.

To Howard and Sally Miller

Contents

Introduction, 3

The Inquisitive Spirit 5

Pensée 139, 7
Blaise Pascal

Of the Division of Labour, 11
Adam Smith

The Dalles to Walla Walla River, 19
Meriwether Lewis and William Clark

The Spirit of Capitalism, 25
Max Weber

Competition, 30
Friedrich Engels

South Wales, 36
George Orwell

The Working Life of Women, 45
B. L. Hutchins

Man and Woman: Images on the Left, 59
Eric Hobsbawm

The Essentials of Leadership, 75
Mary Parker Follett

What I Learned about Business, 86
Henry Ford

Cyrus Hall McCormick: From Farm Boy to Tycoon, 101
Robert Sobel

The Acquisitive Spirit 123

Counting Globally, 125
Roger E. Axtell

The Wherewithal to Spend: The Economic Background
of Renaissance Florence, 127
   Richard Goldthwaite

The Tradesman's Writing Letters, 144
   Daniel Defoe

The Occupational Structure of India, 150
   Dharma Kumar with Meghnad Desai

Foreign Assembly in Haiti, 166
   Joseph Grunwald, Leslie Delatour, and Karl Voltaire

Strategic Intent, 186
   Gary Hamel and C. K. Prahalad

Pouring Ideas into Tin Cans, 206
   Arthur Pound

Black Coats to White Collars: Economic Change, Nonmanual Work,
and the Social Structure of Industrializing America, 214
   Stuart Blumin

Women and the Factory Acts, 228
   Beatrice Webb

Co-Ordination by Committee, 239
   Alfred Pritchard Sloan

Autos: On the Thin Edge, 253
   Paul R. Lawrence and Davis Dyer

America Rediscovers W. Edwards Deming, 275
   Andrea Gabor

Organizations as Physical Structures, 292
   Jeffrey Pfeffer

The Parable of the Sadhu, 303
   Bowen McCoy

The New Product Development Map, 311
   Steven C. Wheelwright and W. Earl Sasser, Jr.

Writing about Business and Industry

Introduction

WRITING ABOUT BUSINESS AND INDUSTRY offers a means for students of business and students of writing to explore the ideas and practices of inventors such as Cyrus McCormick and Henry Ford, economists such as Adam Smith, historians such as Richard Goldthwaite, workers such as B. L. Hutchins, and theorists such as Jeffrey Pfeffer, and Beatrice Webb. The twenty-seven essays collected in this reader are designed to enhance interest in and appreciation of topics related to business and industry by offering a survey of some of the ideas that have shaped business practices in the past and that remain useful in the present. To give insight into some of the forces shaping contemporaneous theories about business and industry, each selection begins with a profile of its author.

Business and industrial practices are constantly evolving because of the effects of the world's economy and governmental concerns as well as public concerns over the environment and equality in pay and job duties. This reader reflects patterns of thought and of concern among working people and theorists that consistently surface in different places and eras. Chief among these concerns are working conditions, methods of organization and leadership, productivity and efficiency, business planning, and the role of competition. The styles of writing are as valuable as the quality of ideas, and both are meant to provide topics for both composition and conversation in the classroom.

Writing about Business and Industry is organized chronologically and is divided into two thematic sections: "The Inquisitive Spirit" and "The Acquisitive Spirit." These labels were chosen as means of uniting what are apparently two disparate types of activities: those associated with business and those associated with industry. But, the two fields share common goals and designs. Part I demonstrates common concerns about the nature of work, the life of the worker, and the character of the workplace, and Part II shows common interests in issues of management style, organizational strategy, and corporate environments. Each selection is followed by questions for discussion and topics for writing. In composing the writing topics, I designed questions that can be adapted for both business writing and English composition; some call for a letter, a memo, a proposal, or a report, and others suggest an essay or a book report.

Writing about Business and Industry may be used by undergraduate and graduate students in introductory and advanced writing classes, as well as by individuals interested in the development of business and industrial thought. Another goal in creating this book was to bring together a diverse group of authors, some of whose ideas were once deemed radical and some of whose ideas, although centuries old, still obtain. Some of these essays are philosophically provocative and others are of a more historical nature, relating the impact of great past economies on the world. In making these selections I have strived, however, to present a unified view of the various shaping ideas of capitalism and industrialism by focusing on how societies have viewed themselves in the world marketplace.

This reader concentrates on nonfictional accounts of invention, commerce, and manufacturing, avoiding fictionalized or poetical views of these same subjects. One need only recall Wordsworth's sonnet "The World is too much with us" or the novels of Upton Sinclair, John Dos Passos, Louisa May Alcott, or E. L. Doctorow, to hear other views. And, although we often ascribe "passion" to the works of literary writers, there is also much passion in the essays of Max Weber, George Orwell, B. L. Hutchins, Henry Ford, and Bowen McCoy as they describe the challenges of their own working lives and the lives of others.

Gathered within these pages are writings that exemplify the human spirit, the resilience of human nature, the capacity to invent and adapt to changing world economies and technology, and the power of success and failure, with candid looks at the costs of both. Reading, thinking, and writing about the works in this book can broaden students' perspectives on business and industry and offer the challenge to put knowledge to work to achieve personal goals and to help others.

The Inquisitive Spirit

Part I of *Writing about Business and Industry* covers a variety of topics ranging from the philosophical description of "ambitious action" to studies of art depicting men and women at work. The twelve essays were written by men and women from the seventeenth through the twentieth centuries. Many of the names are familiar: Blaise Pascal, Meriwether Lewis and William Clark, George Orwell, and Henry Ford, while others, such as the Fabian woman writer B. L. Hutchins, are less widely known. While no particular viewpoint or way of interpreting business and industrial practices is developed, the selections can be seen as addressing such common topics as the happiness of the worker, the meaning and spirit of capitalist pursuits, and the ideas of leadership as they have grown and changed through four centuries.

Beginning with the thoughts of the French mathematician-philosopher Blaise Pascal on "ambition," which establish an important dichotomy between "wants" and "needs," Part I next presents the ideas of Adam Smith on "The Division of Labour" as an example of a pioneering, and enduring, labor practice. The excerpt from *The Journals of Lewis and Clark* profiles trading practices with Native Americans in the Northwest in 1806. These three selections establish the thematic boundaries of Part I: ideas about business and industry, trade practices, and the acceptance of these ideas and efforts over time.

Moving more directly into the industrial age, the German sociologist Max Weber created relevant and necessary definitions of "capitalist" and "capitalism" in 1904, which opened a debate on class structure, opportunity, and individual motivation to succeed, which are echoed in the work of Friedrich Engels. In his essay, *Conditions of the Working Class in England* (1845), Engels addressed the nature of "Competition." Such theories in practice concerned both George Orwell and B. L. Hutchins. In an enlightened study, Hutchins analyzes the pre-World War I treatment of women in factory jobs and their life expectancies, and Orwell meticulously describes the coal miner's life in South Wales in the 1930s. As a change of pace, Eric Hobsbawm's study of paintings and sculptures of men and women at work provides cultural evidence of social attitudes toward industrialism in the period from 1830 to 1914.

M. P. Follett's essay was first prepared as a speech on "The Essentials of Leadership." Her psychological approach to management was well received by both British and American managers in the 1930s and she had a distinguished career as a business theorist. The portraits of Henry Ford and Cyrus McCormick with which Part I closes summarize how ideas became profitable realities for two of the giants of American manufacturing.

The aim of Part I, "The Inquisitive Spirit," is to guide an exploration into some ways in which historians, sociologists, philosophers, and business and industrial leaders have thought about work, profit, the role of labor in society, and the nature of cooperation, both as a leader and as part of a team. The headnotes identify the source of each excerpt and provide a historical and literary context for the work. The topics for discussion and writing follow each selection and are meant to encourage careful writing and creative thinking.

Blaise Pascal
Pensée 139

A seventeenth-century French philosopher, mathematician, theologian, and scientist, Blaise Pascal initially discovered the first twenty-three points of Euclidean geometry at the age of eleven. He patented the first calculator in 1647, and, with his father, conducted experiments with vacuums that led to the invention of the syringe and the barometer. When his father died, Pascal, then 28 years old, entered a Jansenist monastery from which he wrote a series of pamphlets from 1656 to 1657 attacking Jesuit theology.

Seven years after his death, in 1669, the Pensées *were found among his papers, together with an outline for the theory of integral calculus. This passage from his collection of contemplations on life and ideas is representative of Pascal's interest in the impact of decisions on actions. As you read this selection, think about how Pascal would define leadership and about the place of ambition in group deliberations.*

WHEN I HAVE occasionally set myself to consider the different distractions of men, the pains and perils to which they expose themselves at court or in war, whence arise so many quarrels, passions, bold and often bad ventures, etc., I have discovered that all the unhappiness of men arises from one single fact, that they cannot stay quietly in their own chamber. A man who has enough to live on, if he knew how to stay with pleasure at home, would not leave it to go to sea or to besiege a town. A commission in the army would not be bought so dearly, but that it is found insufferable not to budge from the town; and men only seek conversation and entering games, because they cannot remain with pleasure at home.

But on further consideration, when, after finding the cause of all our ills, I have sought to discover the reason of it, I have found that there is one very real reason, namely, the natural poverty of our feeble and mortal condition, so miserable that nothing can comfort us when we think of it closely.

Whatever condition we picture to ourselves, if we muster all the good things which it is possible to possess, royalty is the finest position in the world. Yet, when we imagine a king attended with every pleasure he can feel, if he be without diversion, and be left to consider and reflect on what he is, this feeble happiness will not sustain him; he will necessarily fall into forebodings of dangers, of revolutions which may happen, and, finally, of death and inevitable disease; so that if he be without what is called diversion, he is unhappy, and more unhappy than the least of his subjects who plays and diverts himself.

Hence it comes that play and the society of women, war, and high posts, are so sought after. Not that there is in fact any happiness in them, or that men imagine true bliss to consist in money won at play, or in the hare which they hunt; we would not take these as a gift. We do not seek that easy and peaceful lot which permits us to think of our unhappy condition, nor the dangers of war, nor the labour of office, but the bustle which averts these thoughts of ours, and amuses us.

Reasons why we like the chase better than the quarry.

Hence it comes that men so much love noise and stir; hence it comes that the prison is so horrible a punishment; hence it comes that the pleasure of solitude is a thing incomprehensible. And it is in fact the greatest source of happiness in the condition of kings, that men try incessantly to divert them, and to procure for them all kinds of pleasures.

The king is surrounded by persons whose only thought is to divert the king, and to prevent his thinking of self. For he is unhappy, king though he be, if he think of himself.

This is all that men have been able to discover to make themselves happy. And those who philosophize on the matter, and who think men unreasonable for spending a whole day in chasing a hare which they would not have bought, scarce know our nature. The hare in itself would not screen us from the sight of death and calamities; but the chase which turns away our attention from these, does screen us.

The advice given to Pyrrhus to take the rest which he was about to seek with so much labour, was full of difficulties.

[To bid a man live quietly is to bid him live happily. It is to advise him to be in a state perfectly happy, in which he can think at leisure without finding therein a cause of distress. This is to misunderstand nature.

As men who naturally understand their own condition avoid nothing so much as rest, so there is nothing they leave undone in seeking turmoil. Not that they have an instinctive knowledge of true happiness . . .

So we are wrong in blaming them. Their error does not lie in seeking excitement, if they seek it only as a diversion; the evil is that they seek it as if the possession of the objects of their quest would make them really happy. In this respect it is right to call their quest a vain one. Hence in all this both the censurers and the censured do not understand man's true nature.]

And thus, when we take the exception against them, that what they seek with such fervour cannot satisfy them, if they replied—as they should do if they considered the matter thoroughly—that they sought in it only a violent and impetuous occupation which turned their thoughts from self, and that they therefore chose an attractive object to charm and ardently attract them, they would leave their opponents without a reply. But they do not make this reply, because they do not know themselves. They do not know that it is the chase, and not the quarry, which they seek.

Dancing: we must consider rightly where to place our feet. —A gentleman sincerely believes that hunting is great and royal sport; but a beater is not of this opinion.

They imagine that if they obtained such a post, they would then rest with pleasure, and are insensible of the insatiable nature of their desire. They think they are truly seeking quiet, and they are only seeking excitement.

They have a secret instinct which impels them to seek amusement and occupation abroad, and which arises from the sense of their constant unhappiness. They have another secret instinct, a remnant of the greatness of our original nature, which teaches them that happiness in reality consists only in rest, and not in stir. And of these two contrary instincts they form within themselves a confused idea, which hides itself from their view in the depths of their soul, inciting them to aim at rest through excitement, and always to fancy that the satisfaction which they have not will come to them, if, by surmounting whatever difficulties confront them, they can thereby open the door to rest.

Thus passes away all man's life. Men seek rest in a struggle against difficulties; and when they have conquered these, rest becomes insufferable. For we think either of the misfortunes we have or of those which threaten us. And even if we should see ourselves sufficiently sheltered on all sides, weariness of its own accord would not fail to arise from the depths of the heart wherein it has its natural roots, and to fill the mind with its poison.

Thus so wretched is man that he would weary even without any cause for weariness from the peculiar state of his disposition; and so frivolous is he, that, though full of a thousand reasons for weariness, the least thing, such as playing billiards or hitting a ball, is sufficient to amuse him.

But will you say what object has he in all this? The pleasure of bragging to-morrow among his friends that he has played better than another. So others sweat in their own rooms to show to the learned that they have solved a problem in algebra, which no one had hitherto been able to solve. Many more expose themselves to extreme perils, in my opinion as foolishly, in order to boast afterwards that they have captured a town. Lastly, others wear themselves out in studying all these things, not in order to become wiser, but only in order to prove that they know them; and these are the most senseless of the band, since they are so knowingly, whereas one may suppose of the others, that if they knew it, they would no longer be foolish.

This man spends his life without weariness in playing every day for a small stake. Give him each morning the money he can win each day, on condition he does not play; you make him miserable. It will perhaps be said that he seeks the amusement of play and not the winnings. Make him then play for nothing; he will not become become excited over it, and will feel bored. It is then not the amusement alone that he seeks; a languid and passionless amusement will weary him. He must get excited over it, and deceive himself by the fancy that he will be happy to win what he would not have as a gift on condition of not playing; and he must make for himself an object of passion, and excite over it his desire, his anger, his fear, to obtain his imagined end, as children are frightened at the face they have blackened.

Whence comes it that this man, who lost his only son a few months ago, or who this morning was in such trouble through being distressed by lawsuits and quarrels, now no longer thinks of them? Do not wonder; he is quite taken up in looking out for the boar which his dogs have been hunting so hotly for the last six hours. He requires nothing more. However full of sadness a man may be, he is happy for the time, if you can prevail upon him to enter into some amusement; and however happy a man may be, he will soon be discontented and wretched, if he be not diverted and occupied by some passion or pursuit which prevents weariness from overcoming him. Without amusement there is no joy; with amusement there is no sadness. And this also constitutes the happiness of persons in high position, that they have a number of people to amuse them, and have the power to keep themselves in this state.

Consider this. What is it to be superintendent, chancellor, first president, but to be in a condition wherein from early morning a large number of people come from all quarters to see them, so as not to leave them an hour in the day in which they can think of themselves? And when they are in disgrace and sent back to their country houses, where they lack neither wealth nor servants to help them on occasion, they do not fail to be wretched and desolate, because no one prevents them from thinking of themselves.

Ideas for Discussion

1. What relationship does Pascal believe exists between happiness and possessions?
2. Why is the "chase" more inviting than the "quarry"?
3. What is Pascal's view of man and man's ability to succeed?

Topics for Writing

1. Summarize the main points of this *Pensée* in a memo to your instructor.
2. Analyze Pascal's views on human ambition in a letter to your employer about improving morale.
3. Compose your own essay on any theme the text suggests to you.

Adam Smith
Of the Division of Labour

Adam Smith was an eighteenth-century Scottish economist who is best known for his book An Inquiry into the Nature and Causes of the Wealth of Nations *(1776). Smith was first a professor of logic at Glasgow University (1751), later serving from 1755 to 1764 as professor of moral philosophy there. He published his* Theory of Moral Sentiments *in 1759 and was elected a Fellow of the Royal Society in 1767. In 1776, in London, he published the early chapters of* The Wealth of Nations, *which he saw as the beginning of a major study of society based on legal, economic, political, and philosophical principles. He went to Edinburgh in 1778 as Commissioner of Customs, and by 1787 Smith was appointed Lord Rector of Glasgow University. This selection is sometimes called "The Pin Factory," since it outlines Smith's revolutionary division of labor theory in pin manufacturing. The prose style is representative of eighteenth-century economic and nonfictional composition.*

THE GREATEST improvement in the productive powers of labour, and the greater part of the skill, dexterity, and judgment with which it is anywhere directed, or applied, seem to have been the effects of the division of labour.

The effects of the division of labour, in the general business of society, will be more easily understood by considering in what manner it operates in some particular manufactures. It is commonly supposed to be carried furthest in some very trifling ones; not perhaps that it really is carried further in them than in others of more importance: but in those trifling manufactures which are destined to supply the small wants of but a small number of people, the whole number of workmen must necessarily be small; and those employed in every different branch of the work can often be collected into the same workhouse, and placed at once under the view of the spectator. In those great manufactures, on the contrary, which are destined to supply the great wants of the great body of the people, every different branch of the work employs so great a number of workmen that it is impossible to collect them all into the same workhouse. We can seldom see more, at one time, than those employed in one single branch. Though in such manufactures, therefore, the work may really be divided into a much greater number of parts than in those of a more trifling nature, the division is not near so obvious, and has accordingly been much less observed.

To take an example, therefore, from a very trifling manufacture; but one in which the division of labour has been very often taken notice of, the trade of the pin-maker; a workman not educated to this business (which the division of labour has rendered a distinct trade), nor acquainted with the use of the machinery employed in it (to the invention of which the same division of labour has probably given occasion), could scarce, perhaps, with his utmost industry, make one pin in a day, and certainly could not make twenty. But in the way in which this business is now carried on, not only the whole work is a peculiar trade, but it is divided into a number of branches, of which the greater part are likewise peculiar trades. One man draws out the wire, another straights it, a third cuts it, a fourth points it, a fifth grinds it at the top for receiving the head; to make the head requires two or three distinct operations; to put it on is a peculiar business, to whiten the pins is another; it is even a trade by itself to put them into the paper; and the important business of making a pin is, in this manner, divided into about eighteen distinct operations, which, in some manufactories, are all performed by distinct hands, though in others the same man will sometimes perform two or three of them. I have seen a small manufactory of this kind where ten men only were employed, and where some of them consequently performed two or three distinct operations. But though they were very poor, and therefore but indifferently accommodated with the necessary machinery, they could, when they exerted themselves, make among them about twelve pounds of pins in a day. There are in a pound upwards of four thousand pins of a middling size. Those ten persons, therefore, could make among them upwards of forty-eight thousand pins in a day. Each person, therefore, making a tenth part of forty-eight thousand pins, might be considered as making four thousand eight hundred pins a day. But if they had all wrought separately and independently, and without any of them having been educated to this peculiar business, they certainly could not each of them have made twenty, perhaps not one pin in a day; that is, certainly, not the two hundred and fortieth, perhaps not the four thousand eight hundredth part of what they are at present capable of performing, in consequence of a proper division and combination of their different operations.

In every other art and manufacture, the effects of the division of labour are similar to what they are in this very trifling one; though, in many of them, the labour can neither be so much subdivided, nor reduced to so great a simplicity of operation. The division of labour, however, so far as it can be introduced, occasions, in every art, a proportionable increase of the productive powers of labour. The separation of different trades and employments from one another seems to have taken place in consequence of this advantage. This separation, too, is generally carried furthest in those countries which enjoy the highest degree of industry and improvement; what is the work of one man in a rude state of society being generally that of several in an improved one. In every improved society, the farmer is generally nothing but a farmer; the manufacturer, nothing but a manufacturer. The labour, too, which is necessary to produce any one complete manufacture is almost always divided among a great number of hands. How many different trades are employed in each branch of the linen and woolen manufactures from the growers of the flax and the wool, to the bleachers and smoothers of the linen, or to the dyers and dressers of the cloth! The nature of agriculture, indeed, does not admit of so many subdivisions of labour, nor of so complete a separation of one business from another, as manufactures. It is impossible to separate so entirely the business of the grazier from that of the corn-farmer as the trade of the carpenter is commonly separated from that of the smith. The spinner is almost always a distinct person from the weaver; but the ploughman, the harrower, the sower of the seed, and the reaper of the corn, are often the same. The occasions for those different sorts of labour returning with the different seasons of the year, it is impossible that one man should be constantly employed in any one of them. This impossibility of making so complete and entire a separation of all the different branches of labour employed in agriculture is perhaps the reason why the improvement of the productive powers of labour in this art does not always keep pace with their improvement in manufactures. The most opulent nations, indeed, generally excel all their neighbours in agriculture as well as in manufactures; but they are commonly more distinguished by their superiority in the latter than in the former. Their lands are in general better cultivated, and having more labour and expense bestowed upon them, produce more in proportion to the extent and natural fertility of the ground. But this superiority of produce is seldom much more than in proportion to the superiority of labour and expense. In agriculture, the labour of the rich country is not always much more productive than that of the poor; or, at least, it is never so much more productive as it commonly is in manufactures. The corn of the rich country, therefore, will not always, in the same degree of goodness, come cheaper to market than that of the poor. The corn of Poland, in the same degree of goodness, is as cheap as that of France, notwithstanding the superior opulence and improvement of the latter country. The corn of France is, in the corn provinces, fully as good, and in most years nearly about the same price with the corn of England, though, in opulence and improvement, France is perhaps inferior to England. The corn-lands of England, however, are better cultivated than those of France, and the corn-lands of France are said to be much better cultivated than those of Poland. But though the poor country, notwithstanding the inferiority of its cultivation, can, in some measure, rival the rich in the cheapness and goodness of its corn, it can pretend to no such competition in its manufactures; at least if those manufactures suit the soil, climate, and situation of the rich country. The silks of France are better and cheaper than those of England, because the silk manufacture, at least under the present high duties upon the importation of raw silk, does not so well suit the climate of England as that of France. But the hardware and the coarse woollens of England are beyond all comparison superior to those of France, and much cheaper too in the same degree of goodness. In Poland there are said to be scarce any manufactures of any kind, a few of those coarser household manufactures excepted, without which no country can well subsist.

This great increase of the quantity of work which, in consequence of the division of labour, the same number of people are capable of performing, is owing to three different circumstances; first, to the increase of dexterity in every particular workman; secondly, to the saving of the time which is commonly lost in passing from one species of work to another; and lastly, to the invention of a great number of machines which facilitate and abridge labour, and enable one man to do the work of many.

First, the improvement of the dexterity of the workman necessarily increases the quantity of the work he can perform; and the division of labour, by reducing every man's business to some one simple operation, and by making this operation the sole employment of his life, necessarily increases very much the dexterity of the workman. A common smith, who, though accustomed to handle the hammer, has never been used to make nails, if upon some particular occasion he is obliged to attempt it, will scarce, I am assured, be able to make above two or three hundred nails in a day, and those too very bad ones. A smith who has been accustomed to make nails, but whose sole or principal business has not been that of a nailer, can seldom with his utmost diligence make more than eight hundred or a thousand nails in a day. I have seen several boys under twenty years of age who had never exercised any other trade but that of making nails, and who, when they exerted themselves, could make, each of them, upwards of two thousand three hundred nails in a day. The making of a nail, however, is by no means one of the simplest operations. The same person blows the bellows, stirs or mends the fire as there is occasion, heats the iron, and forges every part of the nail: in forging the head too he is obliged to change his tools. The different operations into which the making of a pin, or of a metal button, is subdivided, are all of them much more simple, and the dexterity of the person, of whose life it has been the sole business to perform them, is usually much greater. The rapidity with which some of the operations of those manufacturers are performed, exceeds what the human hand could, by those who had never seen them, be supposed capable of acquiring.

Secondly, the advantage which is gained by saving the time commonly lost in passing from one sort of work to another is much greater than we should at first view be apt to imagine it. It is impossible to pass very quickly from one kind of work to another that is carried on in a different place and with quite different tools. A country weaver, who cultivates a small farm, must lose a good deal of time in passing from his loom to the field, and from the field to his loom. When the two trades can be carried on in the same workhouse, the loss of time is no doubt much less. It is even in this case, however, very considerable. A man commonly saunters a little in turning his hand from one sort of employment to another. When he first begins the new work he is seldom very keen and hearty; his mind, as they say, does not go to it, and for some time he rather trifles than applies to good purpose. The habit of sauntering and of indolent careless application, which is naturally, or rather necessarily acquired by every country workman who is obliged to change his work and his tools every half hour, and to apply his hand in twenty different ways almost every day of his life, renders him almost always slothful and lazy, and incapable of any vigorous application even on the most pressing occasions. Independent, therefore, of his deficiency in point of dexterity, this cause alone must always reduce considerably the quantity of work which he is capable of performing.

Thirdly, and lastly, everybody must be sensible how much labour is facilitated and abridged by the application of proper machinery. It is unnecessary to give any example. I shall only observe, therefore, that the invention of all those machines by which labour is so much facilitated and abridged seems to have been originally owing to the division of labour. Men are much more likely to discover easier and readier methods of attaining any object when the whole attention of their minds is directed towards that single object than when it is dissipated among a great variety of things. But in consequence of the division of labour, the whole of every man's attention comes naturally to be directed towards some one very simple object. It is naturally to be expected, therefore, that some one or other of those who are employed in each particular branch of labour should soon find out easier and readier methods of performing their own particular work, whereever the nature of it admits of such improvement. A great part of the machines made use of in those manufactures in which labour is most subdivided, were originally the inventions of common workmen, who, being each of them employed in some very simple operation, naturally turned their thoughts towards finding out easier and readier methods of performing it. Whoever has been much accustomed to visit such manufactures must frequently have been shown very pretty machines, which were the inventions of such workmen in order to facilitate and quicken their own particular part of the work. In the first fire-engines, a boy was constantly employed to open and shut alternately the communication between the boiler and the cylinder, according as the piston either ascended or descended. One of those boys, who loved to play with his companions, observed that, by tying a string from the handle of the valve which opened this communication to another part of the machine, the valve would open and shut without his assistance, and leave him at liberty to divert himself with his playfellows. One of the greatest improvements that has been made upon this machine, since it was first invented, was in this manner the discovery of a boy who wanted to save his own labour.

All the improvements in machinery, however, have by no means been the inventions of those who had occasion to use the machines. Many improvements have been made by the ingenuity of the makers of the machines, when to make them became the business of a peculiar trade; and some by that of those who are called philosophers or men of speculation, whose trade it is not to do anything, but to observe everything; and who, upon that account, are often capable of combining together the powers of the most distant and dissimilar objects. In the progress of society, philosophy or speculation becomes, like every other employment, the principal or sole trade and occupation of a particular class of citizens. Like every other employment class of citizens. Like every other employment too, it is subdivided into a great number of different branches, each of which affords occupation to a peculiar tribe or class of philosophers; and this subdivision of employment in philosophy, as well as in every other business, improves dexterity, and saves time. Each individual becomes more expert in his own peculiar branch, more work is done upon the whole, and the quantity of science is considerably increased by it.

It is the great multiplication of the productions of all the different arts, in consequence of the division of labour, which occasions, in a well-governed society, that universal opulence which extends itself to the lowest ranks of the people. Every workman has a great quantity of his own work to dispose of beyond what he himself has occasion for; and every other workman being exactly in the same situation, he is enabled to exchange a great quantity of his own goods for a great quantity, or, what comes to the same thing, for the price of a great quantity of theirs. He supplies them abundantly with what they have occasion for, and they accommodate him as amply with what he has occasion for, and a general plenty diffuses itself through all the different ranks of the society.

Observe the accommodation of the most common artificer or day-labourer in a civilised and thriving country, and you will perceive that the number of people of whose industry a part, though but a small part, has been employed in procuring him this accommodation, exceeds all computation. The woollen coat, for example, which covers the day-labourer, as coarse and rough as it may appear, is the produce of the joint labour of a great multitude of workmen. The shepherd, the sorter of the wool, the wool-comber or carder, the dyer, the scribbler, the spinner, the weaver, the fuller, the dresser, with many others, must all join their different arts in order to complete even this homely production. How many merchants and carriers, besides, must have been employed in transporting the materials from some of those workmen to others who often live in a very distant part of the country! How much commerce and navigation in particular, how many ship-builders, sailors, sail-makers, rope-makers, must have been employed in order to bring together the different drugs made use of by the dyer, which often come from the remotest corners of the world! What a variety of labour, too, is necessary in order to produce the tools of the meanest of those workmen! To say nothing of such complicated machines as the ship of the sailor, the mill of the fuller, or even the loom of the weaver, let us consider only what a variety of labour is requisite in order to form that very simple machine, the shears with which the shepherd clips the wool. The miner, the builder of the furnace for smelting the ore, the seller of the timber, the burner of the charcoal to be made use of in the smelting-house, the brickmaker, the brick-layer, the workmen who attend the furnace, the mill-wright, the forger, the smith, must all of them join their different arts in order to produce them. Were we to examine, in the same manner, all the different parts of his dress and household furniture, the coarse linen shirt which he wears next his skin, the shoes which cover his feet, the bed which he lies on, and all the different parts which compose it, the kitchen-grate at which he prepares his victuals, the coals which he makes use of for that purpose, dug from the bowels of the earth, and brought to him perhaps by a long sea and a long land carriage, all the other utensils of his kitchen, all the furniture of his table, the knives and forks, the earthen or pewter plates upon which he serves up and divides his victuals, the different hands employed in preparing his bread and his beer, the glass window which lets in the heat and the light, and keeps out the wind and the rain, with all the knowledge and art requisite for preparing that beautiful and happy invention, without which these northern parts of the world could scarce have afforded a very comfortable habitation, together with the tools of all the different workmen employed in producing those different workmen employed in producing those different conveniences; if we examine, I say, all these things, and consider what a variety of labour is employed about each of them, we shall be sensible that, without the assistance and co-operation of many thousands, the very meanest person in a civilised country could not be provided, even according to what we very falsely imagine the easy and simple manner in which he is commonly accommodated. Compared, indeed, with the more extravagant luxury of the great, his accommodation must no doubt appear extremely simple and easy; and yet it may be true, perhaps, that the accommodation of a European prince does not always so much exceed that of an industrious and frugal peasant as the accommodation of the latter exceeds that of many an African king, the absolute master of the lives and liberties of ten thousand naked savages.

Ideas for Discussion

1. The chapter reprinted here argues persuasively for the division of labor. Are there drawbacks to this manufacturing system which Smith suppresses?
2. What means does Smith employ to make his essay convincing?
3. What does Smith see as the impact of the division of labor on human integrity?

Topics for Writing

1. Prepare an outline or a diagram of the essay.
2. Illustrate the essay with appropriate visual aids.

3. Imagine you are a newly hired manager wishing to increase productivity in the manner Smith describes. Prepare a proposal for implementing the division of labor concept addressed to either the workers or to the decision maker in the company.
4. Write a short, informal, memo-report on alternative work schedules, include information from Smith's essay. Or, write on the same topic in a letter of inquiry to your company's human resources department to determine the feasibility of such schedules in your company.

Meriwether Lewis and William Clark
The Dalles to Walla Walla River

Meriwether Lewis and William Clark were Virginians who led explorations of lands west of the Mississippi River from 1804 to 1806 for President Thomas Jefferson. The segment from their Journals sheds light on the pre-capitalist economy of America and cultural dealings between Native Americans and explorers of the West. Lewis became governor of Louisiana in 1806 and died in Tennesee in 1809; Clark, the brother of Revolutionary War General George Rogers Clark, became governor of Missouri Territory and later superintendent of Indian Affairs. He died in 1838. Thomas Jefferson's "Instructions To Captain Lewis" appears in Writing about Science, *2nd ed. (Bowen and Schneller, 1991). Original spellings have been retained.*

[Lewis]   SUNDAY APRIL 20TH 1806

This morning I was informed that the natives had pilfered six tommahawks and a knife from the party in the course of the last night. I spoke to the cheif on this subject. he appeared angry with his people and addressed them but the property was not restored. one horse which I had purchased and paid for yesterday and which could not be found when I ordered the horses into close confinement yesterday I was now informed had been gambled away by the rascal who had sold it to me and had been taken away by a man of another nation. I therefore took the goods back from this fellow. I purchased a gun from the cheif for which I gave him 2 Elkskins. in the course of the day I obtained two other indifferent horses for which I gave an extravigant price. I found that I should get no more horses and therefore resolved to proceed tomorrow morning with those which I had and to convey the baggage in two small canoes that the horses could not carry. for this purpose I had a load made up for seven horses, the eighth Bratton was compelled to ride as he was yet unable to walk. I bart[er]ed my Elkskins old irons and 2 canoes for beads. one of the canoes for which they would give us but little I had cut up for fuel. I had the horses graized untill evening and then picquited and hubbled within the limits of our camp. I ordered the indians from our camp this evening and informed them that if I caught them attempting to perloin any article from us I would beat them severely. they went off in reather a bad humour and I directed the party to examine their arms and be on their guard. they stole two spoons from us in the course of the day.

[Clark]   SUNDAY 20TH APRIL 1806

I shewed the Eneshers the articles I had to give for their horses. they without hezitation informed me that they would not sell me any for the articles I had, if I would give them Kittles they would let me have horses, and not without that their horses were at a long ways off in the planes and they would not send for them &c. My offer was a blue robe, a calleco Shirt, a Silk handkerchief, 5 parcels of paint, a knife, a Wampom moon, 8 yards of ribon, several pieces of Brass, a Mockerson awl and 6 braces of yellow beeds; and to that amount for each horse which is more than double what we gave either the Sohsohne or first flatheads we met with on Clarks river I also offered my large blue blanket, my coat sword & plume none of which seamed to entice those people to sell their horses. notwithstanding every exertion not a single horse could be precured of those people in the course of the day.

* * *

[Clark]   MONDAY 21ST APRIL 1806

I found it useless to make any further attempts to trade horses with those unfriendly people who only crowded about me to view and make their remarks and smoke, the latter I did not indulge them with to day. at 12 oClock Capt Lewis and party came up from the Skillutes Village with 9 horses packed and one which bratten who was yet to weak to walk, rode, and soon after the two small canoes also loaded with the residue of the baggage which could not be taken on horses. we had every thing imediately taken above the falls. in the mean time purchased 2 Dogs on which the party dined. whilst I remained at the Enesher Village I subsisted on 2 platters of roots, some pounded fish and sun flower seed pounded which an old man had the politeness to give me in return for which I gave him several small articles.

The man who we had reason to believe had stolen the horse he had given for the Kittle we thretened a little and he produced a very good horse in the place of that one which we cheerfully receved. After dinner we proceeded on about 4 miles to a village of 9 Mat Lodges of the Enesher: one of the canoes joined us, the other not haveing observed us halt continued on. We obtained 2 Dogs and a small quantity of fuel of those people for which we were obliged to give a higher price than usial, our guide continued with us, he appears to be an honest fellow. he tels us that the indians above will treat us with much more hospitallity than those we are now with. we purchased another horse this evening but his back is in such a horrid state that we can put but little on him; we obtained him for a triffle, at least for articles which might be precured in the U. States for 10/ Virginia currency.

* * *

[Lewis]   WEDNESDAY APRIL 23RD 1806

At day light this morning we were informed that the two horses of our Interpreter Charbono were absent; on enquiry it appeared that he had neglected to confine them to picqu[i]ts as had been directed last evening. we immediately dispatched Reubin Feilds and Labuish to assist Charbono in recovering his horses. one of them was found at no great distance and the other was given over as lost. we continued our march along a narrow rocky bottom on the N. side of the river about 12 miles to the Wah-how-pum Village of 12 temperary mat lodges near the Rock rapid. these people appeared much pleased to see us, sold us 4 dogs and some wood for our small articles which we had previously prepared as our only resource to obtain fuel and food through those plains. these articles con[s]isted of pewter buttons, strips of tin iron and brass, twisted wire &c. here we met with a Chopunnish [Nez Percé] man on his return up the river with his family and about 13 head of horses most of them young and unbroken. he offered to hire us some of them to pack as far [as] his nation, but we prefer bying as by hireing his horses we shal have the whole of his family most probably to mentain. at a little distance below this village we passed five lodges of the same people who like those were waiting the arrival of the salmon.

after we had arranged our camp we caused all the old and brave men to set arround and smoke with us. we had the violin played and some of the men danced; after which the natives entertained us with a dance after their method. this dance differed from any I have yet seen. they formed a circle and all sung as well the spectators as the dancers who performed within the circle. these placed their sholders together with their robes tightly drawn about them and danced in a line from side to side, several parties of from 4 to seven will be performing within the circle at the same time. the whole concluded with a premiscuous dance in which most of them sung and danced. these people speak a language very similar to the Chopunnish whome they also resemble in their dress after the dance was ended the indians retired at our request and we retired to rest. we had all our horses side hubbled and turned out to graize; the river is by no means as rapid as when we decended or at least not obstructed with those dangerous rapids the water at present covers most of the rocks in the bed of the river. the natives promised to barter their horses with us in the morning we therefore entertained a hope that we shall be enabled to proceede by land from hence with the whole of our party and baggage.

[Clark]   THURSDAY 24TH APRIL 1806

rose early this morning and sent out after the horses all of which were found except McNeals which I hired an Indian to find and gave him a Tomahawk had 4 pack saddles made ready to pack the horses which we may purchase. we purchased 3 horses, and hired 3 others of the Chopunnish man who accompanies us with his family, and at 1 P.M. set out and proceeded on through a open countrey rugid & sandy between some high lands and the river to a village of 5 Lodges of the Met-cow-we band haveing passed 4 Lodges at 4 miles and 2 Lodges at 6 miles. Great numbers of the nativs pass us on hors back maney meet us and continued with us to the Lodges. we purchased 3 dogs which were pore, but the fattest we could precure, and cooked them with straw and dry willow. we sold our canoes for a fiew strands of beeds.* the nativs had tantelized us with an exchange of horses for our canoes in the first instance, but when they found that we had made our arrangements to travel by land they would give us nothing for them. we sent Drewyer to cut them up, he struck one and split her they discovered that we were deturmined to destroy the canoes and offered us several strans of beeds which were accepted most of the party complain of their feet and legs this evening being very sore. it is no doubt caused by walking over the rough stone and deep sand after being accustomed to a soft soil. my legs and feet give me much pain. I bathed them in cold water from which I experienced considerable relief. we directed that the 3 horses purchased yesterday should be hobbled and confined to pickquets and that the others should be hobbled & spanceled, and strictly attended to by the guard made 12 miles to day.

* * *

[Clark]   MONDAY APRIL 28TH 1806

This morning early the Great Chief Yelleppet brought a very eligant white horse to our camp and presented him to me, signifying his wish to get a kittle but being informed that we had already disposed of every kittle we could possibly spare he said he was content with whatever I thought proper to give him. I gave him my *Swoard,* 100 balls & powder and some small articles of which he appeared perfectly satisfied. it was necessary before we entered on our rout through the plains where we were to meet with no lodges or resident Indians that we should lay in a stock of provisions and not depend altogether on the gun. we derected R. Frazer to whome we have intrusted the duty of makeing the purchases. to lay in as maney fat dogs as he could procure; he soon obtained 10, being anxious to depart we requested the Chief to furnish us with canoes to pass the river, but he insisted on our remaining with him this day at least, that he would be much pleased if we would consent to remain two or 3 days, but he would not let us have canoes to leave him this day. that he had sent for the *Chim-na-pums* his neighbours to come down and join his people this evening and dance for us. We urged the necessity of our proceeding on imediately in order that we might the sooner return to them, with the articles which they wished brought to them but this had no effect, he said that the time he asked could not make any considerable difference. at length urged that there was no wind blowing and that the river was consequently in good order to

---

*From here on travel is by land: to the Nez Percé villages on the Clearwater River, beyond them to Traveller's Rest in the Bitterroot Valley and on to where the dugouts are hidden at the forks of the Jefferson. They detour Celilo Falls, traveling on the high land above the canyon.

pass our horses and if he would furnish us with canoes for that purpose we would remain all night at our present encampment, to this proposition he assented and soon produced a canoe. I saw a man who had his knee contracted who had previously applyed to me for some medisene, that if he would fournish another canoe I would give him some medisene. he readily consented and went himself with his canoe by means of which we passed our horses over the river safely and hobbled them as usial.

We found a *Sho-sho-ne* woman, prisoner among those people by means of whome and *Sah-cah-gah-weah*, Shabono's wife we found means of converceing with the *Wallahwallârs*. we conversed with them for several hours and fully satisfy all their enquiries with respect to our Selves and the Objects of our pursute. they were much pleased. they brought several disordered persons to us for whome they requested some medical aid. to all of whome we administered much to the gratification of those pore wretches, we gave them some eye water which I believe will render them more essential sirvice than any other article in the medical way which we had it in our power to bestow on them sore eyes seam to be a universal complaint among those people; I have no doubt but the fine sands of those plains and the river contribute much to the disorder. [A] man who had his arm broken had it loosely bound in a piece of leather without any thing to surport it. I dressed the arm which was broken short above the wrist & supported it with broad sticks to keep it in place, put [it] in a sling and furnished him with some lint bandages &c. to Dress it in future. a little before sun set the Chimnahpoms arrived; they were about 100 men and a fiew women; they joined the Wallahwallahs who were about 150 men and formed a half circle arround our camp where they waited verry patiently to see our party dance. the fiddle was played and the men amused themselves with danceing about an hour. we then requested the Indians to dance which they very chearfully complyed with; they continued their dance untill 10 at night. the whole assemblage of Indians about 350 men women and children sung and danced at the same time. Most of them danced in the same place they stood and mearly jumped up to the time of their musick. Some of the men who were esteemed most brave entered the space around which the main body were formed in solid column and danced in a circular manner side wise. at 10 P M. the dance ended and the nativs retired; they were much gratified in seeing some of our party join them in their dance. one of their party who made himself the most conspicious charecter in the dance and songs, we were told was a medesene man & could foretell things. that he had told of our comeing into their country and was now about to consult his God the Moon if what we said was the truth &c &c.

Ideas for Discussion

1. What do these journal passages reveal about Lewis and Clark's attitudes toward trading with the Native Americans?

2. Comment on the Native Americans' approaches to trading with Lewis and Clark. How is it a "family affair," for instance?
3. What do these portions of the journal illustrate about the state of American English as a written language in the early nineteenth century?

Topics for Writing

1. Prepare a report on doing business with another country or group of people in your state.
2. For a term paper, research Jefferson's intentions for the Lewis and Clark expeditions and compare them to, say, President Nixon's efforts to open China to Western trade in the 1970s.
3. Compose Lewis and Clark's trip report for President Jefferson.
4. As Lewis or Clark, compose a letter to future traders with Native Americans outlining the steps necessary to be successful in doing business with them.
5. Compose a memo from Lewis and Clark announcing the need for volunteers to serve on this expedition; or, compose a procedural memo summarizing the key points of doing business with Native Americans.

Max Weber
The Spirit of Capitalism

Max Weber was an economist who turned to sociology later in life. Between 1892 and 1897, Weber taught law, political theory, and economics at universities in Berlin, Freiburg, and Heidelberg. From 1897 to 1918, he was on medical leave after suffering a nervous collapse.

A German Calvinist, Weber rejected Marxist views, which is clear from his treatise The Protestant Ethic and the Spirit of Capitalism *(1904/5). He urged the role of strong leaders in guiding the people into the "spirit of capitalism" as an act of will, not determinism.*

In 1918, Weber accepted a position in sociology in Vienna, moving to a similar post in Munich before his death in 1920 at the age of fifty-six. This chapter from The Protestant Ethic *describes "The Spirit of Capitalism," and it marks the importance of Benjamin Franklin in acting out that spirit. Weber's style is characteristic of the blending of political and social philosophy with the asethetics of philosophical writing.*

IN THE TITLE of this study is used the somewhat pretentious phrase, the *spirit* of capitalism. What is to be understood by it? The attempt to give anything like a definition of it brings out certain difficulties which are in the very nature of this type of investigation.

If any object can be found to which this term can be applied with any understandable meaning, it can only be an historical individual, i.e. a complex of elements associated in historical reality which we unite into a conceptual whole from the standpoint of their cultural significance.

Such an historical concept, however, since it refers in its content to a phenomenon significant for its unique individuality, cannot be defined according to the formula *genus proximum, differentia specifica*, but it must be gradually put together out of the individual parts which are taken from historical reality to make it up. Thus the final and definitive concept cannot stand at the beginning of the investigation, but must come at the end. We must, in other words, work out in the course of the discussion, as its most important result, the best conceptual formulation of what we here understand by the spirit of capitalism, that is the best from the point of view which interests us here. This point of view (the one of which we shall speak later) is, further, by no means the only possible one from which the historical phenomena we are investigating can be analysed. Other standpoints would, for this as for every historical phenomenon, yield other characteristics as the essential ones. The result is that it is by no means necessary to understand by the spirit of capitalism only what it will come to mean to *us* for the purposes of our analysis. This is a necessary result of the nature of historical concepts which attempt for their methodological purposes not to grasp historical reality in abstract general formulæ, but in concrete genetic sets of relations which are inevitably of a specifically unique and individual character.

Thus, if we try to determine the object, the analysis and historical explanation of which we are attempting, it cannot be in the form of a conceptual definition, but at least in the beginning only a provisional description of what is here meant by the spirit of capitalism. Such a description is, however, indispensable in order clearly to understand the object of the investigation. For this purpose we turn to a document of that spirit which contains what we are looking for in almost classical purity, and at the same time has the advantage of being free from all direct relationship to religion, being thus, for our purposes, free of preconceptions.

"Remember, that *time* is money. He that can earn ten shillings a day by his labour, and goes abroad, or sits idle, one half of that day, though he spends but sixpence during his diversion or idleness, ought not to reckon *that* the only expense; he has really spent, or rather thrown away, five shillings besides.

"Remember, that *credit* is money. If a man lets his money lie in my hands after it is due, he gives me the interest, or so much as I can make of it during that time. This amounts to a considerable sum where a man has good and large credit, and makes good use of it.

"Remember, that money is of the prolific, generating nature. Money can beget money, and its offspring can beget more, and so on. Five shillings turned is six, turned again it is seven and threepence, and so on, till it becomes a hundred pounds. The more there is of it, the more it produces every turning, so that the profits rise quicker and quicker. He that kills a breeding-sow, destroys all her offspring to the thousandth generation. He that murders a crown, destroys all that it might have produced, even scores of pounds."

* * *

Beware of thinking all your own that you possess, and of living accordingly. It is a mistake that many people who have credit fall into. To prevent this, keep an exact account for some time both of your expenses and your income. If you take the pains at first to mention particulars, it will have this good effect: you will discover how wonderfully small, trifling expenses mount up to large sums, and will discern what might have been, and may for the future be saved, without occasioning any great inconvenience."

* * *

It is Benjamin Franklin who preaches to us in these sentences, the same which Ferdinand Kürnberger satirizes in his clever and malicious *Picture of American Culture* as the supposed confession of faith of the Yankee. That it is the spirit of capitalism which here speaks in characteristic fashion, no one will doubt, however little we may wish to claim that everything which could be understood as pertaining to that spirit is contained in it. . . .

When Jacob Fugger, in speaking to a business associate who had retired and who wanted to persuade him to do the same, since he had made enough money and should let others have a chance, rejected that as pusillanimity and answered that "he (Fugger) thought otherwise, he wanted to make money as long as he could," the spirit of his statement is evidently quite different from that of Franklin. What in the former case was an expression of commercial daring and a personal inclination morally neutral, in the latter takes on the character of an ethically coloured maxim for the conduct of life. The concept spirit of capitalism is here used in this specific sense, it is the spirit of modern capitalism. For that we are here dealing only with Western European and American capitalism is obvious from the way in which the problem was stated. Capitalism existed in China, India, Babylon, in the classic world, and in the Middle Ages. But in all these cases, as we shall see, this particular ethos was lacking.

Now, all Franklin's moral attitudes are coloured with utilitarianism. Honesty is useful, because it assures credit; so are punctuality, industry, frugality, and that is the reason they are virtues. A logical deduction from this would be that where, for instance, the appearance of honesty serves the same purpose, that would suffice, and an unnecessary surplus of this virtue would evidently appear to Franklin's eyes as unproductive waste. And as a matter of fact, the story in his autobiography of his conversion to those virtues, or the discussion of the value of a strict maintenance of the appearance of modesty, the assiduous belittlement of one's own deserts in order to gain general recognition later, confirms this impression. According to Franklin, those virtues, like all others, are only in so far virtues as they are actually useful to the individual, and the surrogate of mere appearance is always sufficient when it accomplishes the end in view. It is a conclusion which is inevitable for strict utilitarianism. The impression of many Germans that the virtues professed by Americanism are pure hypocrisy seems to have been confirmed by this striking case. But in fact the matter is not by any means so simple. Benjamin Franklin's own character, as it appears in the really unusual candidness of his autobiography, belies that suspicion. The circumstance that he ascribes his recognition of the utility of virtue to a divine revelation which was intended to lead him in the path of righteousness, shows that something more than mere garnishing for purely egocentric motives is involved.

* * *

The capitalistic economy of the present day is an immense cosmos into which the individual is born, and which presents itself to him, at least as an individual, as an unalterable order of things in which he must live. It forces the individual, in so far as he is involved in the system of market relationships, to conform to capitalistic rules of action. The manufacturer who in the long run acts counter to these norms, will just as inevitably be eliminated from the economic scene as the worker who cannot or will not adapt himself to them will be thrown into the streets without a job.

* * *

One of the technical means which the modern employer uses in order to secure the greatest possible amount of work from his men is the device of piece-rates. In agriculture, for instance, the gathering of the harvest is a case where the greatest possible intensity of labour is called for, since, the weather being uncertain, the difference between high profit and heavy loss may depend on the speed with which the harvesting can be done. Hence a system of piece-rates is almost universal in this case. And since the interest of the employer in a speeding-up of harvesting increases with the increase of the results and the intensity of the work, the attempt has again and again been made, by increasing the piece-rates of the workmen, thereby giving them an opportunity to earn what is for them a very high wage, to interest them in increasing their own efficiency. But a peculiar difficulty has been met with surprising frequency: raising the piece-rates has often had the result that not more but less has been accomplished in the same time, because the worker reacted to the increase not by increasing but by decreasing the amount of his work.

* * *

Benjamin Franklin was filled with the spirit of capitalism at a time when his printing business did not differ in form from any handicraft enterprise. And we shall see that at the beginning of modern times it was by no means the capitalistic entrepreneurs of the commercial aristocracy, who were either the sole or the predominant bearers of the attitude we have here called the spirit of capitalism. It was much more the rising strata of the lower industrial middle classes. Even in the nineteenth century its classical representatives were not the elegant gentlemen of Liverpool and Hamburg, with their commercial fortunes handed down for generations, but the self-made parvenus of Manchester and Westphalia, who often rose from very modest circumstances. As early as the sixteenth century the situation was similar; the industries which arose at that time were mostly created by parvenus.

* * *

Now, how could activity, which was at best ethically tolerated, turn into a calling in the sense of Benjamin Franklin? The fact to be explained historically is that in the most highly capitalistic centre of that time, in Florence of the fourteenth and fifteenth centuries, the money and capital market of all the great political Powers, this attitude was considered ethically unjustifiable, or at best to be tolerated. But in the backwoods small bourgeois circumstances of Pennsylvania in the eighteenth century, where business threatened for simple lack of money to fall back into barter, where there was hardly a sign of large enterprise, where only the earliest beginnings of banking were to be found, the same thing was considered the essence of moral conduct, even commanded in the name of duty. To speak here of a reflection of material conditions in the ideal superstructure would be patent nonsense. What was the background of ideas which could account for the sort of activity apparently directed toward profit alone as a calling toward which the individual feels himself to have an ethical obligation? For it was this idea which gave the way of life of the new entrepreneur its ethical foundation and justification.

* * *

It might thus seem that the development of the spirit of capitalism is best understood as part of the development of rationalism as a whole, and could be deduced from the fundamental position of rationalism on the basic problems of life. In the process Protestantism would only have to be considered in so far as it had formed a stage prior to the development of a purely rationalistic philosophy. But any serious attempt to carry this thesis through makes it evident that such a simple way of putting the question will not work, simply because of the fact that the history of rationalism shows a development which by no means follows parallel lines in the various departments of life.

Ideas for Discussion

1. What does Weber mean by the phrase "the spirit of capitalism"?
2. How does Weber construct his argument? Be prepared to mention his techniques, style, and key points.
3. What is the difference between precapitalist culture and capitalist culture?
4. Why is Benjamin Franklin central to Weber's view?

Topics for Writing

1. Look up the word "capitalism" and prepare an extended definition, paying special attention to its 1900 meaning.
2. For a longer essay, an oral presentation, or a book report, read either Benjamin Franklin's *Autobiography* or Karl Marx's *Das Kapital*. What impact do these works have on our understanding of Weber's work?
3. In an essay, describe the ideal worker who has "the spirit of capitalism."
4. Write a solicited proposal to the management of a company of your choice, or one you invent, on the subject of making the company more productive. Use appropriate parts of Weber's work in the introduction and conclusion of your work.

Friedrich Engels
Competition

Friedrich Engels, the son of a wealthy German textile manufacturer, emigrated to England at the age of 22 to assume management of a textile factory near Manchester, in 1842. His experience led him to write The Condition of the Working Class in England in 1844, *which was published in 1845. A lifelong friend and intellectual collaborator of Karl Marx, Engels was a cofounder of modern communism. At Marx's death in 1883, Engels took up the task of translating and editing the second and third volumes of* Das Kapital.

This segment on competition, taken from The Condition of the Working Class in England in 1844, *questions the worker's motivation to excel in a class-oriented society. As you read, note passages that reflect Engels's enthusiasm, passion, and interest in his subject, and consider how his style influences your response to the ideas he presents.*

COMPETITION IS THE most extreme expression of that war of all against all which dominates modern middle-class society. This struggle for existence—which in extreme cases is a life and death struggle—is waged not only between different classes of society but also between individuals within these social groups. Everybody competes in some way against everyone else and consequently each individual tries to push aside anyone whose existence is a barrier to his own advancement. The workers compete among themselves, and so do the middle classes. The powerloom weaver competes with the handloom weaver. Among the handloom weavers themselves there is continual rivalry. Those who are unemployed or poorly paid try to undercut and so destroy the livelihood of those who have work and are earning better wages. This competition of workers among themselves is the worst aspect of the present situation as far as the proletariat is concerned. This is the sharpest weapon which the middle classes wield against the working classes. This explains the rise of trade unions, which represent an attempt to eliminate such fratricidal conflict between the workers themselves. It explains, too, the fury of the middle classes against trade unions, and their ill-concealed delight at any setback which the unions suffer.

The worker is helpless; left to himself he cannot survive a single day. The middle classes have secured a monopoly of all the necessities of life.

What the worker needs he can secure only from the middle classes, whose monopoly is protected by the authority of the State. In law and in fact the worker is the slave of the middle classes, who hold the power of life and death over him. The middle classes offer food and shelter to the worker, but only in return for an 'equivalent,' i.e. for his labour. They even disguise the true state of affairs by making it appear that the worker is acting of his own free will, as a truly free agent and as a responsible adult, when he makes his bargain with the middle classes. A fine freedom indeed, when the worker has no choice but to accept the terms offered by the middle classes or go hungry and naked like the wild beasts. A fine "equivalent," when it is the bourgeoisie alone which decides the terms of the bargain. And if a worker is such a fool as to prefer to go hungry rather than accept the "fair" terms of the middle classes who are his "natural superiors"[1]— well, then it is easy enough to find another worker. The working classes are numerous enough in all conscience and not all of them are so stupid as to prefer death to life.

* * *

. . . The "maximum wage" is fixed by the competition among members of the middle classes. It has already been seen that competition also exists within this social group. Members of the middle classes can increase their capital only by engaging in commerce or industry. In either case the services of workers are essential. Even if members of the middle classes invest their capital they still need the services of workpeople in an indirect way, because it is only by using the money in commerce or industry that the borrower is in a position to pay any interest. Although, in a general sense, the middle classes are dependent upon the workers, this does not mean that the middle classes depend upon the workers for their daily needs. After all, at a pinch, the middle classes can live upon their accumulated capital. The middle classes use the workers not to live but to enrich themselves in the same way as they use an article of commerce or a beast of burden. The workers fashion the goods which the middle classes sell at a profit. If the demand for these goods increases to such an extent that there is full employment or even a surplus of jobs for workers who are normally competing for them, competition between the workers themselves ceases and competition among the middle classes begins. The capitalist who is looking for workers knows very well that rising prices, due to an increasing demand for goods, raise profits and so he is prepared to pay somewhat higher wages rather than to lose his profits altogether. He throws out a sprat to catch a mackerel and if he succeeds in landing the mackerel he gladly lets the worker have the sprat. If two or more capitalists are chasing one worker then wages rise. This increase, however, is limited by the increase in the demand [for goods]. The capitalist may well be prepared to sacrifice a little of the increased profits [due to the increased demand for goods] but he takes good care not to give up any part of his "normal" profits and if that danger should arise he is careful not to pay more than "average" wages.

These considerations enable us to determine the "average" rate of wages. Conditions may be regarded as "normal" when there is no intensification of the usual competition either among the workers or among the capitalists. This state of affairs exists when exactly the right number of workers are available for employment to produce precisely the correct quantity of goods to satisfy current demands. In such a state of equilibrium wages will be a little more than the "minimum." The extent to which the level of the "average" wage is above that of the "minimum" wage depends upon the standard of living and the level of culture of the workers. If the workers are accustomed to eat meat several times a week then the capitalists have to face the fact that they must pay wages sufficiently high to make such a diet possible. The capitalist is not in a position to pay less than this, because [in a state of equilibrium] the workers are not competing among themselves and therefore cannot be compelled to reduce their standard of living. On the other hand the capitalist will not pay a higher level of wages, because [in a state of equilibrium] the capitalists are not competing amongst themselves and have no incentive to try to attract labour by offering higher wages or other special inducements.

* * *

The only difference between the old-fashioned slavery and the new is that while the former was openly acknowledged the latter is disguised. The worker *appears* to be free, because he is not bought and sold outright. He is sold piecemeal by the day, the week, or the year. Moreover he is not sold by one owner to another, but he is forced to sell himself in this fashion. He is not the slave of a single individual, but of the whole capitalist class. As far as the worker is concerned, however, there can be no doubt as to his servile status. It is true that the apparent liberty which the worker enjoys does give him some *real* freedom. Even this genuine freedom has the disadvantage that no one is responsible for providing him with food and shelter. His real masters, the middle-class capitalists, can discard him at any moment and leave him to starve, if they have no further use for his services and no further interest in his survival. These arrangements are much more favourable to the middle classes than was the old system of slavery. They can now get rid of their workers whenever it pleases them without losing any of their capital. Adam Smith has consoled the middle-class capitalists by pointing out that they secure their labour at a much cheaper rate than they would do if they were slave owners.[2]

From this it follows that Adam Smith is also quite right when he lays down the principle "that the demand for men, like that for any other commodity, necessarily regulates the production of men; quickens it when it goes on too slowly, and stops it when it advances too fast."[3] *Like that of any other commodity*—if too little labour is available, then its cost, that is to say, wages, rise, and the condition of the worker improves. In such circumstances, more workers marry and the population increases. This goes on until the working-class population has increased sufficiently to meet the demand for labour. On the other hand, if too much labour is available, its cost [i.e. wages] falls, and shortage of food, distress and starvation lead to epidemics carrying off the 'surplus population'. Malthus, who elaborated Adam Smith's observations on the subject of wages, was also correct, according to his lights, when he asserted that there was always a superfluous population in existence. Malthus argued that there were always too many people in the world. He was wrong when he expressed the view that more people existed than could be fed from available resources. The real reason for the existence of the superfluous population is the competition of the workers among themselves. This forces the individual worker to do as much work in a day as is humanly possible. If a factory owner who employs ten workers for nine hours a day finds that nine of them are willing to work for ten hours a day, then he can turn one of his workers on to the street and get the job done by the other nine. At a time when the demand for labour is not very great, the employer can force nine of his workers, on pain of dismissal, to work for an extra hour, i.e. ten hours instead of nine, for the same wages, then he can dismiss the tenth worker, and save the money formerly spent on his wages.[4]

* * *

Consequently English industry must always have a reserve of unemployed workers, except during the short period when the boom is at its height. The existence of such a reserve is essential in order that labour may be available to produce the great quantities of goods which are needed during the few months when the business boom reaches its climax. The size of this reserve varies with the state of trade. When trade is good and more hands are wanted the number of workers in the pool of unemployed declines. . . . This pool of unemployed is the "surplus population" of England. When they are out of work these people eke out a miserable existence by begging and stealing, by sweeping the streets, by collecting horse-dung, by pushing barrows or driving donkey-carts, by hawking and peddling and by turning their hands to anything that will bring in a copper or two. In all the big towns one can see many people of this type, who—as the English say—"keep body and soul together" by doing odd jobs.

* * *

According to the reports of the Poor Law Commissioners the average size of the "surplus population" in England and Wales is 1½ million.[5] The average number of unemployed in Scotland cannot be estimated because of the absence of a poor law[6].

* * *

I need hardly attempt to describe the various degrees of want and misery suffered by the unemployed during such a crisis. Poor rates are quite insufficient to deal with the situation. The charity of the wealthy is a mere drop in the ocean, the effect of which lasts only a moment. When so many are in want begging is of little avail. In times of crisis the petty shopkeepers give the unemployed credit for as long as they can, recouping themselves eventually by charging a high rate of interest on the debts. Moreover, the workers help each other in times of trouble. It is only by these means that large numbers of the surplus population manage to survive a crisis. Were it not for credit from shopkeepers and the self-help of the poor, thousands of the unemployed would die of hunger in times of trade depression.

Notes

1. A favourite expression of English industrialists.
2. Adam Smith, *An Inquiry into the Nature and Causes of the Wealth of Nations* (ed. J. R. McCulloch, 1st edn., vol. I, 1828, p. 134): "The wear and tear of a slave, it has been said, is at the expense of his master; but that of a free servant is at his own expense. The wear and tear of the latter, however, is, in reality, as much at the expense of his master as that of the former. The wages paid to journeymen and servants of every kind must be such as may enable them, one with another, to continue the race of journeymen and servants, according as the increasing, diminishing, or stationary demand of the society may happen to require. But though the wear and tear of a free servant be equally at the expense of his master, it generally costs him much less than that of a slave. The fund destined for replacing or repairing, if I may say so, the wear and tear of the slave, is commonly managed by a negligent master or careless overseer."
3. Adam Smith, op. cit., p. 133.
[4. The difference between the two cases does not seem to be very clear. It may be that in the first instance Engels assumed that the dismissal of the tenth worker is caused by competition between the workers themselves, i.e., the initiative for the dismissal of the tenth worker comes in effect from the other nine. In the second instance the initiative is taken by the employer at a time when the demand for labour is low.]
[5. The number of paupers relieved in England and Wales during the quarter ending Lady-day 1842 was 1,429,356: *Journal of Statistical Society of London*, Vol. 6 (1843), p. 256. See also T. Carlyle, *Past and Present* (1843), p. 2, fn. 1 (Everyman edn., 1912.)]
[6. This is incorrect. For a list of sources on the Scottish poor law in the nineteenth century see, for example W. O. Henderson, "The Cotton Famine in Scotland and the relief of distress": *Scottish Historical Review*, Vol., 30, no. 110, p. 157, n. 5.]

Ideas for Discussion

1. Select key passages in the text that reveal the political undercurrent.
2. How does Engels's definition of the worker compare to Weber's.
3. What recommendations does Engels make to ensure the success of English manufacturing?

4. How would you describe Engels's treatment of his topic: optimistic, realistic, pessimistic? Explain with references to the text.

Topics for Writing

1. Study the development of the textile industry in the country of your choice. Prepare a report on the role of textiles in the economic development of that country. For ideas you may wish to consult Richard Goldthwaite's essay in this volume, "The Wherewithal to Spend" (pp. 127–142).
2. Compose a proposal to a decision maker in manufacturing indicating ways in which working conditions could be improved.
3. Compare and contrast Engels's and Orwell's (pp. 36–45) views of the life of the worker in early modern Britain.
4. Develop an extended definition of "competition" using any readings from this book.

George Orwell
South Wales

George Orwell (the pseudonym of Eric Arthur Blair) was a twentieth-century novelist and writer who died in 1950. The Bengali-born graduate of Eton College, Oxford, served for five years with the Indian imperial police, beginning in 1922. In the 1930s he returned to England, where he pursued a career as a novelist. He fought with the Republicans in the Spanish civil war (1936–1939) and was wounded. During World War II he served as a correspondent, both for the BBC and for newspapers. Orwell is best known for his satirical novels, Animal Farm *(1945) and* Nineteen Eighty-Four *(1949). The description of coal miners' lives in South Wales comes from Orwell's* The Road to Wigan Pier *(1937). This excerpt reflects the power of Orwell's writing, for which he gained respect, both as a novelist and as a social critic.*

OUR CIVILISATION, *pace* Chesterton, is founded on coal, more completely than one realises until one stops to think about it. The machines that keep us alive, and the machines that make the machines, are all directly or indirectly dependent upon coal. In the metabolism of the Western world the coal-miner is second in importance only to the man who ploughs the soil. He is a sort of grimy caryatid upon whose shoulders nearly everything that is *not* grimy is supported. For this reason the actual process by which coal is extracted is well worth watching, if you get the chance and are willing to take the trouble.

When you go down a coal-mine it is important to try and get to the coal face when the "fillers" are at work. This is not easy, because when the mine is working visitors are a nuisance and are not encouraged, but if you go at any other time, it is possible to come away with a totally wrong impression. On a Sunday, for instance, a mine seems almost peaceful. The time to go there is when the machines are roaring and the air is black with coal dust, and when you can actually see what the miners have to do. At those times the place is like hell, or at any rate like my own mental picture of hell. Most of the things one imagines in hell are there—heat, noise, confusion, darkness, foul air, and, above all, unbearably cramped space. Everything except the fire, for there is no fire down there except the feeble beams of Davy lamps and electric torches which scarcely penetrate the clouds of coal dust.

When you have finally got there—and getting there is a job in itself: I will explain that in a moment—you crawl through the last line of pit props and see opposite you a shiny black wall three or four feet high. This is the coal face. Overhead is the smooth ceiling made by the rock from which the coal has been cut; underneath is the rock again, so that the gallery you are in is only as high as the ledge of coal itself, probably not much more than a yard. The first impression of all, overmastering everything else for a while, is the frightful, deafening din from the conveyor belt which carries the coal away. You cannot see very far, because the fog of coal dust throws back the beam of your lamp, but you can see on either side of you the line of half-naked kneeling men, one to every four or five yards, driving their shovels under the fallen coal and flinging it swiftly over their left shoulders. They are feeding it on to the conveyor belt, a moving rubber belt a couple of feet wide which runs a yard or two behind them. Down this belt a glittering river of coal races constantly. In a big mine it is carrying away several tons of coal every minute. It bears it off to some place in the main roads where it is shot into tubs holding half a ton, and thence dragged to the cages and hoisted to the outer air.

It is impossible to watch the "fillers" at work without feeling a pang of envy for their toughness. It is a dreadful job that they do, an almost superhuman job by the standards of an ordinary person. For they are not only shifting monstrous quantities of coal, they are also doing it in a position that doubles or trebles the work. They have got to remain kneeling all the while—they could hardly rise from their knees without hitting the ceiling—and you can easily see by trying it what a tremendous effort this means. Shovelling is comparatively easy when you are standing up, because you can use your knee and thigh to drive the shovel along; kneeling down, the whole of the strain is thrown upon your arm and belly muscles. And the other conditions do not exactly make things easier. There is the heat—it varies, but in some mines it is suffocating—and the coal dust that stuffs up your throat and nostrils and collects along your eyelids, and the unending rattle of the conveyor belt, which in that confined space is rather like the rattle of a machine gun. But the fillers look and work as though they were made of iron. They really do look like iron—hammered iron statues—under the smooth coat of coal dust which clings to them from head to foot. It is only when you see miners down the mine and naked that you realise what splendid men they are. Most of them are small (big men are at a disadvantage in that job) but nearly all of them have the most noble bodies; wide shoulders tapering to slender supple waists, and small pronounced buttocks and sinewy thighs, with not an ounce of waste flesh anywhere. In the hotter mines they wear only a pair of thin drawers, clogs and knee-pads; in the hottest mines of all, only the clogs and knee-pads. You can hardly tell by the look of them whether they are young or old. They may be any age up to sixty or even sixty-five, but when they are black and naked they all look alike. No one could do their work who had not a young man's body, and a figure fit for a guardsman at that; just a few pounds of extra flesh on the waist-line, and the constant bending would be impossible. You can never forget that spectacle once you have seen it—the line of bowed, kneeling figures, sooty black all over, driving their huge shovels under the coal with stupendous force and speed. They are on the job for seven and a half hours, theoretically without a break, for there is no time "off." Actually they snatch a quarter of an hour or so at some time during the shift to eat the food they have brought with them, usually a hunk of bread and dripping and a bottle of cold tea. The first time I was watching the "fillers" at work I put my hand upon some dreadful slimy thing among the coal dust. It was a chewed quid of tobacco. Nearly all the miners chew tobacco, which is said to be good against thirst.

Probably you have to go down several coal-mines before you can get much grasp of the processes that are going on round you. This is chiefly because the mere effort of getting from place to place makes it difficult to notice anything else. In some ways it is even disappointing, or at least is unlike what you have expected. You get into the cage, which is a steel box about as wide as a telephone box and two or three times as long. It holds ten men, but they pack it like pilchards in a tin, and a tall man cannot stand upright in it. The steel door shuts upon you, and somebody working the winding gear above drops you into the void. You have the usual momentary qualm in your belly and a bursting sensation in the ears, but not much sensation of movement till you get near the bottom, when the cage slows down so abruptly that you could swear it is going upwards again. In the middle of the run the cage probably touches sixty miles an hour; in some of the deeper mines it touches even more. When you crawl out at the bottom you are perhaps four hundred yards under ground. That is to say you have a tolerable-sized mountain on top of you; hundreds of yards of solid rock, bones of extinct beasts, subsoil, flints, roots of growing things, green grass and cows grazing on it—all this suspended over your head and held back only by wooden props as thick as the calf of your leg. But because of the speed at which the cage has brought you down, and the complete blackness through which you have travelled, you hardly feel yourself deeper down than you would at the bottom of the Piccadilly tube.

What *is* surprising, on the other hand, is the immense horizontal distances that have to be travelled underground. Before I had been down a mine I had vaguely imagined the miner stepping out of the cage and getting to work on a ledge of coal a few yards away. I had not realised that before he even gets to his work he may have to creep through passages as long as from London Bridge to Oxford Circus. In the beginning, of course, a mine shaft is sunk somewhere near a seam of coal. But as that seam is worked out and fresh seams are followed up, the workings get further and further from the pit bottom. If it is a mile from the pit bottom to the coal face, that is probably an average distance; three miles is a fairly normal one; there are even said to be a few mines where it is as much as five miles.

But these distances bear no relation to distances above ground. For in all that mile or three miles as it may be, there is hardly anywhere outside the main road, and not many places even there, where a man can stand upright.

You do not notice the effect of this till you have gone a few hundred yards. You start off, stooping slightly, down the dim-lit gallery, eight or ten feet wide and about five high, with the walls built up with slabs of shale, like the stone walls in Derbyshire. Every yard or two there are wooden props holding up the beams and girders; some of the girders have buckled into fantastic curves under which you have to duck. Usually it is bad going underfoot—thick dust or jagged chunks of shale, and in some mines where there is water it is as mucky as a farmyard. Also there is the track for the coal tubs, like a miniature railway track with sleepers a foot or two apart, which is tiresome to walk on. Everything is grey with shale dust; there is a dusty fiery smell which seems to be the same in all mines. You see mysterious machines of which you never learn the purpose, and bundles of tools slung together on wires, and sometimes mice darting away from the beam of the lamps. They are surprisingly common, especially in mines where there are or have been horses. It would be interesting to know how they got there in the first place; possibly by falling down the shaft—for they say a mouse can fall any distance uninjured, owing to its surface area being so large relative to its weight. You press yourself against the wall to make way for lines of tubs jolting slowly towards the shaft, drawn by an endless steel cable operated from the surface. You creep through sacking curtains and thick wooden doors which, when they are opened, let out fierce blasts of air. These doors are an important part of the ventilation system. The exhausted air is sucked out of one shaft by means of fans, and the fresh air enters the other of its own accord. But if left to itself the air will take the shortest way round, leaving the deeper workings unventilated; so all short cuts have to be partitioned off.

At the start to walk stooping is rather a joke, but it is a joke that soon wears off. I am handicapped by being exceptionally tall, but when the roof falls to four feet or less it is a tough job for anybody except a dwarf or a child. You have not only got to bend double, you have also got to keep your head up all the while so as to see the beams and girders and dodge them when they come. You have, therefore, a constant crick in the neck, but this is nothing to the pain in your knees and thighs. After half a mile it becomes (I am not exaggerating) an unbearable agony. You begin to wonder whether you will ever get to the end—still more, how on earth you are going to get back. Your pace grows slower and slower. You come to a stretch of a couple of hundred yards where it is all exceptionally low and you have to work yourself along in a squatting position. Then suddenly the roof opens out to a mysterious height—scene of an old fall of rock, probably—and for twenty whole yards you can stand upright. The relief is overwhelming. But after this there is another low stretch of a hundred yards and then a succession of beams which you have to crawl under. You go down on all fours; even this is a relief after the squatting business. But when you come to the end of the beams and try to get up again, you find that your knees have temporarily struck work and refuse to lift you. You call a halt, ignominiously, and say that you would like to rest for a minute or two. Your guide (a miner) is sympathetic. He knows that your muscles are not the same as his. "Only another four hundred yards," he says encouragingly; you feel that he might as well say another four hundred miles. But finally you do somehow creep as far as the coal face. You have gone a mile and taken the best part of an hour; a miner would do it in not much more than twenty minutes. Having got there, you have to sprawl in the coal dust and get your strength back for several minutes before you can even watch the work in progress with any kind of intelligence.

Coming back is worse than going, not only because you are already tired out but because the journey back to the shaft is probably slightly up hill. You get through the low places at the speed of a tortoise, and you have no shame now about calling a halt when your knees give way. Even the lamp you are carrying becomes a nuisance and probably when you stumble you drop it; whereupon, if it is a Davy lamp, it goes out. Ducking the beams becomes more and more of an effort, and sometimes you forget to duck. You try walking head down as the miners do, and then you bang your backbone. Even the miners bang their backbones fairly often. This is the reason why in very hot mines, where it is necessary to go about half naked, most of the miners have what they call "buttons down the back"—that is, a permanent scab on each vertebra. When the track is down hill the miners sometimes fit their clogs, which are hollow underneath, on to the trolley rails and slide down. In mines where the "travelling" is very bad all the miners carry sticks about two and a half feet long, hollowed out below the handle. In normal places you keep your hand on top of the stick and in the low places you slide your hand down into the hollow. These sticks are a great help, and the wooden crash-helmets—a comparatively recent invention—are a godsend. They look like a French or Italian steel helmet, but they are made of some kind of pith and very light, and so strong that you can take a violent blow on the head without feeling it. When finally you get back to the surface you have been perhaps three hours underground and travelled two miles, and you are more exhausted than you would be by a twenty-five-mile walk above ground. For a week afterwards your thighs are so stiff that coming downstairs is quite a difficult feat; you have to work your way down in a peculiar sidelong manner, without bending the knees. Your miner friends notice the stiffness of your walk and chaff you about it. ("How'd ta like to work down pit, eh?" etc.) Yet even a miner who has been long away from work—from illness, for instance—when he comes back to the pit, suffers badly for the first few days.

It may seem that I am exaggerating, though no one who has been down an old-fashioned pit (most of the pits in England are old-fashioned) and actually gone as far as the coal face, is likely to say so. But what I want to emphasise is this. Here is this frightful business of crawling to and fro, which to any normal person is a hard day's work in itself; and it is not part of the miner's work at all, it is merely an extra, like the City man's daily ride in the Tube. The miner does that journey to and fro, and sandwiched in between there are seven and a half hours of savage work. I have never travelled much more than a mile to the coal face; but often it is three miles, in which case I and most people other than coal-miners would never get there at all. This is the kind of point that one is always liable to miss. When you think of a coal-mine you think of depth, heat, darkness, blackened figures hacking at walls of coal; you don't think, necessarily, of those miles of creeping to and fro. There is the question of time, also. A miner's working shift of seven and a half hours does not sound very long, but one has got to add on to it at least an hour a day for "travelling," more often two hours and sometimes three. Of course, the "travelling" is not technically work and the miner is not paid for it; but it is as like work as makes no difference. It is easy to say that miners don't mind all this. Certainly, it is not the same for them as it would be for you or me. They have done it since childhood, they have the right muscles hardened, and they can move to and fro underground with a startling and rather horrible agility. A miner puts his head down and *runs,* with a long swinging stride, through places where I can only stagger. At the workings you see them on all fours, skipping round the pit props almost like dogs. But it is quite a mistake to think that they enjoy it. I have talked about this to scores of miners and they all admit that the "travelling" is hard work; in any case when you hear them discussing a pit among themselves the "travelling" is always one of the things they discuss. It is said that a shift always returns from work faster than it goes; nevertheless the miners all say that it is the coming away, after a hard day's work, that is especially irksome. It is part of their work and they are equal to it, but certainly it is an effort. It is comparable, perhaps, to climbing a smallish mountain before and after your day's work.

When you have been down two or three pits you begin to get some grasp of the processes that are going on underground. (I ought to say, by the way, that I know nothing whatever about the technical side of mining: I am merely describing what I have seen.) Coal lies in thin seams between enormous layers of rock, so that essentially the process of getting it out is like scooping the central layer from a Neapolitan ice. In the old days the miners used to cut straight into the coal with pick and crowbar—a very slow job because coal, when lying in its virgin state, is almost as hard as rock. Nowadays the preliminary work is done by an electrically-driven coal-cutter, which in principle is an immensely tough and powerful band-saw, running horizontally instead of vertically, with teeth a couple of inches long and half an inch or an inch thick. It can move backwards or forwards on its own power, and the men operating it can rotate it this way and that. Incidentally it makes one of the most awful noises I have ever heard, and sends forth clouds of coal dust which make it impossible to see more than two or three feet and almost impossible to breathe. The machine travels along the coal face cutting into the base of the coal and undermining it to the depth of five feet or five feet and a half; after this it is comparatively easy to extract the coal to the depth to which it has been undermined. Where it is "difficult getting," however, it has also to be loosened with explosives. A man with an electric drill, like a rather smaller version of the drills used in street-mending, bores holes at intervals in the coal, inserts blasting powder, plugs it with clay, goes round the corner if there is one handy (he is supposed to retire to twenty-five yards distance) and touches off the charge with an electric current. This is not intended to bring the coal out, only to loosen it. Occasionally, of course, the charge is too powerful, and then it not only brings the coal out but brings the roof down as well.

After the blasting has been done the "fillers" can tumble the coal out, break it up and shovel it on to the conveyor belt. It comes out at first in monstrous boulders which may weigh anything up to twenty tons. The conveyor belt shoots it on to tubs, and the tubs are shoved into the main road and hitched on to an endlessly revolving steel cable which drags them to the cage. Then they are hoisted, and at the surface the coal is sorted by being run over screens, and if necessary is washed as well. As far as possible the "dirt"—the shale, that is—is used for making the roads below. All that cannot be used is sent to the surface and dumped; hence the monstrous "dirt-heaps," like hideous grey mountains, which are the characteristic scenery of the coal areas. When the coal has been extracted to the depth to which the machine has cut, the coal face has advanced by five feet. Fresh props are put in to hold up the newly exposed roof, and during the next shift the conveyor belt is taken to pieces, moved five feet forward and reassembled. As far as possible the three operations of cutting, blasting and extraction are done in three separate shifts, the cutting in the afternoon, the blasting at night (there is a law, not always kept, that forbids its being done when there are other men working near by), and the "filling" in the morning shift, which lasts from six in the morning until half past one.

Even when you watch the process of coal-extraction you probably only watch it for a short time, and it is not until you begin making a few calculations that you realise what a stupendous task the "fillers" are performing. Normally each man has to clear a space four or five yards wide. The cutter has undermined the coal to the depth of five feet, so that if the seam of coal is three or four feet high, each man has to cut out, break up and load on to the belt something between seven and twelve cubic yards of coal. This is to say, taking a cubic yard as weighing twenty-seven hundred-weight, that each man is shifting coal at a speed approaching two tons an hour. I have just enough experience of pick and shovel work to be able to grasp what this means. When I am digging trenches in my garden, if I shift two tons of earth during the afternoon, I feel that I have earned my tea. But earth is tractable stuff compared with coal, and I don't have to work kneeling down, a thousand feet underground, in suffocating heat and swallowing coal dust with every breath I take; nor do I have to walk a mile bent double before I begin. The miner's job would be as much beyond my power as it would be to perform on the flying trapeze or to win the Grand National. I am not a manual labourer and please God I never shall be one, but there are some kinds of manual work that I could do if I had to. At a pitch I could be a tolerable road-sweeper or an inefficient gardener or even a tenth-rate farm hand. But by no conceivable amount of effort or training could I become a coal-miner; the work would kill me in a few weeks.

Watching coal-miners at work, you realise momentarily what different universes different people inhabit. Down there where coal is dug it is a sort of world apart which one can quite easily go through life without ever hearing about. Probably a majority of people would even prefer not to hear about it. Yet it is the absolutely necessary counterpart of our world above. Practically everything we do, from eating an ice to crossing the Atlantic, and from baking a loaf to writing a novel, involves the use of coal, directly or indirectly. For all the arts of peace coal is needed; if war breaks out it is needed all the more. In time of revolution the miner must go on working or the revolution must stop, for revolution as much as reaction needs coal. Whatever may be happening on the surface, the hacking and shovelling have got to continue without a pause, or at any rate without pausing for more than a few weeks at the most. In order that Hitler may march the goosestep, that the Pope may denounce Bolshevism, that the cricket crowds may assemble at Lord's, that the Nancy poets may scratch one another's backs, coal has got to be forthcoming. But on the whole we are not aware of it; we all know that we "must have coal," but we seldom or never remember what coal-getting involves. Here am I, sitting writing in front of my comfortable coal fire. It is April but I still need a fire. Once a fortnight the coal cart drives up to the door and men in leather jerkins carry the coal indoors in stout sacks smelling of tar and shoot it clanking into the coal-hole under the stairs. It is only very rarely, when I make a definite mental effort, that I connect this coal with that far-off labour in the mines. It is just "coal"—something that I have got to have; black stuff that arrives mysteriously from nowhere in particular, like manna except that you have to pay for it. You could quite easily drive a car right across the north of England and never once remember that hundreds of feet below the road you are on the miners are hacking at the coal. Yet in a sense it is the miners who are driving your car forward. Their lamp-lit world down there is as necessary to the daylight world above as the root is to the flower.

It is not long since conditions in the mines were worse than they are now. There are still living a few very old women who in their youth have worked underground, with a harness round their waists and a chain that passed between their legs, crawling on all fours and dragging tubs of coal. They used to go on doing this even when they were pregnant. And even now, if coal could not be produced without pregnant women dragging it to and fro, I fancy we should let them do it rather than deprive ourselves of coal. But most of the time, of course, we should prefer to forget that they were doing it. It is so with all types of manual work; it keeps us alive, and we are oblivious of its existence. More than anyone else, perhaps, the miner can stand as the type of the manual worker, not only because his work is so exaggeratedly awful, but also because it is so vitally necessary and yet so remote from our experience, so invisible, as it were, that we are capable of forgetting it as we forget the blood in our veins. In a way it is even humiliating to watch coal-miners working. It raises in you a momentary doubt about your own status as an "intellectual" and a superior person generally. For it is brought home to you, at least while you are watching, that it is only because miners sweat their guts out that superior persons can remain superior. You and I and the editor of the *Times Lit. Supp.*, and the Nancy poets and the Archbishop of Canterbury and Comrade X, author of *Marxism for Infants*—all of us *really* owe the comparative decency of our lives to poor drudges underground, blackened to the eyes, with their throats full of coal dust, driving their shovels forward with arms and belly muscles of steel.

Ideas for Discussion

1. What is the dominating feature of Orwell's prose style? Support your choice with examples from the text.
2. How does the first-person narration and the address to "you" affect your response to Orwell's piece?
3. Do you sense a political purpose imbedded in Orwell's topics and treatments?

Topics for Writing

1. Profile a day in the life of the worker of your choice. Follow Orwell's style as closely as you can, and write a letter to the worker showing your observations.
2. Look up the laws pertaining to coal mining in Britain or America. Write a report that either summarizes the accomplishments of legislators in improving the working conditions of miners or that recommends ways in which the working conditions can be improved.
3. Write a job description for a coal miner based on Orwell's chapter.
4. Write a proposal for improving working conditions in the mining industry.
5. Write a memo soliciting the report in 2 above to the research department of your firm.

B. L. Hutchins
The Working Life of Women

B. L. Hutchins contributed this essay on women and factory work as a Fabian Society Tract in June 1911. She argues for the fair treatment of women in the workplace and singles out for special comment widows and unmarried women who must support themselves. The Fabian Society was founded in 1884 to promote socialist causes, and it was staunchly opposed to Marxism and violence. Today's British Labour Party grew out of the efforts of such leading Fabians as George Bernard Shaw and Beatrice and Sidney Webb. The text offered here is taken from Sally Alexander's Women's Fabian Tracts *(1988). As you read this essay, it might be useful to consider how Hutchins's style reflects that of Weber, Engels, and Orwell. See if you can list the characteristics of sound writing about social issues, political philosophy, and commercial theories.*

IT IS STILL THE CUSTOM in some quarters to assert that "the proper sphere for women is the home," and to assume that a decree of Providence or a natural law has marked off and separated the duties of men and women. Man, it is said, is the economic support and protector of the family, woman is its watchful guardian and nurse; whence it follows that the wife must be maintained by her husband in order to give her whole time to home and children. The present paper does not attempt to discuss what is in theory the highest life for women; whether the majority of women can ever realize their fullest life outside the family, or whether an intelligent wife and mother has not on the whole, other things equal, more scope for the development of her personality than any single woman can possibly have. The question I am here concerned with relates to the actual position of the women themselves. Is it the lot of all women, or even of a large majority of women, to have their material needs provided for them so that they can reserve themselves for the duties that tend to conserve the home and family?

Let us see what the Census has to tell us on the subject. We find that in 1901 there were in round numbers 15,729,000 men and boys, and 16,799,000 women and girls, in England and Wales. This means that there are 1,070,000 more women than men, and if we omit all children under fifteen there are about 110 women to every 100 men. This surplus of women has increased slowly but steadily in every Census since 1841; that is to say, in 1841 there were in every 1,000 persons 489 males, and 511 females; but in 1901 there were in every 1,000, 484 males, and 516 females.

The disproportionate numbers of women are no doubt partly due to the Imperial needs which compel a large number of men to emigrate to our actual or potential colonies and dependencies. It is impossible to say how many are thus to be accounted for, probably not a very large proportion, save in the upper classes. The Census shows figures for the army, navy, and merchant seamen serving abroad, but if these are added to the population of the United Kingdom the excess of women is still considerable. There seems to be no means of estimating the numbers of men who are absent on private business.

The main cause of the surplus of women seems to be their lower death-rate, and this is popularly accounted for as the advantage resulting to women from their comparatively sheltered life and less exposure to accident and occupational disease. This assumption no doubt accounts for some part of the difference; women do not work on railways or as general laborers, or usually in the most unhealthy processes of trades scheduled as "Dangerous" under the Factory Act. There can be no doubt either that the death-rate of women has been lowered by the operation of the Factory Act in improving conditions of employment. The death-rate of men has also been lowered, but in a less degree, because although men benefit by improved conditions in the factory just as women do, the proportion of men employed in factories and workshops is small comparatively with women, so many men being employed in transport, building, laboring, docks, etc. These latter occupations so far have obtained very little legal protection from the risks and dangers run by the workers, although many of these dangers are notoriously preventible.

Still it is doubtful whether the lower death-rate of women can be entirely accounted for by the greater degree of protection enjoyed. Women often work longer hours even under the Factory Act than most men do under their trade union; much of the work done by women in laundries, jam factories, sack factories, and others, is extremely laborious. Again, the enormous amount of domestic work accomplished by women in their homes, without outside help, in addition to the bearing and caring for infants and young children, must be equal in output of energy to much more than all the industrial work of women, especially when the rough, inconvenient, and inadequate nature of the appliances common in working-class homes is considered, and the still more painful fact is remembered that the very person responsible for all this work is often the one of the family who in case of need is the first to go short of food.

It is true that more men than women die of accidents. But let us add to the accidental deaths the deaths of women from childbirth and other causes peculiar to women. We find that in 1907 10,895 males died from accidents; 4,890 females died from accidents; 4,670 from causes peculiar to women, 9,560 altogether, about 1,300 less than men. But the total deaths of men in

1907 exceeded the deaths of women by 14,297, an excess more than ten times as great.

There is also the question of age, which is important in connection with the death-rate. The number of boys born is larger than the number of girls, about 104 to 100. The death-rate of boy babies is almost always higher than that of girls, and in 1907 the death-rate of boys under four was higher than that of girls, but the death-rate of boys from four to fifteen was lower than that of girls at the same age; then at fifteen the male death-rate again rises above the female and remains higher at all later ages.

Now if the lower death-rate of girls and women is due to their being taken more care of, how inexplicable are these figures. There is little enough difference in the care and shelter given to boys and girls under four, yet the boys die much faster; between four and fifteen, on the other hand, girls usually are a good deal more sheltered and protected than boys, and less likely to run into dangerous places and positions, yet from four to fifteen the male death-rate is slightly lower than the female. At fifteen when, as we shall see, a very large proportion of girls begin industrial work, the death-rates are again reversed, the male death-rate being thenceforward the higher. Nor does it appear that the death-rate of young women is much influenced by the fact of industrial employment. It is true that in Lancashire, where many women and girls work, the death-rate of women is higher than in England and Wales; but in Durham, where comparatively very few women and girls are employed, the death-rate is higher still.

The contrast seems to indicate that it is not the fact of employment, but the conditions, both of life and employment, that are prejudicial to women in these industrial centres, for although death-rates have generally fallen, they are still higher in most of the mining and manufacturing districts, notably in Lancashire and Durham, than the average of England and Wales.

It will be agreed that the greater average duration of life among women is sufficient to account for a large excess number of women over men, over and above the emigration of many young men, which contributes to the same result. The surplus of women is distributed very differently in different districts: it is greater in London and the Home Counties, and also in Lancashire; less in the mining districts and the rural districts; and generally much greater in town than country. In the urban districts women over

Death Rates, 1907, Per 1,000 Living

	Under 1 year per 1,000 births	aged 1	2	3	4	under 5	5	10
Males	130	38.4	15.5	10.1	6.9	44.8	3.3	1.9
Females	104	36.2	14.8	9.7	7.6	37.0	3.4	2.0
	15	20	25	35	45	55	65	all ages
Males	2.9	3.8	5.6	9.5	16.9	33.7	94.1	16.0
Females	2.7	3.2	4.6	7.8	13.1	26.0	85.9	14.1

Percentage of Females Occupied

	LANCASHIRE			DURHAM		
Ages	15	20	25–34	15	20	25–34
Single	78	80	76	40	49	49
Married or widowed	24	25	19	1	2	3
Death-rates, 1907– Male	3.3	4.2	6.1	3.8	4.7	5.6
Female	3.0	3.5	5.4	3.7	4.4	6.3

fifteen number 112, in the rural districts only 102, to every 100 males. This is perhaps partly due to the girls going to towns as domestic servants; for although the percentage of domestic servants is rather higher in the country than in town, the actual numbers are much less, and particular towns and residential urban districts—Bournemouth, Hampstead, and the like—show a very high percentage of servants. But the higher proportion of males in the country must in part be due to the fact that babies born in the country have a better chance of life. Although the number of boys born is greater than the number of girls (it was about 1,037 to 1,000 in 1891–1900, and slightly higher since 1901), the boy babies are on the average more difficult to bring into the world and more delicate for the first few years of life, as is shown by the male infant death-rate being higher than the female. It follows that though boy babies are more numerous at the outset, the girls steadily gain upon them, and at some point in early life the numbers are equal. If infant mortality is high, the surplus boy babies are very soon swept out of existence, and there may be "superfluous women" even under five years old! But in healthy districts, especially in the country, where infant mortality is low, the boys survive in greater numbers, and exceed the girls in numbers up to the age of twenty; thus in later life the disproportion of women is not so great in the country as it is in towns. This fact constitutes one important reason (among others that are better known) for improving the sanitary conditions in towns. A diminution in infant mortality will tend to keep a larger proportion of boys alive, and thus by so much redress the balance of the sexes. To give an instance: in rural districts of Lancashire the boys under five were 1,018 to every 1,000 girls; in the urban districts, which include many towns with a high infant mortality, the boys under five were only 989 to every 1,000 girls. It is impossible here to give many details on this point, but fuller statistics are given in the *Statistical Journal*, June 1909, pp. 211–212.

Marriage and Widowhood

But it is evident that one way or another we must face the fact of a large excess number of women, even though we may hope that improvement in the people's life and health may prevent some of the waste of men and boys' life that occurs at present. How are women provided for? Marriage is still the most important and extensively followed occupation for women. Over 5,700,000 women in England and Wales are married, or 49.6 per cent.; nearly one-half of the female population over fifteen.

```
In every 100 women aged 15-20 ......... 2 are married.
      "         "     "   20-25 ......... 27  "
      "         "     "   25-35 ......... 64  "
      "         "     "   35-45 ......... 75  "
      "         "     "   45-55 ......... 71  "
      "         "     "   55-65 ......... 57  "
      "         "     "   65-75 ......... 37  "
      "         "     "   75    ......... 16  "
```

In middle life—from thirty-five to fifty-five—three-fourths of the women are married. In early life a large proportion are single; in later life a large proportion are widowed. Put it in another way. From twenty to thirty-five, only two out of every four women are married, most of the others being still single; from thirty-five to fifty-five, three in every four women are married; over fifty-five, less than two in every four are married, most of the others being already widowed. It is only for twenty years (between thirty-five and fifty-five) that as many as three-fourths of women can be said to be provided for by marriage, even on the assumption that all wives are provided for by their husbands.

As we have seen, women exceed men in numbers, and not only that, but the age of marriage is usually for economic reasons later for men than women, and some men do not marry at all, consequently it is utterly vain to assume that women *generally* can look to marriage for support, and to talk of the home as "women's true sphere." Mrs. Butler wrote, now many years ago, that, like Pharaoh who commanded the Israelites to make bricks without straw, "these moralisers command this multitude of enquiring women back to homes which are not, and which they have not the material to create." Although about three-fourths of the women in the country do get married some time or other, at any given time fully half the women over fifteen are either single or widowed. Women marry younger and live longer than men, consequently the proportion of widows is considerable, something like one woman in every eight over twenty years old. The largest proportion occurs, as might be expected, at advanced years.

```
In every 100 women aged 35-45 ......... 6 are widows.
      "         "     "   45-55 ......... 16  "
      "         "     "   55-65 ......... 31  '
      "         "     "   65-75 ......... 52  "
      "         "     "   75    ......... 73  "
```

Occupation

The number of women and girls over fifteen returned in 1901 as occupied was 3,970,000, or 34.5. This figure can only be regarded as an approximate one, as there is little information to show how many of the numerous women who work occasionally, but not regularly, do or do not return themselves as occupied, and even if this information were forthcoming, it is difficult to see how any precise line of demarcation could be devised to distinguish the degree of regularity that should constitute an "occupied" woman. The figure is again obviously inadequate in regard to women's *work* (as distinguished from occupation), as no account is taken of the enormous amount of work done at home—cooking, washing, cleaning, mending and making of clothes, tendance of children, and nursing the sick done by women, especially in the working class, who are not returned as belonging to any specific occupation.

It is misleading, however, to take the percentage 34.5 as if it meant that about one-third of all women enter upon a trade or occupation.

In every 100 women aged 15 .... 66 are occupied.
" " " 20 .... 56 "
" " " 25 .... 31 "
" " " 35 .... 23 "
" " " 45 .... 22 "
" " " 55 .... 21 "
" " " 65 .... 16 "
" " " 75 .... 7 "

These figures show what is a very important point to remember, viz., that the majority of women workers are quite young, and this is one great difference in the work of men and women. The Census shows that over 90 per cent. of the men are occupied till fifty-five, and 89 per cent. even from fifty-five to sixty-five. But for women, especially in the industrial classes, the case is different. Their employment is largely an episode of early life. The majority of young working women work for a few years and leave work at marriage, as is shown by the rapid fall in the percentage occupied from the age of twenty-five. It is often stated by social investigators that the prospect of marriage makes working girls slack about trade unions, and indifferent about training. Many girls seem for this reason to fail in some degree to realize their full possibilities or to achieve their full industrial efficiency. In the case of those who do marry, and whose best years will be given to work socially far more important than the episodic employment carried on by them in mill, factory or workroom, this alleged lack of industrial efficiency is not perhaps of much consequence. But although a large proportion of women are married before thirty-five, and as we know, the proportion married is greater in the working classes than among the middle and upper classes, yet it is a mistake to suppose that the mature single woman in industry is so rare as to be a negligible quantity. There are, for instance, nearly a quarter of a million single occupied women between thirty-five and forty-four. They include 88,000 domestic servants, 32,500 professional women (teachers, doctors, etc.), 30,000 textile workers, and 40,000 workers in making clothes and dress. These figures show that self dependence is a necessity for many even at the age when, and in the class where marriage is most frequent. The importance to the single self-supporting woman of a skilled occupation which she can pursue with self-respect and for which she can be decently remunerated, need hardly be emphasized here.

Married and Widowed Women Occupied

The proportion of married or widowed women who are occupied is about 13 per cent., but, unlike the single women, whose percentage of occupation steadily falls as age increases, the percentage of married or widowed occupied is low at first, highest between thirty-five and fifty-five, and then falls to old age.

In every hundred married or widowed women occupied, six are under twenty-five; forty-four are between twenty-five and forty-five; forty are between forty-five and sixty-five; ten are over sixty-five.

The figures in our Census unfortunately do not separate the married or widowed occupied, so it is difficult to estimate from the above figures what proportion falls in to either class, but there can be little doubt that the high percentage of middle-aged women is due to widowhood. Frau Elizabeth Gnauck-Kühne, who has made a very able study of the life and work of German women,[1] tells us that in Germany, of married women only 12 per cent. are occupied, of widowed women as many as 44 per cent. The proportion of occupied widows is probably lower with us, as we have much less small farming, which in Germany is often carried on by women after the husbands' death; but there can be little doubt that the proportion of widows working is higher than the proportion married. In a very interesting passage Madame Gnauck points out the peculiar handicap suffered by a woman who is thus forced to renew industrial activity in middle life. The industrial life of women, she writes, is not continuous, but is split in two. Woman is normally provided for by marriage, let us say, for twenty or thirty years. But marriage is not a life-long provision for the average woman, it is only a provision for the best years of life, those years, in fact, in which a woman is ordinarily most capable of taking care of herself. The husband is, in many cases, swept off in middle life, and in the industrial classes he has usually not had very much chance of saving a competence for his widow. A certain proportion of women, therefore, we cannot say exactly how many, are forced to re-enter the labor market by widowhood, or by other economic causes—illness of the husband, desertion, and so on. Once more the woman appears in the industrial arena, with all the disadvantage of a long period of intermitted employment and loss of industrial experience. Having lost the habit of industrial work, having very usually children to look after and a home to find, she has to compete with girls and young women for wages based on the standard of life of a single unencumbered woman. It may be that the inferior technical skill often attributed to women as compared with men is largely due to this fact, that while a man gives his best years to his work, a woman gives precisely those years to other work, and therefore returns to industry under a considerable handicap. We can hardly doubt that this is a chief cause of pauperism.

The late Mr. Kirkman Gray, in his interesting unfinished work, "Philanthropy and the State," wrote:—"The theory is that the male can earn enough for a family and the female enough for herself. But this theory, even if we accept it as correct, makes no allowance for the fact that every eighth woman is a widow. Here then is the bitter anomaly of the widow's position in the economic sphere. As head of a family, she ought to be able to earn a family wage; as woman she can only gain the customary price of individual subsistence." The Minority Report of the Poor Law Commission recognizes the same anomaly. "It is to the man that is paid the income necessary for the support of the family, on the assumption that the work of the woman is to care for the home and the children. The result is that mothers of young children, if they seek industrial employment, do so under the double disadvantage that the woman's wage is fixed to maintain herself alone, and that even this can be earned only by giving up to work the time that is needed by the care of the children."

Even the Charity Organization Society, which usually inclines to ignore the social aspect of economic hardship and treat every case as merely individual, is forced to recognize the anomaly of the widow's position. "We must look the poor woman's troubles in the face. . . . She has to do the work of two people; she has to be the breadwinner and go out to work, and she must also be the housekeeper. She has to wash, clean, and cook, make and mend clothes, care for and train her children. Can one pair of hands manage all this? And, secondly, when she goes out to work our poor widow will probably only earn low wages . . . about 10s. a week, and she will certainly not be able to support herself and her family on that."[2]

The reflection here occurs that the life of women is inseparably connected with the life of men, and we may well pause to ask whether it is necessary so large a proportion of women should be widows at all. There is an excellent saying, that "we can have as many paupers as we like to pay for." It has an intimate bearing on the toleration of preventible disease and accidents as well as on administrative laxity in the Poor Law. The comparative mortality figure for the general laborer is more than double that of occupied males generally, and it is true the Registrar-General ascribes some of this mortality to confused returns, but even if some allowance, say 25 per cent., be made on this ground, the excess is still great. A pamphlet by Mr. Brockelbank[3] shows that in 1907 one shunter in thirteen was killed or injured at his work on the railway. The same writer gives reasons for supposing that the published returns of fatal accidents to railway servants fall far short of the truth, only those accidents which cause death within twenty-four hours being reported as fatal.

Many other occupations have a deplorably high death-rate, and it would seem that there is still a good deal to be done in improving the conditions of those workers who are not under the Factory Acts or protected by any effective organization. The protection of women by factory regulation has gone on the lines of protecting the individual woman worker at her work. Surely protection is also needed for the woman at home who sees her husband go off daily to some dangerous trade, where, for want of the necessary technical means for the prevention of disease or accident, he may be killed, maimed, or incur disease, and she and her children be left desolate.

It is notorious that a great deal of industrial disease and many accidents are due to causes largely preventible and within control. A very interesting report was issued last year in regard to dangers in building operations, which affect a large number of men—over a million. The report states that laborers are the principal sufferers from accidents, and have the most dangerous part of the work to do. One trade union secretary stated that 9 per cent. of his members had accidents in 1905. On this scale in eleven years each member would have an accident. Another union official said that a large number of accidents were preventible, and asked for more Government inspection. An employer stated that accidents were, in his belief, largely due to the lack of competent foremen and skilled supervision; he had only had three accidents in thirty years' experience, and attributed this immunity to his engagement of a really competent man. He thought the building trade got into bad odor with the public owing to the tendency to save in wages and put incompetent men to work that needs really expert supervision. Another witness complained that accidents were caused by putting unskilled men to skilled work for the sake of cheapness.

Dr. Young stated before the Physical Deterioration Committee in 1903 that factories contributed to the spread of phthisis, and that he considered that while a great deal had been done to combat the special dangers and diseases incidental to special trades in general industrial conditions, a great deal remained to be done, and legislative interference had by no means reached its limit. From the Registrar-General's report we find that very high rates of phthisis occur among men in early manhood and middle life. In 1891–1900 of the total deaths among men twenty-five to thirty-five, nearly one half were due to phthisis and respiratory diseases. The comparative mortality figure for certain occupations in 1900–02 was as follows:—

	Phthisis	Other Respiratory Diseases
All occupied males	175	78
All occupied males in agricultural districts	125	38
Tin miners	838	653
General laborers	567	268
General laborers (industrial districts)	450	171
File makers	375	173
Lead miners	317	187
Dock laborers	291	161

It is in the light of such figures as these, it seems to me, that we have to study the problem of married or widowed women's work and the pauperism of able-bodied widows and their children. As women become better instructed, better organized, able to take more interest in politics, and especially when they obtain the Parliamentary franchise, it is to be hoped that they will agitate for drastic legislation and stringent inspection in the industries carried on by men and unregulated by Factory Law.

In the mining and industrial counties the death-rate is markedly above that of England and Wales as a whole, and it is somewhat curious that while a great deal of attention has been given to the infant mortality of Lancashire, which is usually explained as being due to married women's employment, much less notice has been taken of the fact that the *corrected* death-rate of Lancashire is even more above the average than is the mortality of infants.[4] In 1907, which was an exceptionally healthy year, the death-rates of Lancashire, though diminished, showed themselves still conspicuously above the average; which can be most simply shown by taking the death-rate for the whole country as 100.

Comparative Death-rate, 1907

	General Death-Rate, corrected for age-constitution		
	Infants	Male	Female
England and Wales	100	100	100
Lancashire	117	124	126

A large part of this excess mortality, which is not by any means peculiar to Lancashire but can be paralleled in some mining districts and exceeded in the Potteries, is made up of deaths from phthisis and respiratory diseases, which are now considered to be largely traceable to unhealthy conditions of houses and work places, and in very great measure preventible. It is impossible in the limits of this paper to give full statistics, but those who desire further information are referred to the Reports of the Registrar-General, especially the two parts of the Decennial Supplement, published in 1907 and 1908 respectively, which are an invaluable mine of facts and figures, and also to the *Statistical Journal (loc. cit.)*.

The Woman's Handicap

It is not very easy to summarize briefly the facts of woman's life and employment, which demand a treatment much fuller than is possible within our limits. But there are several points which seem to be of special importance. First, there is the curious fact that women, though physically weaker than men, seem to have a greater stability of nerves, a greater power of resistance to disease, and a stronger hold of life altogether. It is notorious that there are more male lunatics, and very many more male criminals than female, and much fewer women die from alcoholism, nervous diseases, suicide, and various complaints that indicate mental and physical instability, while more women than men die of old age. On the other hand, there are more female paupers and more female old-age pensioners than male, and these facts seem to indicate that women on the whole are handicapped rather by their economic position than by physical disability. We have seen that in this country women are more numerous than men, and that for various reasons they cannot all be maintained by men, even if it were theoretically desirable that they should be so maintained, a point which I am not here discussing. It follows that (quite apart from the question of economic independence as an ideal) economic self dependence is in a vast number of cases a necessity. It is impossible to estimate in how many cases this occurs, but it is safe to say that many women do in fact support themselves and others, and that many more would do so if they could.

Normally working women seem to pass from one plane of social development to another, not once only but in many cases twice or thrice in their lives. We might distinguish these planes as status and contract, or value-in-use or value-in-exchange. All children, it is evident, are born into a world of value-in-use; they are not, for some years at all events, valued at what their services will fetch in the market. At an age varying somewhere between eight and eighteen or twenty the working girl, like the boy, starts on an excursion into the world of competition and exchange; she sells her work for what it will fetch. This stage, the stage of the cash nexus, lasts for the majority of girls a few years only. If she marries and leaves work, she returns at once into the world of value-in-use: the work she does for husband, home, and children is not paid at so much per unit, but is done for its own sake. This accounts on an average for say twenty-five years; then she, in numbers at present unknown, is forced again to enter competitive industry on widowhood. This is what Madame Gnauck has called the "cleft" (*Spalte*) in the woman's industrial career. The lower death-rate of women is actually a source of weakness to them, in so far as it leaves a disproportionate number of women without partners at the very time when owing to the care of young children they are least capable of self-support, and it increases the competition of women for employment. Their use-value in the home, however great, will not fetch bread and shelter for their children. Professor Thomas Jones, in his deeply interesting report to the Poor Law Commission (Appendix XVII., Out-Relief and Wages) has been impressed by the pitiful fact that outside work should be forced on women whose whole desire is usually to be at home. He writes in reference to the well-intentioned efforts made by the Charity Organization Society to train widows for self support, efforts which, unfortunately, have not met with much success: "The widow whom it is sought to train is no longer young. It is rather late to begin. . . . Further, many women are domestic by instinct, and dislike factory life. More important still in explaining failure . . . is the conflict between the bread-winner and the housemother. Many a mother is distracted during the training time with anxiety for the children at home who may or may not be properly cared for."[5]

Many serious discourses and amiable sermons are delivered in public and in private on the supreme beauty and importance of woman's influence, the necessity of maintaining a high standard of home life, and the integrity of the family. All this may be true, but for many women it is singularly irrelevant. *Il faut vivre.*

A woman may possess all the domestic virtues in the highest possible degree, but she cannot live by them. Value-in-use is subordinated to value-in-exchange. Mrs. Brown may be much more useful, from the point of view of her family and the community, when she is engaged in keeping her little home clean and tidy and caring more or less efficiently for the fatherless little Browns' bodily and spiritual needs, than she is when fruit-picking, sackmaking, or washing for an employer's profit. But the point is that these kinds of work do at worst bring her in a few shillings a week, and the former—nothing at all. In the face of such facts it is absurd to tell women that their work as mothers is of the highest importance to the State. We may hope, however, that public opinion will ere long be convinced that the present system of dealing with indigent widows, as described in Professor Jones's Report, is wasteful of child life, destructive of the home, and cruelly burdensome to the most conscientious and tender-hearted mothers. The truly statesmanlike course will be to grant widows with young children a pension sufficient for family maintenance, on the condition that the home should be under some form of efficient inspection or control to ensure the money being properly laid out and the children cared for.[6] In the case of those women who are not naturally adapted to an entirely domestic life and prefer to work for themselves, it might be arranged that some portion of the pension should be diverted to pay a substitute. These cases would probably not be numerous, but it is as well to recognize that some such do exist.[7]

Socialists will not fail to realize that the case of the mother of small children forced under a competitive system to do unskilful and ill-remunerated work and neglect the work that is all-important for the State, viz., the care and nurture of its future citizens, is only an extreme instance of the anomaly of the whole position of woman in an individualist industrial community. This is not a place to enter on a discussion of the lines on which the economic position of women may be expected to develop under Socialism. I desire here merely to emphasize the importance of the distinction between value-in-use and value-in-exchange which seems to me to lie at the root of the whole social question; but most especially so as regards women. Our present industrial system, and therewith largely our social system also, is continually balanced perilously on the possibility of profit. Production is directed, not towards satisfying the needs and building up of the character of the nation's citizens, but merely towards what will yield most profit to the individuals who control the process. Except to the extent of the regulations of the Factory, Public Health, and Adulteration Acts (often inadequate and imperfectly enforced), it makes no difference at all whether the objects produced are useful or poisonous, beautiful or hideous, whether the conditions are healthy or dangerous, ennobling or degrading; profit is the only test. The special anomaly of the woman's position is that while the pressure of social tradition is continually used to induce her to cultivate qualities that, so far from helping, are a positive hindrance to success in competitive industry, yet when circumstances throw her out into the struggle there is little or no social attempt made to compensate her for her deficiencies. Her very virtues are often her weakness.

No sane person can argue that adaptability to the conditions of profit-making industry can afford any test of a woman's merit *quâ* woman, yet it is all that many women have to depend on for their own and their children's living. The position ought at once to be frankly faced that women's work at home is service to the State, and it may be hoped that ere long some practical step may be taken to put in force the Minority Report suggestions regarding allowances to widows with young children.

Notes

1. "Die Deutsche Frau."
2. "How to Help Widows," by A. M. Humphrey, p. 1. (Published by the Charity Organization Society.)
3. "A Question of National Importance." (Hapworth and Co., 1909.)
4. See Corrected Death-Rates in Counties. Registrar-General's Report for 1907, pp. 12–20, cf. p. 14.
5. Poor Law Commission, Appendix, Vol. XVII.
6. See Minority Report Poor Law Commission, Part I., p. 184 (Longmans' edition).
7. I am not here alluding to cruel, depraved, or drunken mothers. In those cases children should obviously be entirely removed from the mother, and she herself dealt with penally or curatively, as may be deemed advisable.

Ideas for Discussion

1. Hutchins's essay first appeared in 1911. Is her understanding of women workers dated?
2. What functions do the tables and statistics serve in giving Hutchins credibility?

Topics for Writing

1. Research and prepare a comprehensive, formal report on current issues facing women workers. Topics to consider are the glass ceiling, wages, childcare, or sexual harassment.

2. Read Beatrice Webb's essay on "Women and the Factory Acts" in this volume pp. 228–239 for a report on the common concerns of these women writers.
3. Prepare an unsolicited proposal on one of the issues itemized in 1 for the company of your choice.
4. Visit a day-care center in a business or interview workers on issues such as wage equity, sexual harassment, and the glass ceiling. Write a trip report or a summary of your findings in the form of a short memo.

Eric Hobsbawm
Man and Woman: Images on the Left

Eric Hobsbawm was a graduate of Birkbeck College, Cambridge, where he was in residence until his death in 1994. Hobsbawm had a distinguished career as a social historian, author, and educator. He taught in Great Britain, the United States, and Italy, and his publications include Primitive Rebels (1959), The Age of Capitalism (1973), The Worlds of Labour (1984), *and* Nations and Nationalism since 1780 (1990). *This essay was first published in 1978 as part of* The Worlds of Labour *(retitled in the United States as* Workers: The Worlds of Labor). *This piece is characteristic of the late professor's interest in a variety of fields and his ability to weave them into seamless coherence.*

WOMEN HAVE OFTEN pointed out that male historians in the past, including Marxists, have grossly neglected the female half of the human race. The criticism is just; the present writer accepts that it applies to his own work. Yet if this deficiency is to be remedied, it cannot be simply by developing a specialized branch of history which deals exclusively with women, for in human society the two sexes are inseparable. What we need also to study is the changing forms of the relations between the sexes, both in social reality and in the image which both sexes have of one another. The present paper is a preliminary attempt to do this for the revolutionary and socialist movements of the nineteenth and early twentieth centuries by means of the ideology expressed in the images and emblems associated with these movements. Since these were overwhelmingly designed by men, it is of course impossible to assume that the sex-roles they represent express the views of most women. However, it is possible to compare these images of roles and relationships with the social realities of the period, and with the more specifically formulated ideologies of revolutionary and socialist movements.

That such a comparison is possible, is the assumption which underlies this paper. It is not suggested that the images here analysed directly reflect social realities, except where they were specifically designed to do so, as in pictures intended to have documentary value, and even then they clearly did not only reflect reality. My assumption is merely that in images designed to be seen by and to have an impact upon a wide public, e.g. of workers, the public's experience of reality sets limits to the degree to which they may diverge from that experience. If the capitalist in socialist cartoons of the *Belle Epoque* were to have been *habitually* presented not as a fat man smoking a cigar and in a top hat, but as a fat woman, these permissible limits would have been exceeded, and the caricatures would have been less effective; for most bosses were not only conceived as males but were males. It does not follow that all capitalists were fat with top hats and cigars, though these attributes were readily understood as indicating wealth in a bourgeois society, and had to be understood as specifying one particular form of wealth and privilege as distinct from others, e.g. the nobleman's. Such a correspondence with reality was evidently less necessary in purely symbolic and allegorical images, and yet even here they were not completely absent; if the deity of war had been presented as a woman, it would have been with the intention to shock. To interpret iconography in this manner is naturally not to make a serious analysis of image and symbol. My purpose is more modest.

Let us begin with perhaps the most famous of revolutionary paintings, though one not created by a revolutionary: Delacroix' *Liberty on the Barricades* in 1830. The picture will be familiar to many: a bare-breasted girl in Phrygian bonnet with a banner, stepping over the fallen, followed by armed men in characteristic costumes. The sources of the picture have been much investigated. Whatever they are, its contemporary interpretation is not in doubt. Liberty was seen not as an allegorical figure, but as a real woman (inspired no doubt by the heroic Marie Deschamps, whose feats suggested the picture). She was seen as a woman of the people, belonging to the people, at ease among the people:

> C'est une forte femme aux puissantes mamelles,
> à la voix rauque, aux durs appas qui . . .
> Agile et marchant à grands pas
> Se plaît aux cris du peuple. . . .
> <div align="right">Barbier, <i>La Curée</i></div>
>
> (A strong woman, stout bosom'd,
> With raucous voice and rough charm . . .
> She strides forward with confidence,
> Rejoicing in the clamour of the people. . . .
> <div align="right"><i>The Bandwagon</i>)</div>

She was for Balzac, of peasant stock: "dark-skinned and ardent, the very image of the people." She was proud, even insolent (Balzac's words), and thus the very opposite of the public image of women in bourgeois society. And, as the contemporaries stress, she was sexually emancipated. Barbier, whose *La Curée* is certainly one of Delacroix' sources, invents an entire history of sexual emancipation and initiative for her:

qui ne prend ses amours que dans la populace,
qui ne prête son large flanc
qu'à des gens forts comme elle

(who takes her lovers only from among the masses,
who gives her sturdy body only to men as strong as herself)

after having, *enfant de la Bastille* ("child of the Bastille"), spread universal sexual excitement around her, tired of her early lovers and followed Napoleon's banners and a *capitaine de vingt ans* ("20-year-old captain"). Now she returned,

toujours belle et *nue* [my emphasis, EJH]
avec l'écharpe aux trois couleurs
(still beautiful and *naked* with the tricolour sash)

to win the "Trois Glorieuses" (the July Revolution) for her people.

Heine, who comments on the picture itself, pushes the image even further towards another ambiguous stereotype of the independent and sexually emancipated woman, the courtesan: "a strange mixture of Phyrne, fishwife and goddess of freedom." The theme is recognizable: Flaubert in *Education Sentimentale* returns to it in the context of 1848, with his image of Liberty as a common prostitute in the ransacked Tuileries (though operating the habitual bourgeois transition from the equation liberty = good to that of license = bad): "In the ante-chamber, bolt upright on a pile of clothes, stood a woman of the streets posing as a statue of liberty." The same note is hinted at by the reactionary Félicien Rops, who had actually represented "the Commune personified by a naked woman, a soldier's cap on her head and sword at her side," an image which came not only to his own mind. His powerful *Peuple* is a naked young woman, in the posture of a whore dressed only in stockings and a night cap, possibly hinting at the Phrygian bonnet, her legs opening on her sex.

The novelty of Delacroix' *Liberty* therefore lies in the identification of the nude female figure with a real woman of the people, an emancipated woman, and one playing an active—indeed a leading—role in the movement of men. How far back this revolutionary image can be traced is a question which must be left to art historians to answer. Here we can only note two things. *First*, its concreteness removes it from the usual allegorical role of females, though she maintains the nakedness of such figures, and this nudity is indeed stressed by painter and observers. She does not inspire or represent: she *acts*. *Second*, she seems clearly distinct from the traditional iconographic image of woman as an active freedom-fighter, notably Judith, who, with David, so often represents the successful struggle of the weak against the strong. Unlike David and Judith, Delacroix' *Liberty* is not alone, nor does she represent weakness. On the contrary, she represents the concentrated force of the invincible people. Since "the people" consists of a collection of different classes and occupations, and is presented as such, a general symbol not identified with any of them is desirable. For traditional iconographic reasons this was likely to be female. But the woman chosen represents "the people."

The Revolution of 1830 seems to represent the high point of this image of Liberty as an active, emancipated girl accepted as leader by men, though the theme continues to be popular in 1848, doubtless because of Delacroix' influence on other painters. She remains naked in Phrygian cap in Millet's *Liberty on the Barricades*, but her context is now vague. She remains a leader-figure in Daumier's draft of *The Uprising* but, once again, her context is shadowy. On the other hand, though there are not many representations of the Commune and of Liberty in 1871, they tended to be naked (as in the design of Rops mentioned above) or bare-breasted. Perhaps the notably active part played by women in the Commune also accounts for the symbolization of this revolution by a non-allegorical (i.e. clothed) and obviously militant woman in at least one foreign illustration.

The revolutionary concept of republic or liberty thus still tended to be a naked, or more likely bare-breasted, female. The Communard Dalou's celebrated statue of the Republic on the Place de la Nation still has at least one breast bared. Only research could show how far the revelation of the breast retains this rebellious or at least polemical association, as perhaps in the cartoon from the Dreyfus period (January 1898) in which a young and virginal Marianne, one breast exposed, is protected against a monster by a matronly and armed Justice over the line: "Justice: Have no fear of the monster! I am here." On the other hand the institutionalized Republic, Marianne, in spite of her revolutionary origins, is now normally though lightly, clothed. The reign of decency has been re-established. Perhaps also the reign of lies, since it is characteristic of the allegorical female figure of Truth—she still appears frequently, notably in the caricatures of the Dreyfus period—that she should be naked. And indeed, even in the iconography of the respectable British labour movement of Victorian England, she remains naked, as on the emblem of the Amalgamated Society of Carpenters and Joiners, 1860, until late Victorian morality prevails.

Generally, the role of the female figure, naked or clothed, diminishes sharply with the transition from the democratic-plebeian revolutions of the nineteenth century to the proletarian and socialist movements of the twentieth. In a sense, the main problem of this paper consists in this masculinization of the imagery of the labour and socialist movement.

For obvious reasons the working woman proletarian is not much represented by artists, outside the few industries which were predominantly female. This was certainly not due to prejudice. Constantin Meunier, the Belgian who pioneered the typical idealization of the male worker, painted—and to a lesser extent sculpted—women wage-workers as readily as men; sometimes, as in his *Le retour des mines* (Coming back from the mines) (1905) working together with men—as women still did in Belgian mines. However, it is probable that the image of woman as a wage-worker and an active participant together with males in political activity was largely due to socialist influence. In Britain it does not become noticeable in the trade union iconography until this influence is felt. In the emblems of pre-socialist British trade unions, uninfluenced by intellectuals, real women appear mainly in those small images by which unions advertised their fraternal help to members in distress: sickness, accident and funeral benefit. They stand by the bedside of the sick husband as his mates come to visit him adorned with the sash of their union. Surrounded by children, they shake hands with the union representative who hands them money after the death of the breadwinner.

Of course women are still present in the form of symbol and allegory, though towards the end of the century in Britain union emblems are to be found without any female figures, especially in such purely masculine industries as coal-mining, steel-smelting and the like. Still, the allegories of liberal self-help continue to be largely female, because they had always been. Prudence, Industry (= diligence) Fortitude, Temperance, Truth and Justice presided over the Stone Masons' Friendly Society in 1868; Art, Industry, Truth and Justice over the Amalgamated Society of Carpenters and Joiners. From the 1880s on one has the impression that only Justice and Truth, possibly supplemented by Faith and Hope, survive among these traditional figures. However as socialism advances, other female persons enter the iconography of the left, though they are in no sense supposed to represent real women. They are goddesses or muses.

Thus on a banner of the (left-wing) Workers' Union, 1898–1929, a sweet-young lady in white drapery and sandals points to a rising sun labelled "A better life" for the benefit of a number of realistically painted workers in working dress. She is Faith, as the text below the picture makes clear. A militant figure, also in white draperies and sandals, but with sword and buckler marked "Justice & Equality," not a hair out of place on her well-styled head, stands before a muscular worker in an open shirt who has evidently just defeated a beast labelled "Capitalism" which lies dead on the ground before him. The banner is labelled "The Triumph of Labour," and represents the Southend-on-Sea branch of the National Union of General Workers, another socialist union. The Tottenham branch of the same union has the same young lady, this time with flowing hair, her dress marked "Light, Education, Industrial Organisation, Political Action and Real International," pointing out the promised land in the shape of a children's playground to the usual group of workers. The promised land is labelled "gain the Cooperative Commonwealth," and the entire banner illustrates the slogan "Producers of the Nation's Wealth, Unite! And have your share of the world."

These images are all the more significant because they are obviously linked to the new socialist movement, which develops its own iconography, and because (unlike the old allegorical vocabulary) this new iconography is in part inspired by the tradition of French revolutionary imagery, from which Delacroix' *Liberty* is also derived. Stylistically, in Britain at least, it belongs to the progressive arts-and-crafts movement and its offshoot, *art nouveau*, which provided British socialism with its chief artists and illustrators, William Morris and Walter Crane. Yet Walter Crane's widely popular image of humanity advancing to socialism—a couple in loose summery clothes, the man carrying a child on his shoulder—like so many of his designs still reflects the debt to 1789 in the presence of the Phrygian bonnet. The earliest of the First-of-May badges of the Austrian social-democrats make the connection even more obvious. They represent a female figure with the motto: Fraternity, Equality, Liberty and the Eight-Hour Day.

Yet what is the role of the women in this new socialist iconography? They inspire. The emblem of the *Labour Annual*, published from 1895, is T. A. West's *Light and Life*. A lady in flowing robes, half-visible behind an escutcheon, blows a ritual trumpet for the benefit of a handsome boy with open-necked shirt and sleeves rolled up beyond the elbow, carrying a basket from which he sows the seed of, presumably, socialist propaganda; rays, stars and waves form the background to the design. Insofar as human women appear in this iconography, they are part of an idealized couple, with or without children. Insofar as each is symbolically identified with some activity, it is the man who represents industrial labour. In Crane's couple he has beside him a pick and a shovel, while she, carrying a basket of corn, and with a rake by her side, represents nature or at most agriculture. Curiously enough, the same division occurs in Mukhina's famous sculpture of the (male) worker and the (female) *kolkhoz* peasant on the Soviet Pavilion at the Paris International Exposition of 1937: he the hammer, she the sickle.

Of course actual women of the working classes also occur in the new socialist iconography, and embody a symbolic meaning, at least by implication. Yet they are quite different from the militant girls of the Paris Commune. They are figures of suffering and endurance. Meunier, that great pioneer of proletarian art and socialist realism—both as realism and as idealisation—anticipates them, as usual. His *Femme du Peuple* (Woman of the People) (1893) is old, thin, her hair drawn back so tightly as to suggest little more than a naked skull, her withered flat chest suggested by the very (and untypical) nakedness of her shoulders. His even better-known *Le Grisou* (Firedamp) has the female figure, swathed in shawls, grieving over the corpse of the dead miner. These are the suffering proletarian mothers best known from Gorki's novel or Kaethe Kollwitz' tragic drawings. And it is perhaps not insignificant that their bodies become invisible under shawls and headcloths. The typical image of the proletarian woman has been desexualized and hides behind the clothes of poverty. She is spirit, not body. (In real life this image of the suffering wife and mother turned militant is perhaps exemplified by the blackclad eloquence of *La Pasionaria* in the days of the Spanish Civil War.)

Yet while the female body in socialist iconography is increasingly dressed, if not concealed, a curious thing is happening to the male body. It is increasingly revealed for symbolic purposes. The image which increasingly symbolizes the working class is the exact counterpart to Delacroix'

Liberty, namely a topless young man: the powerful figure of a masculine labourer, swinging hammer or pick and *naked to the waist*. This image is unrealistic in two ways. In the first place, it was by no means easy to find many nineteenth-century male workers in the countries with strong labour movements labouring with a naked torso. This, as Van Gogh recognized, was one of the difficulties of an era of artistic realism. He would have liked to paint the naked bodies of peasants, but in real life they did not go naked. The numerous pictures representing industrial labour, even under conditions when it would today seem reasonable to take off one's shirt, as in the heat and glow of ironworks or gasworks, almost universally show them clothed, however lightly. This includes not merely what might be called broad evocations of the world of labour such as Madox Ford's *Work*, or Alfred Roll's *Le Travail* (1881)—a scene of open-air building work—but realistic paintings or graphic reporting. Naturally bare torsoed workers could be seen—for instance among some, but by no means all, British coal-face workers. In such cases workers could be realistically presented as semi-nudes, as in G. Caillebotte's *Raboteurs de Parquet* (Floor-polishers), or in the figure of a coal-hewer on the emblem of the Ironfounders' Union (1857). In real life, however, these were all special cases. In the second place, the image of nakedness is unrealistic because it almost certainly excluded the vast body of skilled and factory workers, who would not have dreamed of working without their shirts at any time, and who, incidentally, in general formed the bulk of the organized labour movement.

When the bare-torsoed worker first appears in art is uncertain. Certainly what must be one of the earliest sculptured proletarians, Westmacott's slate-worker on the Penrhyn monument, Bangor (1821), is dressed, while the peasant girl near him is, perhaps semi-allegorically, rather decolletée. At all events from the 1880s on he was familiar in sculpture in the work of the Belgian, Constantin Meunier, perhaps the first artist to devote himself wholeheartedly to the presentation of the manual worker; possibly also of the Communard Dalou, whose unfinished monument to labour contains similar motifs. Obviously he was much more prominent in sculpture, which had, by long tradition, a much stronger tendency to present the human figure nude than painting. In fact, Meunier's drawings and paintings are much more often realistically clothed, and, as has been shown for at least one of his themes, dockers unloading a ship, were only undressed in the three-dimensional design for a monument of labour. Perhaps this is one reason why the semi-nude figure is less prominent in the period of the Second International, when the socialist movement was not in a position to commission many public monuments as yet, and comes into his own after 1917 in Soviet Russia, where it was. Yet, though a direct comparison between painted and sculptured image is therefore misleading, the bare male torso may already be found here and there on two-dimensional emblems, banners and other pictures of the labour movement even in the nineteenth century. Still, in sculpture he triumphed after 1917 in Soviet Russia, under such titles as *Worker, The Weapons of the Proletariat, Memorial of Bloody*

Sunday 1905 etc. The theme is not yet exhausted, since a statue called "Friendship of the Peoples" of the 1970s still presents the familiar topless Hercules swinging a hammer.

Painting and graphics still found it harder to break the links with realism. It is not easy to find any bare-torsoed workers in the heroic age of the Russian revolutionary poster. Even the symbolic painting *Trud* (Toil) presents a design of an idealized young man *in working clothes*, surrounded by the tools of a skilled artisan, rather than the heavy-muscled and basically unskilled titan of the more familiar kind. The powerful hammer-swinger engaged in breaking the chains binding the globe, who symbolized the *Communist International* on the covers of its periodical from 1920, wore clothes on his torso, though only sketchy ones. The symbolic decorations of this review in its early numbers were non-human: five pointed stars, rays, hammers, sickles, ears of grain, beehives, cornucopias, roses, thorns, crossed torches and chains. While there were more modern images such as stylizations of smoking factory chimneys in the art-nouveau fashion* and driving bands of transmission-belts, there were no bare-chested workers. Propaganda photographs of such men do not become common, if they occur at all, before the first Five-Year Plan. Nevertheless, though the progress of the two-dimensional bare torso was slower than might be thought, the image was familiar. Thus it is the symbol decorating the cover of the French edition of the *Compte Rendu Analytique* of the 5th Congress of Comintern (Paris, 1924).

Why the bare body? The question can only be briefly discussed, but takes us back both to the language of idealized and symbolic presentation and to the need to develop such a language for the socialist revolutionary movement. There is no doubt that eighteenth-century aesthetic theory linked the naked body and the idealization of the human being; often quite consciously as in Winckelmann. An idealized person (as distinct from an allegorical figure) could not be clothed in the garments of real life, and—as in the nude statues of Napoleon—should if possible be presented without garments. Realism had no place in such a presentation. When Stendhal criticized the painter David, because it would have been suicidal for his warriors of antiquity to go into battle naked, armed only with helmet, sword and shield, he was simply drawing attention, in his usual role as provocator, to the incompatibility of symbolic and realistic statement in art. But the socialist movement, in spite of its profound attachment in principle to realism in art—an attachment which goes back to the Saint-Simonians—required a language of symbolic statement, in which to state its ideals. As we have seen, the emblems and banners of the British trade unions—rightly described by Klingender as "the true folk-art of nineteenth century Britain"—are a combination of realism, allegory and symbol. They are probably the last flourishing form of the allegorical and symbolic language outside public monumental sculpture. An idealized presentation of the subject

---

* In Russia this motif occurs as early as 1905–7.

of the movement, the struggling working class itself, must sooner or later involve the use of the nude—as on the banner of the Export Branch of the Dockers' Union in the 1890s, where a naked muscular figure, his loins lightly draped, kneels on a rock wrestling with a large green serpent, surrounded by suitable mottoes. In short, though the tension between realism and symbolism remained, it was still difficult to devise a complete vocabulary of symbol and ideal without the nude. On the other hand, it may be suggested that the total nude was no longer acceptable. It cannot have been easy to overlook the absurdity of the 1927 "Group: October" which consists of three muscular men, naked except for the Red-army cap worn by one of them, with hammers and other suitable paraphernalia. Let us conjecture that the bare-torsoed image expressed a compromise between symbolism and realism. There were after all *real* workers who could be so presented.

We are left with a final, but crucial question. Why is the struggling working class symbolized exclusively by a *male* torso? Here we can only speculate. Two lines of speculation may be suggested.

The first concerns the changes in the actual sexual division of labour in the capitalist period, both productive and political. It is a paradox of nineteenth-century industrialization that it tended to increase and sharpen the sexual division of labour between (unpaid) household work and (paid) work outside, by depriving the producer of control over the means of production. In the pre-industrial or proto-industrial economy (peasant farming, artisanal production, small shopkeeping, cottage industry, putting-out, etc.) household and production were generally a single or combined unit, and though this normally meant that women were grossly overworked—since they did most of the housework and shared in the rest of the work—they were not confined to one type of work. Indeed, in the great expansion of "proto-industrialism" (cottage industry) which has recently been investigated the actual productive processes attenuated or even abolished the differences in work between men and women, with far-reaching effects on the social and sexual roles and conventions of the sexes.

On the other hand in the increasingly common situation of the worker who laboured for an employer in a workplace belonging to the employer, home and work were separate. Typically it was the male who had to leave home every day to work for wages and the woman who did not. Typically women worked outside the home (where they did so at all) only before or, if widowed or separated, after marriage, or where the husband was unable to earn sufficient to maintain wife and family, and very likely only so long as he was unable to do so. Conversely, an occupation in which an adult man was normally unable to earn a family wage was—very understandably—regarded as underpaid. Hence, the labour movement quite logically developed the tendency to calculate the desirable minimum wage in terms of the earnings of a single (i.e. in practice male) breadwinner, and to regard a wage-working wife as a symptom of an undesirable economic situation. In fact the situation was often undesirable, and the number of married women obliged to work for wages or their equivalent was substantial, though a very large proportion of them did so at home—i.e. outside the effective range of labour movements. Moreover, even in industries in which the work of married women was traditionally well established—as in the Lancashire textile region—its scope can be exaggerated. In 1901 38 per cent of married and widowed women in Blackburn were employed for wages, but only 15 per cent of those in Bolton.

In short, conventionally women aimed to stop working for wages outside the house once they got married. Britain, where in 1911 only 11 per cent of wage-working women had husbands and only 10 per cent of married women worked, was perhaps an extreme case; but even in Germany (1907) where 30 per cent of wage-working women had husbands the sex-difference was striking. For every wife at wage-work in the age-groups from 25 to 40 years, there were four wage-working husbands. The situation of the married woman was not substantially changed as yet by the tendency—rather marked after 1900—for women to enter industry in larger numbers, and by the growing variety of occupations and leisure activities open to unmarried girls. "The trend towards a larger number of married women having a specified occupation had not been firmly established at the turn of the century." The point is worth stressing, since some feminist historians, for reasons difficult to understand, have attempted to deny it. Nineteenth-century industrialization (unlike twentieth-century industrialization) tended to make marriage and the family the major career of the working class woman who was not obliged by sheer poverty to take other work. Insofar as she worked for wages before marriage, she saw wage-work as a temporary, though no doubt desirable, phase in her life. Once married, she belonged to the proletariat not as a worker, but as the wife, mother and housekeeper of workers.

Politically the pre-industrial struggle of the poor not only produced ample room for women to take part beside men—neither sex had such political rights as the right to vote—but in some respects a specific and leading role for them. The commonest form of struggle was that to assert social justice, i.e. the maintenance of what E. P. Thompson has called "the moral economy of the crowd" through direct action to control prices. In the form of action, which could be politically decisive—we recall the march of the women on Versailles in 1789—women not only took the lead, but were conventionally expected to. As Luisa Accati rightly states: "in a large number of cases (I would almost say in practically all cases) women have the decisive role, whether because it is they who take the initiative, or because they form a very large part of the crowd." We need not here consider the well-known pre-industrial practice in which rebellious men take action disguised as women, as in the so-called Rebecca Riots of Wales (1843).

Furthermore, the characteristic urban revolution of the pre-industrial period was not proletarian but plebeian. Within the *menu peuple,* a socially heterogeneous coalition of elements, united by common "littleness" and poverty rather than by occupational or class criteria, women could play a political role, provided only they could come out on the streets. They could and did help to build barricades. They could assist those who fought behind them. They could even fight or bear arms themselves. Even the image of the modern "people's revolution" in a large non-industrial metropolis contains them, as anyone who recalls the street scenes of Havana after the triumph of Fidel Castro will testify.

On the other hand the specific form of struggle of the proletariat, the trade union and the strike, largely excluded the women, or greatly reduced their visible role as active participants, except in the few industries in which they were heavily concentrated. Thus in 1896 the total number of women in British trade unions (excluding teachers) was 142,000 or something like 8 per cent; but 60 per cent of these were in the extremely strongly organized cotton industry. By 1910 it was above 10 per cent but though there had been some growth in trade unionism among white-collar and shopworkers, the great bulk of the expansion in industry was still in textiles. Elsewhere their role was indeed crucial, but distinct, even in small industrial and mining centres where place and work and community were inseparable. Yet if in such places their role in strikes was public, visible and essential, it was nevertheless not that of strikers themselves.

Moreover, where men's work and women's work were not so separate and distinct that no question of intermixture could arise, the normal attitude of male trade unionists towards women seeking to enter their occupation was, in the words of S. and B. Webb, "resentment and abhorrence." The reason was simple: since their wages were so much lower, they represented a threat to the rates and conditions of men. They were—to quote the Webbs again—"as a class, the most dangerous enemies of the artisan's Standard of Life," though the men's attitude was also—in spite of the growing influence on the left—strongly influenced by what would today be called "sexism": "the respectable artisan has an instinctive distaste for the promiscuous mixing of men and women in daily intercourse, whether this be in the workshop or in a social club." Consequently the policy of all unions capable of doing so was to exclude women from their work, and the policy even of those unions incapable of doing so (e.g. the cotton weavers) was to segregate the sexes or at least to avoid women and girls working "in conjunction with men, especially if (they are) removed from constant association with other female workers." Thus both the fear of the economic competition of women workers and the maintenance of "morality" combined to keep women outside or on the margins of the labour movement—except in the conventional role of family members.

The paradox of the labour movement was thus that it encouraged an ideology of sexual equality and emancipation, while in practice discouraging the actual joint participation of men and women in the process of labour as workers. For the minority of emancipated women of all classes, including workers, it provided the best opportunities to develop as human beings, indeed as leaders and public figures. Probably it provided the only environment in the nineteenth century which gave them such opportunities. Nor should we underestimate the effect on the ordinary, even the married, working-class women of a movement passionately committed to female emancipation. Unlike the petty-bourgeois "progressive" movement which, as among the French Radical Socialists, virtually flaunted its male chauvinism, the socialist labour movement tried to overcome the tendencies within the proletariat and elsewhere to maintain sexual inequality, even if it failed to achieve as much as it would have wished. It is not insignificant that the major work by the charismatic leader of the German socialists, August Bebel—and by far the most popular work of socialist propaganda in Germany at that period—was his *Woman and Socialism*. Yet at the same time the labour movement unconsciously tightened the bonds which kept the majority of (non-wage-earning) married women of the working class in their assigned and subordinate social role. The more powerful it became as a mass movement, the more effective these brakes on its own emancipatory theory and practice became; at least until the economic transformations destroyed the nineteenth-century industrial phase of the sexual division of labour. In a sense the iconography of the movement reflects this unconscious reinforcement of the sexual division of labour. In spite of and against the movement's conscious intentions, its image expressed the essential "maleness" of the proletarian struggle in its elementary form before 1914, the trade-union struggle.

It should now be clear why, paradoxically, the historical change from an era of plebeian and democratic to one of proletarian-socialist movements should have led, iconographically, to a decline in the role of the female. However, there may be another factor which reinforced this masculinization of the movement: the decline of classical pre-industrial millennialism. This is an even more speculative question, and I touch on it with caution and hesitation.

As has already been suggested, in the iconography of the left, the female figure maintained herself best as an image of utopia: the goddess of freedom, the symbol of victory, the figure who pointed towards the perfect society of the future. And indeed the imagery of the socialist utopia was essentially one of nature, of fertility and growth, of blossoming, for which the female metaphor came naturally:

> Les générations écloses
> Verront fleurir leurs bébés roses
> Comme églantiers en Floréal
> Ce sera la saison des roses . . .
> Voilà l'avenir social
> E. Pottier

> (The budding generations
> Will see their rosy babies flower
> Like briars in the spring.
> It will be the season of roses . . .
> That's the people's future.)

Engène Pottier, the Fourierist author of the *Internationale*, is full of such images of femaleness, even in its literal sense of the maternal breast:

> pour tes enfants longtemps sevrés
> reprends le rôle du mamelle
>
> *(L'Age d'Or)*
>
> Ah, chassons-la. Dans l'or des blés
> Mère apparais, les seins gonflées
> à nos phalanges collectives
>
> *(La fille du Thermidor)*
>
> Du sein de la nourrice, il coule ce beau jour
> Une inondation d'existence et d'amour.
> Tout est fécondité, tout pullule et foisonne
>
> *(Abondance)*
>
> Nature—toi qui gonfles ton sein
> pour ta famille entière
>
> *(La Cremaillère)*
>
> (To your children, though weaned long ago,
> Give once again your breast.
>
> *The Golden Age*
>
> In the golden meadows come to us Mother,
> Your breasts full for the collective hosts.
>
> *Daughter of Thermidor*
>
> This beautiful day flows from the nurse's breast,
> A flood of life and love.
> All is fruitfulness, everything swarms and abounds.
>
> *Abundance*
>
> Nature—you whose breast has swell'd
> To feed your entire family . . .
>
> *Celebration*)

So, in a less explicitly physical way, is Walter Crane who, as we have seen, was largely responsible for the themes of socialist imagery in Britain from the 1880s on. It was an imagery of spring and flowers, of harvest (as in the well-known "The Triumph of Labour" designed for the 1891 May Day demonstration), of girls in light flowing dresses and Phrygian bonnets. Ceres was the goddess of communism.

It is not surprising that the period of socialist ideology most deeply imbued with feminism, and most inclined to assign a crucial, indeed sometimes a dominant, role to women, was the romantic-utopian era before 1848. Of course at this period we can hardly speak of a socialist "movement" at all, but only of small and atypical groups. Moreover, the actual number and prominence of women in leading positions in such groups was far smaller than in the years of the non-utopian Second International. There is nothing to compare in the Britain of Owenism and Chartism with the role of women as writers, public speakers and leaders in the 1880s and 1890s, not only in the middle-class ambiance of the Fabian Society, but in the much more working-class atmosphere of the Independent Labour Party, not to mention such figures as Eleanor Marx in the trade-union movement.

Moreover, the women who then became prominent, like Beatrice Webb or Rosa Luxemburg, did not make their reputation because they were women, but because they were outstanding irrespective of sex. Nevertheless, the role of women's emancipation in socialist ideology has never been more obvious and central than in the period of "utopian socialism."

This was partly due to the crucial role assigned to the destruction of the traditional family in the socialism of that period; a role which is still very clear in *The Communist Manifesto*. The family was seen as the prison-house not only of the women, who were not on the whole very active in politics, or indeed as a mass very enthusiastic about the abolition of marriage, but also of young people, who were much more attracted to revolutionary ideologies. Moreover, as J. F. C. Harrison has rightly pointed out, even on empirical grounds the new proletarians might well conclude that "their rude little homes were a restrictive and circumscribing influence, and that in community they would have a means of breaking out of this: 'we can afford to live in palaces as well as the rich . . . were we only to adopt the principle of combination, the patriarchal principle of large families, such as that of Abraham.' " It has been the consumer-society, combined—paradoxically—with the replacement of mutual aid by state welfare, which has weakened this argument against the privatized nuclear-family household.

Yet utopian socialism also assigned another role to women, which was basically similar to the female role in the chiliastic religious movements with which the utopians had much in common. Here women were not only—perhaps not even primarily—equal, but superior. Their specific role was that of prophets, like Joanna Southcott, founder of an influential millennial movement in early nineteenth-century England, or the *"femme-mère-messie"* (woman-mother-messiah) of the Saint-Simonian religion. This role incidentally provided opportunities for a public career in a masculine world for a small number of women. The foundresses of Christian Science and Theosophy come to mind. However, the tendency of the socialist and labour movements to move away from chiliasm towards rationalist theory and organization ("scientific socialism") made this social role for women in the movement increasingly marginal. Able women, whose talents lay in filling it, were pushed out of the centre of the movement into fringe religions which provided more scope for them. Thus Annie Besant, secularist and socialist, found fulfilment and her major political role after 1890 as high priestess of Theosophy and—through Theosophy—an inspirer of the Indian national liberation movement.

All that remained of the utopian/messianic role of women in socialism was the image of the female as inspiration and symbol of the better world. But paradoxically this image by itself was hardly distinguishable from Goethe's *"das ewig weibliche zieht uns hinan"* ("the eternal feminine raises us to the heavens"). In actuality it could be no different from the bourgeois-masculine idealization of the female in theory, which was only too readily compatible with her inferiority in practice. At most the female image of the inspirer became the image of a Joan of Arc, easily recognizable in Walter Crane's designs. Joan of Arc was indeed an icon of women's militancy, but she did not represent either political or personal emancipation, or indeed activism, in any sense that could become a model for real women. Even if we forget that she excluded the majority of women who were no longer virgins—i.e. women as sexual beings—there was, by historic definition, room for only a very few Joans of Arc in the world at any given moment. And, incidentally, as the increasingly enthusiastic adoption of Joan of Arc by the French right-wing demonstrates, her image was ideologically and politically undetermined. She might or might not represent Liberty. She might be on the barricades, but she did not—unlike Delacroix' girl—necessarily belong there.

Unfortunately it is at present impossible to continue the iconographic analysis of the socialist movement beyond a point of history which is already fairly remote. The traditional language of symbol and allegory is no longer much spoken or understood, and with its decline women as goddesses and muses, as personifications of virtue and ideals, even as Joans of Arc, have lost their specific place in political imagery. Even the famous international symbol of peace in the 1950s was no longer a woman, as it would almost certainly have been in the nineteenth-century, but Picasso's dove. The same is probably true of masculine images, though the hammer-wielding Promethean man survived longer as the personification of movement and struggle. The iconography of the movement since, say, World War II, is non-traditional. We do not at present have the analytical tools to interpret it, e.g. to make symbolic readings of the main modern iconographic medium, which is ostensibly naturalistic, the photograph or film.

Iconography can therefore not throw significant light at present on the relations between men and women in the mid-twentieth-century socialist movement, as it can for the nineteenth-century. Still, it can make one final suggestion about the masculine image. This, as has already been suggested, is in some senses paradoxical, since it typifies not so much the worker as sheer muscular effort; not intelligence, skill and experience, but brute strength. Even, as in Meunier's famous Iron-puddler, physical effort which virtually excludes and exhausts the mind. One can see artistic reasons for this. As Brandt points out, in Meunier "the proletariat is transformed into a Greek athlete," and for this form of idealization the expression of intelligence is irrelevant. One can also see historical reasons for it. The period 1870–1914 was above all the period in which industry relied on a massive influx of inexperienced but physically strong labour to perform the very large proportion of labour-intensive and relatively unskilled tasks; and when the dramatic environment of darkness, flame and smoke typified the revolution in man's capacity to produce by steampowered industry.

And yet, as we know, the bulk of the militants of organized labour in this period consisted, if we leave aside the admittedly important contingent of miners, essentially of skilled men. How is it that an image which omits all the characteristics of their kind of labour, established itself as the expression of the working class? Three explanations may be suggested. The first, and perhaps psychologically the most convincing, is that for most workers, whatever their skill, the criterion of belonging to their class was precisely the performance of manual, physical labour. The instincts of genuine labour movements were *ouvrieriste:* a distrust of those who did not get their hands dirty. This the image represented. The second is that the movement wished to stress precisely its inclusive character. It comprised all proletarians, not merely printers, skilled mechanics and their like. The third, which probably prevailed in the period of the Third International, was that in some sense the relatively unskilled, purely manual labourer, the miner or docker, was considered more revolutionary, since he did not belong to the labour aristocracy with its penchant for reformism and social-democracy. He represented "the masses" to whom revolutionaries appealed over the heads of the social-democrats. The image was reality, insofar as it represented the fundamental distinction between manual and non-manual work; aspiration insofar as it implied a programme or a strategy. How realistic it was in the second respect is a question which does not belong to the present paper. But it is nevertheless not insignificant that, as an image, it omitted much that was most characteristic about the working class and its labour movement.

Ideas for Discussion

1. What does an essay such as Hobsbawm's contribute to contemporary views of gender-oriented communication in the workplace?
2. What role did the visual arts play in defining the Socialist movement?
3. What kinds of definitions of workers do these art works create?

Topics for Writing

1. Examine some modern representations of working people, perhaps as found in advertisements for products in business and industry periodicals. Do they reflect capitalist or democratic ideologies to you? Write a memo or letter to your instructor or class, as directed.
2. Write an essay on the sense of belonging at work or at home, which Hobsbawm describes as part of the labor movement's visualizing of the worker's life choices.
3. In an essay compare the realities of a worker's life as described in any selection in this book and the art work idealizing it.

Mary Parker Follett
The Essentials of Leadership

Mary Parker Follett was a graduate of what is now known as Radcliffe College, where she studied economics, government, and philosophy. Her best-known book, Creative Experience *(1924), describes her work with the Roxbury (Massachusetts) Neighborhood House, which brought together workers from diverse backgrounds and occupations to discuss their differences in the spirit of inquiry and understanding. Follett, who became a member of the Boston school board's subcommittee on vocational guidance and a member of the Minimum Wage Board of the Women's Municipal League, was devoted to creating a theory of management based on psychological principles. She moved to England in 1924 where she studied the working conditions in British industry. She returned to Boston in 1933 where she died after an illness. Although never in business herself, Follett was a respected theorist who spent her career lecturing in Britain and America on ideas such as those found in her book* Freedom and Co-Ordination, *a posthumous 1949 publication of a lecture first delivered at the London School of Economics and Political Science in February 1933.*

I HAVE TRIED to show you certain changes which are creeping into our thinking on business management. As I have said, in the more progressively managed businesses an order was no longer an arbitrary command but—the law of the situation. A week ago I defined authority as something which could not really be conferred on someone, but as a power which inhered in the job. Now I want to show the difference between the theory of leadership long accepted and a conception which is being forced on our attention by the way in which business is today conducted.

What I call the old-fashioned theory of leadership is well illustrated by a study made a few years ago by two psychologists. They worked out a list of questions by which to test leadership ability. Here are some of the questions:

At a reception or tea do you try to meet the important persons present?
At a lecture or entertainment do you go forward and take a front seat?
At a hairdresser's are you persuaded to try a new shampoo or are you able to resist?

If you make purchases at Woolworth's, are you ashamed to have your friends know it?

If you are at a stupid party, do you try to inject life into it?

If you hold an opinion the reverse of which the lecturer has expressed, do you usually volunteer your opinion?

Do you find it difficult to say No when a salesman is trying to sell you something?

What do you do when someone tries to push in ahead of you in a line at the box office?

When you see someone in a public place whom you think you have met, do you go up to him and enquire whether you have met?

And so on—there were a good many more. But what on earth has all this to do with leadership? I think nothing whatever. These psychologists were making tests, they said, for aggressiveness, assuming that aggressiveness and leadership are synonymous, assuming that you cannot be a good leader unless you are aggressive, masterful, dominating. But I think, not only that these characteristics are not the qualities essential to leadership, but, on the contrary, that they often militate directly against leadership. I knew a boy who was very decidedly the boss of his gang all through his youthful days. That boy is now forty-eight years old. He has not risen in his business or shown any power of leadership in his community. And I do not think that this has been in spite of his dominating traits, but because of them.

But I cannot blame the psychologists too much, for in the business world too there has been an idea long prevalent that self-assertion, pugnacity even, are necessary to leadership. Or at any rate, the leader is usually supposed to be one who has a compelling personality, who can impose his own will on others, can make others do what he wants done.

One writer says that running a business is like managing an unruly horse, a simile I particularly dislike. Another writer says, "The successful business man feels at his best in giving orders. . . . The business man tends to lay down the law—he feels himself to be an individual source of energy." While this is undoubtedly true of many business men, yet there are many today of whom it is not true. It is no longer the universally accepted type of administrative leadership. We saw two weeks ago that in scientifically managed plants, with their planning departments, their experts, their staff officials, their trained managers of the line, few "orders" are given in the old sense of the word. When therefore we are told that large-scale ability means masterfulness and autocratic will, some of us wish to reply: But that is the theory of the past, it is not what we find today in the best-managed industries.

This does not however denote that less leadership is required than formerly, but a different kind. Let me take two illustrations, one of the foreman, one of the salesman. We find in those plants where there is little order-giving of the old kind, where the right order is found by research, that the foreman is not only as important but more important than formerly. He is by no means less of a leader; indeed he has more opportunities for leadership in the sense of that word which is now coming to be accepted by many. This is because his time is freed for more constructive work. With the more explicitly defined requirements made upon him—requirements in regard to time, quality of work and methods—he has a greater responsibility for group accomplishment. In order to meet the standards set for group accomplishment, he is developing a technique very different from the old foreman technique. The foreman today does not merely deal with trouble, he forestalls trouble. In fact we don't think much of a foreman who is always dealing with trouble; we feel that if he were doing his job properly, there wouldn't be so much trouble. The job of the head of any unit—foreman or head of department—is to see that conditions (machines, materials, etc.) are right, to see that instructions are understood, and to see that workers are trained to carry out the instructions, trained to use the methods which have been decided on as best. The test of a foreman now is not how good he is at bossing, but how little bossing he has to do, because of the training of his men and the organisation of their work.

Now take the salesman. If the foreman was supposed to dominate by aggressiveness, the salesman was supposed to dominate by persuasiveness. Consider the different demand made on salesmen today. Salesmen are being chosen less and less for their powers of persuasion, but for their general intelligence, for their knowledge of the goods handled, and for their ability to teach prospective customers the best way to use the goods. A business man said to me: "The training of salesmen . . . is being carried on with increasing elaboration, and always with more emphasis on knowledge of the product and its uses, and distinctly less on the technique of persuasion. . . ." For the firms who sell production equipment, this means sending men who sometimes act as consulting engineers for their customers.

We find this same doctrine taught in the salesmanship classes held in the big shops. The shop assistants are told not to overpersuade a customer, else when the customer gets home she may be sorry she has bought that article and may not come to that shop again. The saleswoman may have made one sale and lost a dozen by her persuasiveness. Her job is to know her goods and to study the needs of her customer.

To dominate, either by a masterful or a persuasive personality, is going out of fashion. People advertise courses in what they call "applied psychology" and promise that they will teach you how to develop your personality and thus become leaders, but wiser teachers say to their students, "Forget your personality, learn your job."

What then are the requisites of leadership? First, a thorough knowledge of your job. And this fact is keenly appreciated today as business is becoming a profession and business management a science. Men train themselves to become heads of departments or staff officials by learning all that goes with the particular position they wish to attain.

Consider the influence which it is possible for the cost accountant to exercise because of his special knowledge. Where there is cost-accounting and unit budgeting, the cost accountant is in a position to know more about the effect of a change in price than anyone else. His analyses and his interpretations may dictate policy to the chief executive.

Moreover, we find leadership in many places besides these more obvious ones, and this is just because men are learning special techniques and therefore naturally lead in those situations. The chairman of a committee may not occupy a high official position or be a man of forceful personality, but he may know how to guide discussion effectively, that is, he may know the technique of *his* job. Or consider the industrial-relations-man now maintained in so many industries. This man is an adept at conciliation. He has a large and elaborate technique for that at his command.

When it is a case of instruction, the teacher is the leader. Yet a good instructor may be a poor foreman. Again, some men can make people produce, and some are good at following up quality who could never make people produce.

There is also individual leadership which may come to the fore irrespective of any particular position; of two girls on a machine, one may be the leader. We often see individual leadership, that is, leadership irrespective of position, springing up in a committee. There was an instance of this in a sales committee. The chairman of the committee was the sales manager, Smith. Smith was narrow but not obstinate. Not being obstinate, Jones was able to get Smith to soften his opinion on the particular matter in question, and there was then an integration of the opinion of that committee around Jones' leadership.

I think it is of great importance to recognise that leadership is sometimes in one place and sometimes in another. For it tends to prevent apathy among under-executives. It makes them much more alert if they realise that they have many chances of leadership before they are advanced to positions which carry with them definitely, officially, leadership. Moreover, if such occasional leadership is exercised with moderation without claiming too much for oneself, without encroaching on anyone's official position, it may mean that person will be advanced to an official position of leadership.

But let us look further at the essentials of leaderships. Of the greatest importance is the ability to grasp a total situation. The chief mistake in thinking of leadership as resting wholly on personality lies probably in the fact that the executive leader is not a leader of men only but of something we are learning to call the total situation. This includes facts, present and potential, aims and purposes and men. Out of a welter of facts, experience, desires, aims, the leader must find the unifying thread. He must see the relation between all the different factors in a situation. The higher up you go, the more ability you have to have of this kind, because you have a wider range of facts from which to seize the relations. The foreman has a certain range—a comparatively small number of facts and small number of people. The head of a sub-department has a wider range; the head of a department a wider still, the general manager the widest of all. One of the principal functions of the general manager is to organise all the scattered forces of the business. The higher railway officials may not understand railway accounting, design of rolling stock, and assignment of rates as well as their expert assistants, but they know how to use their knowledge, how to relate it, how to make a total situation.

The leader then is one who can organise the experience of the group—whether it be the small group of the foreman, the larger group of the department, or the whole plant—can organise the experience of the group and thus get the full power of the group. The leader makes the team. This is pre-eminently the leadership quality—the ability to organise all the forces there are in an enterprise and make them serve a common purpose. Men with this ability create a group power rather than express a personal power. They penetrate to the subtlest connections of the forces at their command, and make all these forces available and most effectively available for the accomplishment of their purpose.

Some writers tell us that the leader should represent the accumulated knowledge and experience of his particular group, but I think he should go far beyond this. It is true that the able executive learns from everyone around him, but it is also true that he is far more than the depository where the wisdom of the group collects. When leadership rises to genius it has the power of transforming, of transforming experience into power. And that is what experience is for, to be made into power. The great leader creates as well as directs power. The essence of leadership is to create control, and that is what the world needs today, control of small situations or of our world situation.

I have said that the leader must understand the situation, must see it as a whole, must see the inter-relation of all the parts. He must do more than this. He must see the evolving situation, the developing situation. His wisdom, his judgment, is used, not on a situation that is stationary, but on one that is changing all the time. The ablest administrators do not merely draw logical conclusions from the array of facts of the past which their expert assistants bring to them, they have a vision of the future. To be sure, business estimates are always, or should be, based on the probable future conditions. Sales policy, for instance, is guided not only by past sales but by probable future sales. The leader, however, must see all the future trends and unite them. Business is always developing. Decisions have to anticipate the development. You remember how Alice in Wonderland had to run as fast as she could in order to stand still. That is a commonplace to every business man. And it is up to the general manager to see that his executives are running as fast as they can. Not, you understand, working as hard as they can—that is taken for granted—but anticipating as far as they can.

This insight into the future we usually call in business anticipating. But anticipating means more than forecasting or predicting. It means far more than meeting the next situation, it means making the next situation. If you will watch decisions, you will find that the highest grade decision does not have to do merely with the situation with which it is directly concerned. It is always the sign of the second-rate man when the decision merely meets the present situation. It is the left-over in a decision which gives it the greatest value. It is the carry-over in a decision which helps develop the situation in the way we wish it to be developed. In business we are always passing from one significant moment to another significant moment, and the leader's task is pre-eminently to understand the moment of passing. The leader sees one situation melting into another and has learned the mastery of that moment. We usually have the situation we make—no one sentence is more pregnant with meaning for business success. This is why the leader's task is so difficult, why the great leader requires great qualities— the most delicate and sensitive perceptions, imagination and insight, and at the same time courage and faith.

The leader should have the spirit of adventure, but the spirit of adventure need not mean the temperament of the gambler. It should be the pioneer spirit which blazes new trails. The insight to see possible new paths, the courage to try them, the judgment to measure results—these are the qualifications of the leader.

And now let me speak to you for a moment of something which seems to me of the utmost importance, but which has been far too little considered, and that is the part of the followers in the leadership situation. Their part is not merely to follow, they have a very active part to play and that is to keep the leader in control of a situation. Let us not think that we are either leaders or—nothing of much importance. As one of those led we have a part in leadership. In no aspect of our subject do we see a greater discrepancy between theory and practice than here. The definition given over and over again of the leader is one who can induce others to follow him. Or that meaning is taken for granted and the question is asked: "What is the technique by which a leader keeps his followers in line?" Some political scientists discuss why men obey or do not obey, why they tend to lead or to follow, as if leading and following were the essence of leadership. I think that following is a very small part of what the other members of a group have to do. I think that these authors are writing of theory, of words, of stereotypes of the past, that they are, at any rate, not noticing the changes that are going on in business thinking and business practice. If we want to treat these questions realistically, we shall watch what is actually happening, and what I see happening in some places is that the members of a group are not so much following a leader as helping to keep him in control of a situation.

How do we see this being done? For one thing, in looking at almost any business we see many suggestions coming up from below. We find sub-executives trying to get upper executives to instal mechanical improvements, to try a new chemical process, to adopt a plan for increasing incentives for workers, and so on. The upper executives try to persuade the general manager and the general manager the board of directors. We have heard a good deal in the past about the consent of the governed; we have now in modern business much that might be called the consent of the governing, the suggestions coming from below and those at the top consenting. I am not trying to imitate Shaw and Chesterton and being paradoxical; there is actually a change going on in business practice in this respect which I want to emphasise to you at every point.

How else may a man help to keep those above him in control? He may, instead of trying to "get by" on something, instead of covering up his difficulties so that no one will know he is having any, inform his chief of his problems, tell him the things he is not succeeding in as well as all his wonderful achievements. His chief will respect him just as much for his failures as for his successes if he himself takes the right attitude towards them.

Another way is to take a wrong order back for correction. It may have been an error, or it may be that it was all right once, but that it must be changed to meet changing conditions. The worker has not met his responsibility by merely obeying. Many a worker thinks that the pointing out of a wrong order is a gratuitous thing on his part, a favour he generously confers but which he need not because it is not really his job, his job is to obey. As a matter of fact, however, obeying is only a small part of his job. One general manager told me that what they disliked in his factory was what they called there the Yes, yes man. The intelligent leader, this man said, does not want the kind of follower who thinks of his job only in terms of passive obedience.

But there is following. Leader and followers are both following the invisible leader—the common purpose. The best executives put this common purpose clearly before their group. While leadership depends on depth of conviction and the power coming therefrom, there must also be the ability to share that conviction with others, the ability to make purpose articulate. And then that common purpose becomes the leader. And I believe that we are coming more and more to act, whatever our theories, on our faith in the power of this invisible leader. Loyalty to the invisible leader gives us the strongest possible bond of union, establishes a sympathy which is not a sentimental but a dynamic sympathy.

Moreover, when both leader and followers are obeying the same demand, you have instead of a passive, an active, self-willed obedience. The men on a fishing smack are all good fellows together, call each other by their first names, yet one is captain and the others obey him; but it is an intelligent, alert, self-willed obedience.

The best leaders get their orders obeyed because they too are obeying. Sincerity more than aggressiveness is a quality of leadership.

If the leader should teach his followers their part in the leadership situation, how to help keep their chief in control, he has another duty equally important. He has to teach them how to control the situations for which they are specifically responsible. This is an essential part of leadership and a part recognised today. We have a good illustration of this in the relation between upper executives and heads of departments in those firms where the Budget is used as a tool of control. Suppose an upper executive is dissatisfied with the work of a department. When this happens it is either because quality is too poor or costs are too high. The old method of procedure was for the upper executive simply to blame the head of the department. But in a plant where the departments are budgeted, an upper executive can ask the head of a department to sit down with him and consider the matter. The Budget objectifies the whole situation. It is possible for an upper executive to get the head of the department to find out himself where the difficulty lies and to make him give himself the necessary orders to meet the situation.

Many are coming to think that the job of a man higher up is not to make decisions for his subordinates but to teach them how to handle their problems themselves, teach them how to make their own decisions. The best leader does not persuade men to follow his will. He shows them what it is necessary for them to do in order to meet their responsibility, a responsibility which has been explicitly defined to them. Such a leader is not one who wishes to do people's thinking for them, but one who trains them to think for themselves.

Indeed the best leaders try to train their followers themselves to become leaders. A second-rate executive will often try to suppress leadership because he fears it may rival his own. I have seen several instances of this. But the first-rate executive tries to develop leadership in those under him. He does not want men who are subservient to him, men who render him an unthinking obedience. While therefore there are still men who try to surround themselves with docile servants—you all know that type—the ablest men today have a larger aim, they wish to be leaders of leaders. This does not mean that they abandon one iota of power. But the great leader tries also to develop power wherever he can among those who work with him, and then he gathers all this power and uses it as the energising force of a progressing enterprise.

If any of you think I have underestimated the personal side of leadership, let me point out that I have spoken against only that conception which emphasises the dominating, the masterful man. I most certainly believe that many personal qualities enter into leadership—tenacity, sincerity, fair dealings with all, steadfastness of purpose, depth of conviction, control of temper, tact, steadiness in stormy periods, ability to meet emergencies, power to draw forth and develop the latent possibilities of others, and so on. There are many more. There is, for instance, the force of example on which we cannot lay too great stress. If workers have to work overtime, their head should be willing to do the same. In every way he must show that he is willing to do what he urges on others.

One winter I went yachting with some friends in the inland waterways of the southern part of the United States. On one occasion our pilot led us astray and we found ourselves one night aground in a Carolina swamp. Obviously the only thing to do was to try to push the boat off, but the crew refused, saying that the swamps in that region were infested with rattlesnakes. The owner of the yacht offered not a word of remonstrance, but turned instantly and jumped overboard. Every member of the crew followed.

So please remember that I do not underestimate what is called the personal side of leadership, indeed there is much in this paper, by implication, on that side. And do not think that I underestimate the importance of the man at the top. No one could put more importance on top leadership than I do, as I shall try to show you next week when we consider the part of the chief executive in that intricate system of human relationship which business has now become.

I might say as a summary of this talk that we have three kinds of leadership: the leadership of position, the leadership of personality and the leadership of function. My claim for modern industry is that in the best managed plants the leadership of function is tending to have more weight and the leadership of mere position or of mere personality less.

Please note that I say only a tendency. I am aware how often a situation is controlled by a man either because his position gives him the whip hand and he uses it, or because he knows how to play politics. My only thesis is that in the more progressively managed businesses there is a tendency for the control of a particular situation to go to the man with the largest knowledge of that situation, to him who can grasp and organise its essential elements, who understands its total significance, who can see it through—who can see length as well as breadth—rather than to one with merely a dominating personality or in virtue of his official position.

And that thought brings me to my conclusion. The chief thing I have wanted to do in this hour is to explode a long-held superstition. We have heard repeated again and again in the past, "Leaders are born, not made." I read the other day, "Leadership is a capacity that cannot be acquired." I believe that leadership can, in part, be learned. I hope you will not let anyone persuade you that it cannot be. The man who thinks leadership cannot be learned will probably remain in a subordinate position. The man who believes it can be, will go to work and learn it. He may not ever be president of the company, but he can rise from where he is.

Moreover, if leadership could not be learned, our large, complex businesses would not have much chance of success, for they require able leadership in many places, not only in the president's chair.

Leadership is a part of business management and there is a rapidly developing technique for every aspect of the administration and management of a business.

I urge you then, instead of accepting the idea that there is something mysterious about leadership, to analyse it. I think that then you cannot fail to see that there are many aspects of it which can be acquired. For instance, a part of leadership is all that makes you get on most successfully in your direct contacts with people—how and when to praise, how and when to point out mistakes, what attitude to take toward failures. All this can of course be learned. The first thing to do is to discover what is necessary for leadership and then to try to acquire by various methods those essentials. Even those personal characteristics with which we were endowed by birth can often be changed. For instance, vitality, energy, physical endurance, are usually necessary for leadership, but even this is not always beyond us. Theodore Roosevelt was a delicate lad and yet became an explorer, a Rough Rider, a fighter, and by his own determined efforts. You have seen timid boys become self-confident. You have seen bumptious little boys have all that taken out of them by their schoolmasters.

Leadership is not the "intangible," the "incalculable" thing we have often seen it described. It is capable of being analysed into its different elements, and many of these elements can be acquired and become part of one's equipment.

My paper has been concerned with functional leadership and with multiple leadership. Our present historians and biographers are strengthening the conception of multiple leadership by showing us that in order to understand any epoch we must take into account the lesser leaders. They tell us also that the number of these lesser leaders has been so steadily increasing that one of the most outstanding facts of our life today is a widely diffused leadership. Wells goes further and says that his hope for the future depends on a still more widely diffused leadership. In the past, he says, we depended on a single great leader . . . today many men and women must help to lead. In the past, he says, Aristotle led the world in science, today there are thousands of scientists each making his contribution.

Industry gives to men and women the chance for leadership, the chance to make their contribution to what all agree is the thing most needed in the world today.

Business used to be thought of as trading, managing as manipulating. Both ideas are now changing. Business is becoming a profession and management a science and an art. This means that men must prepare themselves for business as seriously as for any other profession. They must realise that they, as all professional men, are assuming grave responsibilities, that they are to take a creative part in one of the large functions of society, a part which, I believe, only trained and disciplined men can in the future hope to take with success.

Ideas for Discussion

1. This essay was originally a speech. What features of an oral presentation do you detect in its style, structure, and content?
2. What is the definition of leadership that Follett promotes? Is it a valid definition today?
3. What do Follett's ideas share with psychologists who specialize in industrial relations?

Topics for Writing

1. Write a letter inviting Follett to speak at an organizational meeting that you will chair. To complete the cycle, write her reply and your own follow-up letter.
2. Write a memo to the management staff announcing Follett's lecture.
3. Develop Follett's distinctions between multiple and functional leadership into a proposal for the human resources manager.
4. Write an essay or letter (to Follett) in which you support or refute the last paragraph of her talk.

Henry Ford
What I Learned about Business

Henry Ford, the son of a Michigan farm family, left home in 1879, when he was fifteen years old to find work as an apprentice machinist in Detroit. He soon returned home, where he continued to experiment with machinery, and in 1890 he went back to Detroit where he worked for the power company as a machinist and engineer. Two years later, in 1892, he completed the development of the automobile and formed the Detroit Automobile Company. He and his partners disagreed on how to operate the company, and he formed a new partnership with James Couzins and the Dodge Brothers, among others. Together they created the Ford Motor Company in 1903, which pioneered mass production with the Model T in 1908 and 1909. In 1919, Ford's son, Edsel, succeeded him. At Edsel Ford's death in 1943, Henry Ford resumed leadership of the company until his grandson, Henry Ford II took over in 1945. Ford's Life and Works *(1922), from which the selection that follows is taken, offers an autobiographical account of his start as a leading American industrialist.*

ON MAY 31, 1921, the Ford Motor Company turned out Car No. 5,000,000. It is out in my museum along with the gasoline buggy that I began work on thirty years before and which first ran satisfactorily along in the spring of 1893. I was running it when the bobolinks came to Dearborn and they always come on April 2nd. There is all the difference in the world in the appearance of the two vehicles and almost as much difference in construction and materials, but in fundamentals the two are curiously alike—except that the old buggy has on it a few wrinkles that we have not yet quite adopted in our modern car. For that first car or buggy, even though it had but two cylinders, would make twenty miles an hour and run sixty miles on the three gallons of gas the little tank held and is as good to-day as the day it was built. The development in methods of manufacture and in materials has been greater than the development in basic design. The whole design has been refined; the present Ford car, which is the "Model T," has four cylinders and a self starter—it is in every way a more convenient and an easier riding car. It is simpler than the first car. But almost every point in it may be found also in the first car. The changes have been brought about through experience in the making and not through any change in the basic principle—which I take to be an important fact demonstrating that, given a good idea to start with, it is better to concentrate on perfecting it than to hunt around for a new idea. One idea at a time is about as much as any one can handle.

It was life on the farm that drove me into devising ways and means to better transportation. I was born on July 30, 1863, on a farm at Dearborn, Michigan, and my earliest recollection is that, considering the results, there was too much work on the place. That is the way I still feel about farming. There is a legend that my parents were very poor and that the early days were hard ones. Certainly they were not rich, but neither were they poor. As Michigan farmers went, we were prosperous. The house in which I was born is still standing, and it and the farm are part of my present holding.

There was too much hard hand labour on our own and all other farms of the time. Even when very young I suspected that much might somehow be done in a better way. That is what took me into mechanics—although my mother always said that I was born a mechanic. I had a kind of workshop with odds and ends of metal for tools before I had anything else. In those days we did not have the toys of to-day; what we had were home made. My toys were all tools—they still are! And every fragment of machinery was a treasure.

The biggest event of those early years was meeting with a road engine about eight miles out of Detroit one day when we were driving to town. I was then twelve years old. The second biggest event was getting a watch—which happened in the same year. I remember that engine as though I had seen it only yesterday, for it was the first vehicle other than horse-drawn that I had ever seen. It was intended primarily for driving threshing machines and sawmills and was simply a portable engine and boiler mounted on wheels with a water tank and coal cart trailing behind. I had seen plenty of these engines hauled around by horses, but this one had a chain that made a connection between the engine and the rear wheels of the wagon-like frame on which the boiler was mounted. The engine was placed over the boiler and one man standing on the platform behind the boiler shovelled coal, managed the throttle, and did the steering. It had been made by Nichols, Shepard & Company of Battle Creek. I found that out at once. The engine had stopped to let us pass with our horses and I was off the wagon and talking to the engineer before my father, who was driving, knew what I was up to. The engineer was very glad to explain the whole affair. He was proud of it. He showed me how the chain was disconnected from the propelling wheel and a belt put on to drive other machinery. He told me that the engine made two hundred revolutions a minute and that the chain pinion could be shifted to let the wagon stop while the engine was still running. This last is a feature which, although in different fashion, is incorporated into modern automobiles. It was not important with steam engines, which are easily stopped and started, but it became very important with the gasoline engine. It was that engine which took me into automotive transportation. I tried to make models of it, and some years later I did make one that ran very well, but from the time I saw that road engine as a boy of twelve right forward to to-day, my great interest has been in making a machine that would travel the roads. Driving to town I always had a pocket full of trinkets—nuts, washers, and odds and ends of machinery. Often I took a broken watch and tried to put it together. When I was thirteen I managed for the first time to put a watch together so that it would keep time. By the time I was fifteen I could do almost anything in watch repairing—although my tools were of the crudest. There is an immense amount to be learned simply by tinkering with things. It is not possible to learn from books how everything is made—and a real mechanic ought to know how nearly everything is made. Machines are to a mechanic what books are to a writer. He gets ideas from them, and if he has any brains he will apply those ideas.

From the beginning I never could work up much interest in the labour of farming. I wanted to have something to do with machinery. My father was not entirely in sympathy with my bent toward mechanics. He thought that I ought to be a farmer. When I left school at seventeen and became an apprentice in the machine shop of the Drydock Engine Works I was all but given up for lost. I passed my apprenticeship without trouble—that is, I was qualified to be a machinist long before my three-year term had expired—and having a liking for fine work and a leaning toward watches I worked nights at repairing in a jewellery shop. At one period of those early days I think that I must have had fully three hundred watches. I thought that I could build a serviceable watch for around thirty cents and nearly started in the business. But I did not because I figured out that watches were not universal necessities, and therefore people generally would not buy them. Just how I reached that surprising conclusion I am unable to state. I did not like the ordinary jewellery and watchmaking work excepting where the job was hard to do. Even then I wanted to make something in quantity. It was just about the time when the standard railroad time was being arranged. We had formerly been on sun time and for quite a while, just as in our present daylight-saving days, the railroad time differed from the local time. That bothered me a good deal and so I succeeded in making a watch that kept both times. It had two dials and it was quite a curiosity in the neighbourhood.

In 1879—that is, about four years after I first saw that Nichols-Shepard machine—I managed to get a chance to run one and when my apprenticeship was over I worked with a local representative of the Westinghouse Company of Schenectady as an expert in the setting up and repair of their road engines. The engine they put out was much the same as the Nichols-Shepard engine excepting that the engine was up in front, the boiler in the rear, and the power was applied to the back wheels by a belt. They could make twelve miles an hour on the road even though the self-propelling feature was only an incident of the construction. They were sometimes used as tractors to pull heavy loads and, if the owner also happened to be in the threshing-machine business, he hitched his threshing machine and other paraphernalia to the engine in moving from farm to farm. What bothered me was the weight and the cost. They weighed a couple of tons and were far too expensive to be owned by other than a farmer with a great deal of land. They were mostly employed by people who went into threshing as a business or who had sawmills or some other line that required portable power.

Even before that time I had the idea of making some kind of a light steam car that would take the place of horses—more especially, however, as a tractor to attend to the excessively hard labour of ploughing. It occurred to me, as I remember somewhat vaguely, that precisely the same idea might be applied to a carriage or a wagon on the road. A horseless carriage was a common idea. People had been talking about carriages without horses for many years back—in fact, ever since the steam engine was invented—but the idea of the carriage at first did not seem so practical to me as the idea of an engine to do the harder farm work, and of all the work on the farm ploughing was the hardest. Our roads were poor and we had not the habit of getting around. One of the most remarkable features of the automobile on the farm is the way that it has broadened the farmer's life. We simply took for granted that unless the errand were urgent we would not go to town, and I think we rarely made more than a trip a week. In bad weather we did not go even that often.

Being a full-fledged machinist and with a very fair workshop on the farm it was not difficult for me to build a steam wagon or tractor. In the building of it came the idea that perhaps it might be made for road use. I felt perfectly certain that horses, considering all the bother of attending them and the expense of feeding, did not earn their keep. The obvious thing to do was to design and build a steam engine that would be light enough to run an ordinary wagon or to pull a plough. I thought it more important first to develop the tractor. To lift farm drudgery off flesh and blood and lay it on steel and motors has been my most constant ambition. It was circumstances that took me first into the actual manufacture of road cars. I found eventually that people were more interested in something that would travel on the road than in something that would do the work on the farms. In fact, I doubt that the light farm tractor could have been introduced on the farm had not the farmer had his eyes opened slowly but surely by the automobile. But that is getting ahead of the story. I thought the farmer would be more interested in the tractor.

I built a steam car that ran. It had a kerosene-heated boiler and it developed plenty of power and a neat control—which is so easy with a steam throttle. But the boiler was dangerous. To get the requisite power without too big and heavy a power plant required that the engine work under high pressure; sitting on a high-pressure steam boiler is not altogether pleasant. To make it even reasonably safe required an excess of weight that nullified the economy of the high pressure. For two years I kept experimenting with various sorts of boilers—the engine and control problems were simple enough—and then I definitely abandoned the whole idea of running a road vehicle by steam. I knew that in England they had what amounted to locomotives running on the roads hauling lines of trailers and also there was no difficulty in designing a big steam tractor for use on a large farm. But ours were not then English roads; they would have stalled or racked to pieces the strongest and heaviest road tractor. And anyway the manufacturing of a big tractor which only a few wealthy farmers could buy did not seem to me worth while.

But I did not give up the idea of a horseless carriage. The work with the Westinghouse representative only served to confirm the opinion I had formed that steam was not suitable for light vehicles. That is why I stayed only a year with that company. There was nothing more that the big steam tractors and engines could teach me and I did not want to waste time on something that would lead nowhere. A few years before—it was while I was an apprentice—I read in the *World of Science,* an English publication, of the "silent gas engine" which was then coming out in England. I think it was the Otto engine. It ran with illuminating gas, had a single large cylinder, and the power impulses being thus intermittent required an extremely heavy fly-wheel. As far as weight was concerned it gave nothing like the power per pound of metal that a steam engine gave, and the use of illuminating gas seemed to dismiss it as even a possibility for road use. It was interesting to me only as all machinery was interesting. I followed in the English and American magazines which we got in the shop the development of the engine and most particularly the hints of the possible replacement of the illuminating gas fuel by a gas formed by the vaporization of gasoline. The idea of gas engines was by no means new, but this was the first time that a really serious effort had been made to put them on the market. They were received with interest rather than enthusiasm and I do not recall any one who thought that the internal combustion engine could ever have more than a limited use. All the wise people demonstrated conclusively that the engine could not compete with steam. They never thought that it might carve out a career for itself. That is the way with wise people—they are so wise and practical that they always know to a dot just why something cannot be done; they always know the limitations. That is why I never employ an expert in full bloom. If ever I wanted to kill opposition by unfair means I would endow the opposition with experts. They would have so much good advice that I could be sure they would do little work.

The gas engine interested me and I followed its progress, but only from curiosity, until about 1885 or 1886 when, the steam engine being discarded as the motive power for the carriage that I intended some day to build, I had to look around for another sort of motive power. In 1885 I repaired an Otto engine at the Eagle Iron Works in Detroit. No one in town knew anything about them. There was a rumour that I did and, although I had never before been in contact with one, I undertook and carried through the job. That gave me a chance to study the new engine at first hand and in 1887 I built one on the Otto four-cycle model just to see if I understood the principles. "Four cycle" means that the piston traverses the cylinder four times to get one power impulse. The first stroke draws in the gas, the second compresses it, and the third is the explosion or power stroke, while the fourth stroke exhausts the waste gas. The little model worked well enough; it had a one-inch bore and a three-inch stroke, operated with gasoline, and while it did not develop much power, it was slightly lighter in proportion than the engines being offered commercially. I gave it away later to a young man who wanted it for something or other and whose name I have forgotten; it was eventually destroyed. That was the beginning of the work with the internal combustion engine.

I was then on the farm to which I had returned, more because I wanted to experiment than because I wanted to farm, and, now being an all-around machinist, I had a first-class workshop to replace the toy shop of earlier days. My father offered me forty acres of timber land, provided I gave up being a machinist. I agreed in a provisional way, for cutting the timber gave me a chance to get married. I fitted out a sawmill and a portable engine and started to cut out and saw up the timber on the tract. Some of the first of that lumber went into a cottage on my new farm and in it we began our married life. It was not a big house—thirty-one feet square and only a story and a half high—but it was a comfortable place. I added to it my workshop, and when I was not cutting timber I was working on the gas engines—learning what they were and how they acted. I read everything I could find, but the greatest knowledge came from the work. A gas engine is a mysterious sort of thing—it will not always go the way it should. You can imagine how those first engines acted!

It was in 1890 that I began on a double-cylinder engine. It was quite impractical to consider the single cylinder for transportation purposes—the fly-wheel had to be entirely too heavy. Between making the first four-cycle engine of the Otto type and the start on a double cylinder I had made a great many experimental engines out of tubing. I fairly knew my way about. The double cylinder I thought could be applied to a road vehicle and my original idea was to put it on a bicycle with a direct connection to the crankshaft and allowing for the rear wheel of the bicycle to act as the balance wheel. The speed was going to be varied only by the throttle. I never carried out this plan because it soon became apparent that the engine, gasoline tank, and the various necessary controls would be entirely too heavy for a bicycle. The plan of the two opposed cylinders was that, while one would be delivering power the other would be exhausting. This naturally would not require so heavy a fly-wheel to even the application of power. The work started in my shop on the farm. Then I was offered a job with the Detroit Electric Company as an engineer and machinist at forty-five dollars a month. I took it because that was more money than the farm was bringing me and I had decided to get away from farm life anyway. The timber had all been cut. We rented a house on Bagley Avenue, Detroit. The workshop came along and I set it up in a brick shed at the back of the house. During the first several months I was in the night shift at the electric-light plant—which gave me very little time for experimenting—but after that I was in the day shift and every night and all of every Saturday night I worked on the new motor. I cannot say that it was hard work. No work with interest is ever hard. I always am certain of results. They always come if you work hard enough. But it was a very great thing to have my wife even more confident than I was. She has always been that way.

I had to work from the ground up—that is, although I knew that a number of people were working on horseless carriages, I could not know what they were doing. The hardest problems to overcome were in the making and breaking of the spark and in the avoidance of excess weight. For the transmission, the steering gear, and the general construction, I could draw on my experience with the steam tractors. In 1892 I completed my first motor car, but it was not until the spring of the following year that it ran to my satisfaction. This first car had something of the appearance of a buggy. There were two cylinders with a two-and-a-half-inch bore and a six-inch stroke set side by side and over the rear axle. I made them out of the exhaust pipe of a steam engine that I had bought. They developed about four horsepower. The power was transmitted from the motor to the countershaft by a belt and from the countershaft to the rear wheel by a chain. The car would hold two people, the seat being suspended on posts and the body on elliptical springs. There were two speeds—one of ten and the other of twenty miles per hour—obtained by shifting the belt, which was done by a clutch lever in front of the driving seat. Thrown forward, the lever put in the high speed; thrown back, the low speed; with the lever upright the engine could run free. To start the car it was necessary to turn the motor over by hand with the clutch free. To stop the car one simply released the clutch and applied the foot brake. There was no reverse, and speeds other than those of the belt were obtained by the throttle. I bought the iron work for the frame of the carriage and also the seat and the springs. The wheels were twenty-eight-inch wire bicycle wheels with rubber tires. The balance wheel I had cast from a pattern that I made and all of the more delicate mechanism I made myself. One of the features that I discovered necessary was a compensating gear that permitted the same power to be applied to each of the rear wheels when turning corners. The machine altogether weighed about five hundred pounds. A tank under the seat held three gallons of gasoline which was fed to the motor through a small pipe and a mixing valve. The ignition was by electric spark. The original machine was air-cooled—or to be more accurate, the motor simply was not cooled at all. I found that on a run of an hour or more the motor heated up, and so I very shortly put a water jacket around the cylinders and piped it to a tank in the rear of the car over the cylinders.

Nearly all of these various features had been planned in advance. That is the way I have always worked. I draw a plan and work out every detail on the plan before starting to build. For otherwise one will waste a great deal of time in makeshifts as the work goes on and the finished article will not have coherence. It will not be rightly proportioned. Many inventors fail because they do not distinguish between planning and experimenting. The largest building difficulties that I had were in obtaining the proper materials. The next were with tools. There had to be some adjustments and changes in details of the design, but what held me up most was that I had neither the time nor the money to search for the best material for each part. But in the spring of 1893 the machine was running to my partial satisfaction and giving an opportunity further to test out the design and material on the road.

II

My "gasoline buggy" was the first and for a long time the only automobile in Detroit. It was considered to be something of a nuisance, for it made a racket and it scared horses. Also it blocked traffic. For if I stopped my machine anywhere in town a crowd was around it before I could start up again. If I left it alone even for a minute some inquisitive person always tried to run it. Finally, I had to carry a chain and chain it to a lamp post whenever I left it anywhere. And then there was trouble with the police. I do not know quite why, for my impression is that there were no speed-limit laws in those days. Anyway, I had to get a special permit from the mayor and thus for a time enjoyed the distinction of being the only licensed chauffeur in America. I ran that machine about one thousand miles through 1895 and 1896 and then sold it to Charles Ainsley of Detroit for two hundred dollars. That was my first sale. I had built the car not to sell but only to experiment with. I wanted to start another car. Ainsley wanted to buy. I could use the money and we had no trouble in agreeing upon a price.

It was not at all my idea to make cars in any such petty fashion. I was looking ahead to production, but before that could come I had to have something to produce. It does not pay to hurry. I started a second car in 1896; it was much like the first but a little lighter. It also had the belt drive which I did not give up until some time later; the belts were all right excepting in hot weather. That is why I later adopted gears. I learned a great deal from that car. Others in this country and abroad were building cars by that time, and in 1895 I heard that a Benz car from Germany was on exhibition in Macy's store in New York. I travelled down to look at it but it had no features that seemed worth while. It also had the belt drive, but it was much heavier than my car. I was working for lightness; the foreign makers have never seemed to appreciate what light weight means. I built three cars in all in my home shop and all of them ran for years in Detroit. I still have the first car; I bought it back a few years later from a man to whom Mr. Ainsley had sold it. I paid one hundred dollars for it.

During all this time I kept my position with the electric company and gradually advanced to chief engineer at a salary of one hundred and twenty-five dollars a month. But my gas-engine experiments were no more popular with the president of the company than my first mechanical leanings were with my father. It was not that my employer objected to experiments—only to experiments with a gas engine. I can still hear him say:

"Electricity, yes, that's the coming thing. But gas—no."

He had ample grounds for his skepticism—to use the mildest terms. Practically no one had the remotest notion of the future of the internal combustion engine, while we were just on the edge of the great electrical development. As with every comparatively new idea, electricity was expected to do much more than we even now have any indication that it can do. I did not see the use of experimenting with electricity for my purposes. A road car could not run on a trolley even if trolley wires had been less expensive; no storage battery was in sight of a weight that was practical. An electrical car had of necessity to be limited in radius and to contain a large amount of motive machinery in proportion to the power exerted. That is not to say that I held or now hold electricity cheaply; we have not yet begun to use electricity. But it has its place, and the internal combustion engine has its place. Neither can substitute for the other—which is exceedingly fortunate.

I have the dynamo that I first had charge of at the Detroit Edison Company. When I started our Canadian plant I bought it from an office building to which it had been sold by the electric company, had it revamped a little, and for several years it gave excellent service in the Canadian plant. When we had to build a new power plant, owing to the increase in business, I had the old motor taken out to my museum—a room out at Dearborn that holds a great number of my mechanical treasures.

The Edison Company offered me the general superintendency of the company but only on condition that I would give up my gas engine and devote myself to something really useful. I had to choose between my job and my automobile. I chose the automobile, or rather I gave up the job—there was really nothing in the way of a choice. For already I knew that the car was bound to be a success. I quit my job on August 15, 1899, and went into the automobile business.

It might be thought something of a step, for I had no personal funds. What money was left over from living was all used in experimenting. But my wife agreed that the automobile could not be given up—that we had to make or break. There was no "demand" for automobiles—there never is for a new article. They were accepted in much the fashion as was more recently the airplane. At first the "horseless carriage" was considered merely a freak notion and many wise people explained with particularity why it could never be more than a toy. No man of money even thought of it as a commercial possibility. I cannot imagine why each new means of transportation meets with such opposition. There are even those to-day who shake their heads and talk about the luxury of the automobile and only grudgingly admit that perhaps the motor truck is of some use. But in the beginning there was hardly any one who sensed that the automobile could be a large factor in industry. The most optimistic hoped only for a development akin to that of the bicycle. When it was found that an automobile really could go and several makers started to put out cars, the immediate query was as to which would go fastest. It was a curious but natural development—that racing idea. I never thought anything of racing, but the public refused to consider the automobile in any light other than as a fast toy. Therefore later we had to race. The industry was held back by this initial racing slant, for the attention of the makers was diverted to making fast rather than good cars. It was a business for speculators.

A group of men of speculative turn of mind organized, as soon as I left the electric company, the Detroit Automobile Company to exploit my car. I was the chief engineer and held a small amount of the stock. For three years we continued making cars more or less on the model of my first car. We sold very few of them; I could get no support at all toward making better cars to be sold to the public at large. The whole thought was to make to order and to get the largest price possible for each car. The main idea seemed to be to get the money. And being without authority other than my engineering position gave me, I found that the new company was not a vehicle for realizing my ideas but merely a money-making concern—that did not make much money. In March, 1902, I resigned, determined never again to put myself under orders. The Detroit Automobile Company later became the Cadillac Company under the ownership of the Lelands, who came in subsequently.

I rented a shop—a one-story brick shed—at 81 Park Place to continue my experiments and to find out what business really was. I thought that it must be something different from what it had proved to be in my first adventure.

The year from 1902 until the formation of the Ford Motor Company was practically one of investigation. In my little one-room brick shop I worked on the development of a four-cylinder motor and on the outside I tried to find out what business really was and whether it needed to be quite so selfish a scramble for money as it seemed to be from my first short experience. From the period of the first car, which I have described, until the formation of my present company I built in all about twenty-five cars, of which nineteen or twenty were built with the Detroit Automobile Company. The automobile had passed from the initial stage where the fact that it could run at all was enough, to the stage where it had to show speed. Alexander Winton of Cleveland, the founder of the Winton car, was then the track champion of the country and willing to meet all comers. I designed a two-cylinder enclosed engine of a more compact type than I had before used, fitted it into a skeleton chassis, found that I could make speed, and arranged a race with Winton. We met on the Grosse Point track at

Detroit. I beat him. That was my first race, and it brought advertising of the only kind that people cared to read.

The public thought nothing of a car unless it made speed—unless it beat other racing cars. My ambition to build the fastest car in the world led me to plan a four-cylinder motor. But of that more later.

The most surprising feature of business as it was conducted was the large attention given to finance and the small attention to service. That seemed to me to be reversing the natural process which is that the money should come as the result of work and not before the work. The second feature was the general indifference to better methods of manufacture as long as whatever was done got by and took the money. In other words, an article apparently was not built with reference to how greatly it could serve the public but with reference solely to how much money could be had for it—and that without any particular care whether the customer was satisfied. To sell him was enough. A dissatisfied customer was regarded not as a man whose trust had been violated, but either as a nuisance or as a possible source of more money in fixing up the work which ought to have been done correctly in the first place. For instance, in automobiles there was not much concern as to what happened to the car once it had been sold. How much gasoline it used per mile was of no great moment; how much service it actually gave did not matter; and if it broke down and had to have parts replaced, then that was just hard luck for the owner. It was considered good business to sell parts at the highest possible price on the theory that, since the man had already bought the car, he simply had to have the part and would be willing to pay for it.

The automobile business was not on what I would call an honest basis, to say nothing of being, from a manufacturing standpoint, on a scientific basis, but it was no worse than business in general. That was the period, it may be remembered, in which many corporations were being floated and financed. The bankers, who before then had confined themselves to the railroads, got into industry. My idea was then and still is that if a man did his work well, the price he would get for that work, the profits and all financial matters, would care for themselves and that a business ought to start small and build itself up and out of its earnings. If there are no earnings then that is a signal to the owner that he is wasting his time and does not belong in that business. I have never found it necessary to change those ideas, but I discovered that this simple formula of doing good work and getting paid for it was supposed to be slow for modern business. The plan at that time most in favour was to start off with the largest possible capitalization and then sell all the stock and all the bonds that could be sold. Whatever money happened to be left over after all the stock and bond-selling expenses and promoters, charges and all that, went grudgingly into the foundation of the business. A good business was not one that did good work and earned a fair profit. A good business was one that would give the opportunity for the floating of a large amount of stocks and bonds at high prices. It was the stocks and bonds, not the work, that mattered. I could not see how a new business or an old business could be expected to be able to charge into its product a great big bond interest and then sell the product at a fair price. I have never been able to see that.

I have never been able to understand on what theory the original investment of money can be charged against a business. Those men in business who call themselves financiers say that money is "worth" 6 per cent. or 5 per cent. or some other per cent., and that if a business has one hundred thousand dollars invested in it, the man who made the investment is entitled to charge an interest payment on the money, because, if instead of putting that money into the business he had put it into a savings bank or into certain securities, he could have a certain fixed return. Therefore they say that a proper charge against the operating expenses of a business is the interest on this money. This idea is at the root of many business failures and most service failures. Money is not worth a particular amount. As money it is not worth anything, for it will do nothing of itself. The only use of money is to buy tools to work with or the product of tools. Therefore money is worth what it will help you to produce or buy and no more. If a man thinks that his money will earn 5 per cent. or 6 per cent. he ought to place it where he can get that return, but money placed in a business is not a charge on the business—or, rather, should not be. It ceases to be money and becomes, or should become, an engine of production, and it is therefore worth what it produces—and not a fixed sum according to some scale that has no bearing upon the particular business in which the money has been placed. Any return should come after it has produced, not before.

Business men believed that you could do anything by "financing" it. If it did not go through on the first financing then the idea was to "refinance." The process of "refinancing" was simply the game of sending good money after bad. In the majority of cases the need of refinancing arises from bad management, and the effect of refinancing is simply to pay the poor managers to keep up their bad management a little longer. It is merely a postponement of the day of judgment. This makeshift of refinancing is a device of speculative financiers. Their money is no good to them unless they can connect it up with a place where real work is being done, and that they cannot do unless, somehow, that place is poorly managed. Thus, the speculative financiers delude themselves that they are putting their money out to use. They are not; they are putting it out to waste.

I determined absolutely that never would I join a company in which finance came before the work or in which bankers or financiers had a part. And further that, if there were no way to get started in the kind of business that I thought could be managed in the interest of the public, then I simply would not get started at all. For my own short experience, together with what I saw going on around me, was quite enough proof that business as a mere money-making game was not worth giving much thought to and was distinctly no place for a man who wanted to accomplish anything. Also it did not seem to me to be the way to make money. I have yet to have it demonstrated that it is the way. For the only foundation of real business is service.

A manufacturer is not through with his customer when a sale is completed. He has then only started with his customer. In the case of an automobile the sale of the machine is only something in the nature of an introduction. If the machine does not give service, then it is better for the manufacturer if he never had the introduction, for he will have the worst of all advertisements—a dissatisfied customer. There was something more than a tendency in the early days of the automobile to regard the selling of a machine as the real accomplishment and that thereafter it did not matter what happened to the buyer. That is the shortsighted salesman-on-commission attitude. If a salesman is paid only for what he sells, it is not to be expected that he is going to exert any great effort on a customer out of whom no more commission is to be made. And it is right on this point that we later made the largest selling argument for the Ford. The price and the quality of the car would undoubtedly have made a market, and a large market. We went beyond that. A man who bought one of our cars was in my opinion entitled to continuous use of that car, and therefore if he had a breakdown of any kind it was our duty to see that his machine was put into shape again at the earliest possible moment. In the success of the Ford car the early provision of service was an outstanding element. Most of the expensive cars of that period were ill provided with service stations. If your car broke down you had to depend on the local repair man—when you were entitled to depend upon the manufacturer. If the local repair man were a forehanded sort of a person, keeping on hand a good stock of parts (although on many of the cars the parts were not interchangeable), the owner was lucky. But if the repair man were a shiftless person, with an inadequate knowledge of automobiles and an inordinate desire to make a good thing out of every car that came into his place for repairs, then even a slight breakdown meant weeks of laying up and a whopping big repair bill that had to be paid before the car could be taken away. The repair men were for a time the largest menace to the automobile industry. Even as late as 1910 and 1911 the owner of an automobile was regarded as essentially a rich man whose money ought to be taken away from him. We met that situation squarely and at the very beginning. We would not have our distribution blocked by stupid, greedy men.

That is getting some years ahead of the story, but it is control by finance that breaks up service because it looks to the immediate dollar. If the first consideration is to earn a certain amount of money, then, unless by some stroke of luck matters are going especially well and there is a surplus over for service so that the operating men may have a chance, future business has to be sacrificed for the dollar of to-day.

And also I noticed a tendency among many men in business to feel that their lot was hard—they worked against a day when they might retire and live on an income—get out of the strife. Life to them was a battle to be ended as soon as possible. That was another point I could not understand, for as I reasoned, life is not a battle except with our own tendency to sag with the downpull of "getting settled." If to petrify is success, all one has to do is to humour the lazy side of the mind; but if to grow is success, then one must wake up anew every morning and keep awake all day. I saw great businesses become but the ghost of a name because someone thought they could be managed just as they were always managed, and though the management may have been most excellent in its day, its excellence consisted in its alertness to its day, and not in slavish following of its yesterdays. Life, as I see it, is not a location, but a journey. even the man who most feels himself "settled" is not settled—he is probably sagging back. Everything is in flux, and was meant to be. Life flows. We may live at the same number of the street, but it is never the same man who lives there.

And out of the delusion that life is a battle that may be lost by a false move grows, I have noticed, a great love for regularity. Men fall into the half-alive habit. Seldom does the cobbler take up with the new-fangled way of soling shoes, and seldom does the artisan willingly take up with new methods in his trade. Habit conduces to a certain inertia, and any disturbance of it affects the mind like trouble. It will be recalled that when a study was made of shop methods, so that the workmen might be taught to produce with less useless motion and fatigue, it was most opposed by the workmen themselves. Though they suspected that it was simply a game to get more out of them, what most irked them was that it interfered with the well-worn grooves in which they had become accustomed to move. Business men go down with their businesses because they like the old way so well they cannot bring themselves to change. One sees them all about—men who do not know that yesterday is past, and who woke up this morning with their last year's ideas. It could almost be written down as a formula that when a man begins to think that he has at last found his method he had better begin a most searching examination of himself to see whether some part of his brain has not gone to sleep. There is a subtle danger in a man thinking that he is "fixed" for life. It indicates that the next jolt of the wheel of progress is going to fling him off.

There is also the great fear of being thought a fool. So many men are afraid of being considered fools. I grant that public opinion is a powerful police influence for those who need it. Perhaps it is true that the majority of men need the restraint of public opinion. Public opinion may keep a man better than he would otherwise be—if not better morally, at least better as far as his social desirability is concerned. But it is not a bad thing to be a fool for righteousness' sake. The best of it is that such fools usually live long enough to prove that they were not fools—or the work they have begun lives long enough to prove they were not foolish.

The money influence—the pressing to make a profit on an "investment"—and its consequent neglect of or skimping of work and hence of service showed itself to me in many ways. It seemed to be at the bottom of most troubles. It was the cause of low wages—for without well-directed work high wages cannot be paid. And if the whole attention is not given to the work it cannot be well directed. Most men want to be free to work; under the system in use they could not be free to work. During my first experience I was not free—I could not give full play to my ideas. Everything had to be planned to make money; the last consideration was the work. And the most curious part of it all was the insistence that it was the money and not the work that counted. It did not seem to strike any one as illogical that money should be put ahead of work—even though everyone had to admit that the profit had to come from the work. The desire seemed to be to find a short cut to money and to pass over the obvious short cut—which is through the work.

Take competition; I found that competition was supposed to be a menace and that a good manager circumvented his competitors by getting a monopoly through artificial means. The idea was that there were only a certain number of people who could buy and that it was necessary to get their trade ahead of someone else. Some will remember that later many of the automobile manufacturers entered into an association under the Selden Patent just so that it might be legally possible to control the price and the output of automobiles. They had the same idea that so many trades unions have—the ridiculous notion that more profit can be had doing less work than more. The plan, I believe, is a very antiquated one. I could not see then and am still unable to see that there is not always enough for the man who does his work; time spent in fighting competition is wasted; it had better be spent in doing the work. There are always enough people ready and anxious to buy, provided you supply what they want and at the proper price—and this applies to personal services as well as to goods.

Ideas for Discussion

1. What problems did Ford encounter as he began auto manufacturing? How did he address them?
2. Provide a summary of Ford's "business principles."
3. Has Ford romanticized his early life in any way? Cite examples.

Topics for Writing

1. Design an advertising campaign for Ford's new Model T and his company.
2. Prepare a report on the current condition of Ford Motor Company. Is there any lingering evidence of its founder in its current status?
3. As a decision maker, prepare a new product development report suggesting something new Ford could manufacture.

Robert Sobel
Cyrus Hall McCormick: From Farm Boy to Tycoon

Robert Sobel teaches history at the New College of Hofstra University in New York. Since 1966, he has served Greenwood Press as an editor in business history. He is the author of eight books on Wall Street and related subjects in American business. This biographical profile of Cyrus McCormick is taken from Sobel's The Entrepreneurs *(1974), which profiles major figures in American business and industry. Written for a mass market, Sobel's style is at once interesting and informative.*

THE UNITED STATES was an agrarian nation at the time of the founding of the Boston Manufacturing Company, and it would remain essentially agrarian to the end of the century.[1] Lowell's factory at Waltham, the Almy and Brown operation in Providence, and the dozens of others that dotted the landscape in the 1820s and 1830s were viewed in somewhat the same way as space launchers are today—as intrusions on the countryside, on a land clearly meant for farmers. America had excellent soils, inadequate capital resources, and was chronically short of labor. Such transportation as had been developed was geared to the agrarian sector, used by farmers to bring goods to market, and by merchants to carry supplies to the farms. Nature and economics appeared to have destined the land for grains and cotton, not for factories. So Jefferson appeared correct in appealing for a nation of farmers—or at least he would appear to have been accurate in his projections during his lifetime. Even Hamilton, with his bias toward manufacturing, asked little more than self-sufficiency for the nation in all goods. Neither man spoke of an industrial America. How could they, given the technology of the period in which they lived, the mood of the populace, and the heritage of European civilization they shared?

Of the merchant class and himself one of the precursors of an age of industry, Francis Lowell might be viewed as little more than an appendage of the cotton culture, existing to process ginned cotton into cloth and so serve the plantations of the South. This, it would appear, was the view of men like Calhoun, who visited Waltham and surrounding towns in the 1820s and 1830s to see what the Yankees were doing with their cotton. Just as a previous generation of the Boston aristocrats had carried and sold the products of agrarian America overseas, and then imported goods for sale to farmers, so the men of the 1820s and 1830s—their sons and grandsons—would process the raw materials of the farms into finished goods, and most of these would be consumed by farmers. Industry depended upon and existed to serve agriculture; the American business tradition had its roots in the soil.

Two northerners, Eli Whitney and Francis Cabot Lowell, had pioneered in developments that helped make cotton king in the South. A Virginian, Cyrus McCormick, developed the technology that created emperor wheat, which a generation later would triumph in the battle between the two crops. In the process McCormick accomplished much more. If Lowell was the father of the American factory system, McCormick was the precursor of and model for the big businessmen who followed him, men whose ties to the soil were indirect and in some cases nonexistent. Yet McCormick, like Lowell, had his origins in products of the soil.

McCormick is generally credited with being the inventor of the reaper, a device that cut wheat in such a way that the grain would not fall from the stalks, leaving it in the field for a binder, who would then gather it and arrange the stalks in orderly sheaves. Later on, McCormick helped develop a machine that would bind the stalks as well, and his name is associated with a variety of other devices used in the wheat fields. But even McCormick's most ardent champions concede that had he never lived, effective reapers would have been invented, while lingering doubt as to his inventive genius remains to this day.

In Chicago he is still considered one of the founders of the modern city that emerged from the great fire that leveled the wooden town in 1871. McCormick's role in rebuilding Chicago was significant, but not central, and in any case Chicago did more for him than he for it, while the fire proved a blessing in disguise in that it enabled McCormick to create a modern factory instead of modifying an already obsolete facility.

Some historians view the Civil War as an economic contest between cotton and wheat, and so credit McCormick with a central role in the North's victory. The war was far more than that, however, and McCormick's loyalty to the Union cause was suspect at the time.

The popular literature of a half century ago pictures McCormick as the man who unveiled American inventiveness to the Europeans, with the reapers of his design changing the face of agricultural Europe, causing social changes of a major magnitude, and creating relative prosperity for the masses of the old continent. But the Europeans had developed reapers and other farm equipment too, some of which were superior to the McCormick machines and in most ways competitive in price, while McCormick's American competitors did better in the European markets than he for a while. The reaper would have been invented, the North would have won the Civil War, European as well as American wheat fields would have been mechanized, and Chicago rebuilt, all without the aid of Cyrus Hall McCormick.

Having said all this is not to deny that McCormick remains a major figure in the American business tradition. It is not McCormick the inventor, however, who made the major contribution. Instead, McCormick's most significant work came in the area of sales and distribution, servicing and credit, popularization and education. Without these accomplishments and armed only with a superior machine, McCormick might yet have failed. But with his innovations in distribution and popularization, together with a set of machines that in some ways were inferior to those of his competitors, McCormick triumphed.

McCormick's grandfather arrived in America in 1735 and settled in Pennsylvania. His fifth son, Robert, was born three years later, and grew up to become a farmer and part-time weaver. Robert McCormick moved south with his family during the Revolution, arriving in Virginia in 1779. Soon after, his first wife died, and Robert married Mary Ann Hall, the daughter of a local farmer. Their son, Cyrus Hall McCormick, was born in 1809.

In most respects Cyrus McCormick's childhood was like that of the majority of farm boys of that time and place. He managed to survive diseases, received some schooling, attended the local Presbyterian church, and courted the local girls. At an early age he went into the fields to work alongside his father, and learned to tell whether the soils were good or worn out, whether the crops that year would be successful or fail, and how to use horse and slave power effectively. By all accounts Robert McCormick was a good farmer, a man who learned his skills not only from practice but by reading agricultural journals and swapping information with his fellows. In one respect, however, he was different from most. A half century later, Cyrus McCormick wrote:

> My father was both mechanical and inventive, and could and did at that time, use the tools of his shops in making any piece of machinery he wanted. He invented, made and patented several more or less valuable agricultural implements, but, with perhaps less inventive speculation than some others, most of his inventions dropped into disuse after the lapse of some years. Among these were a threshing machine, a hydraulic machine, a hemp-breaking machine, with a peculiar horsepower adapted to it, and others.[2]

All of the machines Robert McCormick worked on were related to wheat farming in one way or another, with the sole exception of a method or device to aid in the teaching of "performing on the violin." Later on, McCormick's son recalled that he was always more interested in working with his father in the shops than in the fields. By the time Cyrus McCormick was in his early teens he and Robert had become co-inventors in the shop, trading information and seeking improvements as they went. None of Robert McCormick's devices ever succeeded financially. He would use them on his own farm and on occasion sell one or another to a neighbor, all of whom appeared interested in what he was doing. In the process, however, he was training his son and educating him in the arts of farming and machines. This too was not unusual in the Virginia of the early nineteenth century. By working with and observing his father, Cyrus McCormick learned of machines, wheat, patents, and the frustrations that accompanied farming, inventing, manufacturing, and selling.

Robert McCormick was particularly interested in his thresher, and so were his neighbors. He claimed it was superior in design and construction to others, built a few in his shop, and sold them to other Virginia farmers for $70 apiece. Actually, this thresher was not unusual in any way, and probably no better in performance than others turned out by rival farmer-inventors. Nor was the idea of such implements novel. Throughout history wheat farming had been seasonal. A busy planting period was followed by the long, tedious, and steady work of caring for the crop. Then came the harvest, a few days during which the wheat was cut, gathered, and threshed. Afterward the farmers would settle down for several months, preparing for the next planting season.

The lone farmer had little difficulty caring for growing wheat or preparing for planting. He might need help at these times, and this usually came from members of his family. The harvest was the critical time. The wheat had to be cut and gathered at the right time, for otherwise it would break down and decay. It could not be left in its cut state in the field, but had quickly to be taken to the barn for threshing. On small farms the entire family would work around the clock during the four-to-ten day harvest season, hoping to gather and thresh as much of the crop as possible. Sometimes this labor was insufficient, and then the hogs and cattle would be turned loose to feed on the already rotting grain. This constituted waste, the fattening of livestock on expensive fodder. The short harvest season could bring disaster to larger farmers whose holdings could not be serviced by their immediate families. At such times as harvest they would seek occasional labor, paying whatever wages were then current. These farmers could not rely upon friends and neighbors for assistance, since all were busy with their own harvests and storage of crops. Thus, from the first, the wheat farmers recognized the need for farm machinery. These machines would be most useful at the time of planting. They would be vital at harvests.

The lack of help during harvests was one of the major factors holding down the size of farms in wheat areas. Slaves were not the answer: it would hardly pay to own a slave for those few days and not be able to use him effectively the rest of the year. Nor would occasional labor do except in areas with large numbers of unemployed men and women, and such was not the case in America. Clearly American farmers needed a well-designed reaper, one that would cut the wheat cleanly and not break down at critical moments. Many farmers were at work on just such a device. Robert McCormick joined them, and worked on a reaper as well as a thresher.

Medieval farmers must have dreamed of such machines. They would thresh wheat by stamping on it and then blowing the chaff away, or they would wait for the wind to do the job for them. These same farmers used scythes and sickles to cut the grain stalks. They would stumble through the wheat fields with large, ungainly scythes in hand and flail at the wheat. Or they would go more slowly, cutting clumps of stalks with the smaller but more effective sickles, taking more time in the process. Both hand tools were ineffective, and they knew it. At first they tried to design better sickles and scythes. Then they turned to the development of other hand tools. Later on they would try to invent reaping machines.

Actually, the Romans made stabs at developing reapers—they are mentioned in Pliny and Palladius. Crude "harvesting carts" were used by some English farmers in the late sixteenth century. With the coming of the agricultural revolution in England in the late eighteenth century, several models appeared, none of them very effective. By the early nineteenth century, dozens of inventors in England were experimenting with reaping devices.[3] Some produced working models, but few of them were able to sell many to skeptical farmers. It was one thing to advertise the miracle machines but quite another to convince customers they could do the job as claimed. Farmers were asked to purchase the machines at prices they really couldn't afford, with no guarantee as to performance and no recourse if they failed. Some of the inventors would take their devices into wheat fields, invite local farmers to watch, and then give demonstrations. Most of these ended in disaster, with the grain uncut or butchered; and more often than not, the machine broke down. One inventor tried a novel approach in 1814, when he advertised:

> J. Dobbs most respectfully informs his Friends and the Public, that having invented a Machine to expedite the Reaping of Corn, etc. but having been unable to obtain the Patent till too late to give it a general inspection in the field with safety, he is induced to take advantage of his Theatrical Profession and make it known to his Friends, who have been anxious to see it, through that medium. . . . To conclude [the performance] will be presented the celebrated farce of Fortune's Frolic. The part of Robin Roughhead will be taken by Mr. Dobbs, in which he will work the Machine in character, in an Artificial Field of Wheat, planted as near as possible in the manner it grows.[4]

As far as we can tell, the demonstration was unimpressive. In any case, that was the last the world heard of the Dobbs machine. English farmers continued to use sickles and scythes, the larger farmers employing jobless men and women, usually Irish.

A similar situation existed in America, compounded by the labor shortage and the resultant higher wages paid to occasional workers. Working with a scythe, an experienced farmer—followed by men or boys who would bind the wheat—might cut two acres in a good day. At prevailing wages, the farmer who hired workers could expect to spend three dollars an acre for his labor, which would be about 10 percent of the money he would receive for his wheat. Little wonder, then, that American wheat farmers sought methods to lower the cost. Increased labor was not the answer, so it would have to come from new tools, animal power, or machines. In 1823, Jonathan Roberts of the Pennsylvania Agricultural Society put it this way:

> In practical husbandry the expense of labor is a cardinal consideration. Since the year 1818, farmers have very sensibly felt that labour has been much dearer than produce. We can not speedily look for their equalization; a mitigation of this effect may be sought in some degree by improved implements. . . . Nothing is more wanted than the application of animal labour in the cutting of grain. It is the business on the farm which requires the most expedition, and it is always the most expensive labour. Such an invention can be no easy task, or the ingenuity of our fellow citizens would, ere this, have effected it. But we have no right to despair where there is not a physical impossibility. A liberal premium might well be employed to obtain such an object.[5]

The farmers and inventors of America were well aware of this. Some knew of Jeremiah Bailey's reaper, developed the previous year, and Peter Gaillard's earlier model. But these and others proved unworkable or unreliable, and in any case the inventors were unable to market them in numbers. Alexis de Tocqueville, the perceptive French visitor to America who traveled through the land in the early 1830s and wrote of his observations in 1835, did not seem to believe one would be forthcoming, at least not until Americans had become an urbanized people. "To cultivate the ground promises an almost certain reward for his efforts, but a slow one," he wrote in *Democracy in America*. "In that way you only grow rich little by little and with toil. Agriculture only suits the wealthy, who already have a great superfluity, or the poor, who only want to live." As for the rest, "his choice is made; he sells his field, moves from his house, and takes up some risky but lucrative profession."[6] In other words, only those who could afford to hire gangs of workers in harvesting crops, and the subsistence farmer, working with his family at harvest, would survive.

In the summer of 1831, Cyrus McCormick, then twenty-two years old, was working on an improved scythe, while his father attempted to develop a mechanical reaper. Robert McCormick had produced a reaper in 1816 but abandoned it for other projects. The new device appeared more promising, though like its predecessors it had flaws and probably was unmarketable. Robert McCormick was one of many farmers and inventors at work on reapers that summer. William Manning of New Jersey had his own machine, developed independently of the McCormick reaper, and Obed Hussey, a retired sailor, was developing the machine he would demonstrate two years later. Enoch Ambler and Alexander Wilson of New York and Samuel Lane of Hallowell, Maine, were preparing their reapers, as were others who are forgotten today.

Robert McCormick's reaper was not completed in time for the 1831 harvest, but another model, this one fashioned with the help of Cyrus, was given field trials the following year. About a hundred people attended the first demonstration, which started out poorly. Then a neighbor, William Taylor, offered to permit Cyrus to harvest one of his fields. The McCormicks accepted, and cut six acres of wheat that day. One observer, a teacher at a local school, proclaimed the machine was worth at least $100,000. Another, a farmer, noted that the machine handled the stalks roughly, shaking the grain off and onto the field. He wanted his wheat reaped, not threshed, he said, and would have nothing to do with the reaper.[7] On the other hand, a second farmer, James McDowell, Sr., ordered one to be built for him to be used in the harvest of the following year.[8]

In September 1833, McCormick advertised his reapers for sale at $50 each, but none were taken. By then the basic machine devised by Robert McCormick had been changed in so many ways that it could be called a Cyrus McCormick machine. Later accounts indicate it was as good as any then being produced, but no better. And like his fellows, Cyrus McCormick hadn't the ability to sell his devices yet. The construction of the reaper was an impressive but not difficult feat; farmer-inventors elsewhere had already shown it could be done. Selling farmers on the idea of using the devices was another matter, while making it possible for them to obtain them at reasonable rates taxed the imagination of all producers.

At one time it was common to talk of the conservatism of farmers generally, including the Americans in this catchall. Today we know such generalizations are meaningless or deceiving. The introduction of farm machinery in England in the eighteenth century provided the impetus for what later came to be called the industrial revolution, and this would not have been possible unless the English farmers were willing to accept modifications of their old methods. The libraries of men like George Washington and Thomas Jefferson reveal that large farmers of the late eighteenth and early nineteenth centuries were interested in new methods of planting, cultivating, and harvesting their crops, and several of the founding fathers invented new plows and scythes or experimented with fertilizers and labor usage. In most cases these inventions and experiments involved the perfection of old techniques rather than the introduction of new ones, and this was to be expected; the same developments were taking place in textile production, commercial dealings, and other areas of business.

The acceptance of complex farm machinery such as the reaper, thresher, binder, and seeder was something else. It involved the introduction of a new technology to farming, not the further development of the old. Convincing wheat farmers to accept this new technology—to reorient their thinking, as it were—would be difficult. But it had been done in cotton with the gin (a lesser change, to be sure, but a change that affected an entire society, and more so than would any of the wheat-related inventions) and it could be done with other crops. The fact that most of the inventors of reapers and other wheat machines were farmers themselves, like McCormick, indicates that many farmers were prepared to accept the changes, and even welcome them. Some argued that the machines frightened their horses, others that they broke down and couldn't be repaired easily, and most were troubled by the costs. Perhaps they would have accepted reapers more easily if they could be rented. Or they might have been willing to hire "reaping teams," which would sell a service rather than a device. Just as wheat farmers would not keep slaves for effective use during only two or three weeks of the year, they balked at the purchase of a reaper, which would be stored in the barn for fifty or so weeks, and then used intensively the other two. If the machine broke down during the harvest and couldn't be repaired in time to continue the harvest, then it might be said it caused more harm than good, offering farmers a promise of help when little was really forthcoming.

There is no indication that McCormick or any other farm machine manufacturer considered the sale of the service rather than the machine at this time. Even had they done so, they lacked the capital for such an approach. On the other hand, individual farmers in the 1850s did sell their land, purchase machines and hire labor, and then go from farm to farm selling their services. Other wheat farmers, in the Midwest in particular, would join together to purchase two, three, and more machines, and then work together to bring in the crops of the township. Some would run the machines, others would become experts at repair, and the rest would organize teams for binding and threshing. These organizational innovations came from the farmers, not the manufacturers. McCormick and other businessmen, who were really farmers-inventors-manufacturers-distributors combined, understood what was happening and adjusted their policies accordingly. Throughout the early history of the reaper industry, the feedback from salesmen to the factories was more important than the orders that came from the factory and went into the field. McCormick and others like him would organize the industry, but they would do so in response to the market. Those who learned to do this—McCormick being one of them—would succeed. The others would fail.

The wheat farmers of the late 1830s were skeptical of the machines but did not reject them out of hand. As de Tocqueville indicated, farming was becoming less attractive to young men of that period. "Every farmer's son and daughter are in pursuit of some genteel mode of living," wrote a contributor to the *New England Farmer* in 1838. "After consuming the farm in the expenses of a fashionable, flashy, fanciful education, they leave the honorable profession of their fathers to become doctors, lawyers, merchants, or ministers or something of the kind. . . ." Others went to the mills, such as those begun by Lowell and his followers, to earn more in a week than they could obtain on the farm in a month. A New Yorker, writing in the same period, complained:

> Thousands of young men do annually forsake the plough, and the honest profession of their fathers, if not to win the fair, at least from an opinion, too often confirmed by mistaken parents, that agriculture is not the road to wealth, to honor, nor to happiness. And such will continue to be the

case, until our agriculturalists become qualified to assume that rank in society to which the importance of their calling, and their numbers, entitle them, and which intelligence and self-respect can alone give them.[9]

There was some hope that the native-born farmers who were fleeing to the cities or assuming new occupations in rural areas would be replaced by immigrants from Ireland, England, and the continent. Many of these did go to the farm areas seeking land, and more for employment, but their talents did not impress local farmers, who complained of their lack of skills. As a result of this labor shortage, the costs of farm labor rose. In Massachusetts, for example, the average wage of a farm worker, without board, went from $0.78 a day in 1811–20 to $0.88 a day in 1831–40. The prices for wheat declined irregularly from 1815 to 1836, to the point where some wheat farmers wondered whether it paid to have their crops harvested. In such a circumstance, machinery—even if the machines had to be purchased— would not only be beneficial but, in some cases, obligatory. In writing of the introduction of new reapers, ploughs, drills, threshers, etc., Jesse Buell, one of the more astute observers of the agricultural scene, said: "A farm may now be worked with half the expense of labor that was wont to be worked forty years ago, and may be better worked withal."[10]

Tragedy struck the wheat fields in the second half of the 1830s. First there was the failure of the crops in 1836 due to bad weather. Late in the year wheat sold for $2.00 a bushel and flour for $12.00, twice the price of the year before. Some farmers did well, but most suffered and many went bankrupt. It was the culmination of a series of below-average harvests.[11] Successful and unsuccessful farmers alike strove to increase their productivity, to take advantage of higher prices and at the same time earn sufficient money to repay their debts. To some this meant the purchase of machinery. The editor of the *Farmers' Cabinet*, writing in December 1836, said:

> We have long been firm in the faith that the time would come when most of the operations carried on in the growth of corn and grain, would be done by machinery. . . . We have no doubt that ploughing will be done successfully by steam, and that mowing and reaping will be done by the same Herculean power. For a long time our farmers were opposed to the threshing machines,—this opposition arose from imperfect machinery, but still this very opposition retarded the perfection of the very machines it opposed. So in reaping and mowing, some imperfect attempts have been made which were not perfectly successful; and hence the whole scheme has been condemned.[12]

The crop failures coincided with, and in part caused, the financial panic of 1837, which was followed by a depression that deeply wounded the young industrial sector of the nation's economy.[13] The increasing demand for farm machines pulled McCormick and others into the business.

But not immediately. After obtaining patents on his machine, McCormick's interests were diverted by a venture into iron manufacturing. For a while it seemed the business would prosper, and had this happened, McCormick might have remained in that field. Then the depression struck the iron business with a hammer blow. The iron facility lost money, and eventually went out of business. McCormick turned his attention to reapers once more, determined to recoup his losses through sales to farmers.

While McCormick attempted to make a go of it in iron, Obed Hussey concentrated on the reaper. He traveled to New York in 1834, where he found a manufacturer for his device, appointed agents, and met with editors of agricultural newspapers to tell them of his work. For a while Hussey's business languished. He sold several machines from his Maryland factory, conducted field demonstrations for skeptical farmers, and placed advertisements in farmers' newspapers. "My next year's machine will be much superior to any which I have before made," claimed Hussey in 1839, "and to which I apprehend but little improvement can be subsequently made." Hussey later claimed that his machine could cut twenty acres a day, rather high a figure and one he rarely could attain. Often Hussey oversold what amounted to experimental models, and so would lose credibility among some of his customers. His approach was that of a manufacturer. Hussey would produce machines and then try to sell them, using the money obtained from the sales to turn out slightly better models. He was the first man to enter the business as a would-be mass-producer, and for years claimed to have been the inventor of the reaper, challenging McCormick and others on this ground. There is reason to believe that his early machines were the best in America, superior to McCormick's in most ways. McCormick's early success came not from a better design or construction but from his approach to the market.

McCormick did not sell a machine outside his immediate neighborhood until 1840. Up to then he spent so much time on the iron business that he had little to spare for the reaper. He would construct a model, give the machine a field trial, make improvements, and all the while consult with his neighbors, speaking as one farmer to another. Even then McCormick seemed to feel he would have to sell the service rather than the hardware, and take account of farmer psychology. When one of his neighbors decided to take a machine, McCormick would build it, make the delivery himself, and then teach the farmer and his hired hands how to operate the machine, make simple repairs, and service it. By 1843, McCormick was able to advertise, "They all give satisfaction, allowance being made for defects which I had afterwards to correct," and later on—only after the machine had been thoroughly tested—would he warrant it to cut from fifteen to twenty acres a day.[14] His methodology and business strategy were those of a farmer-inventor rather than that of a sailor turned businessman. In order to function successfully in the industry, McCormick had to teach himself about business. Hussey, who already knew something of business, had to learn more about farming and farmers. Both men applied themselves to these tasks in the 1840s, and each in his own way succeeded. Still, their basic approaches differed, with McCormick's being derived from the field to the factory, and Hussey's the opposite. It may be an oversimplification to say that McCormick tried to help the farmers and then allow them to transform the face of American agriculture, while Hussey was developing a vision of a new kind of agriculture that he tried to "sell" to the farmers. Often they came to the same conclusion in practice, and neither man articulated his beliefs in that way during his business career. But the difference in approach remained nonetheless, and McCormick's was more palatable to the wheat farmers than Hussey's, even when his machines proved inferior to those of his rival.

In 1842, after selling some machines outside of his neighborhood, McCormick decided to expand operations. He advertised that:

> for some time to come, he [McCormick] intends to devote his attention exclusively to introducing his machines in different parts of the country, by establishing agencies, selling rights (which he now offers for the first time), or otherwise; and will continue to have them *manufactured in the best manner*, on the same terms as heretofore . . . guaranteeing their performance in every respect; and if they perform as *warranted* to do, it will be seen, as stated also by others, that they will pay for themselves in one year's use . . . and if so, what *tolerable* farmer can hesitate to purchase?

Hussey saw this and other McCormick advertisements and correctly viewed them as challenges. In a letter to the *Richmond Southern Planter* he noted "an account of another reaper in your State, which is attracting some attention."

> It shall be my endeavor to meet the machine in the field in the next harvest. I think it but justice to give this public notice that the parties concerned may not be taken unawares, but have the opportunity to prepare themselves for such a contest, that no advantage may be taken. Those gentlemen who have become prudently cautious, by being often deceived by humbugs, will then have an opportunity to judge for themselves.[15]

McCormick accepted the challenge, and a contest was arranged for July in Virginia. Hussey had difficulties in transporting his machines to the site, and in the end had to use the smaller and inferior of his two models. As a result, McCormick won the first round. The judges reported "great reluctance in deciding between them, but on the whole prefer McCormick's." On the other hand, they found the large Hussey machine "heavier, stronger, and more efficient."[16]

Other trials followed, with machines constructed by several manufacturers competing. No conclusive verdict was arrived at, but the trials of the early 1840s served to publicize and popularize the machines. Within a few years the question no longer was whether to purchase a reaper, but when, under what terms, and which model.

McCormick did well in the competition for sales, so much so that he had to license others, including A. C. Brown of Cincinnati and other small firms in New York and Missouri, to produce his machines for local distribution. Furthermore, he began to job out parts production to others, since his small installation in Virginia could not handle the demand for machines. Too, he no longer was able to fabricate machines on demand but rather went into full production after the autumn harvests, worked through the year, and then turned his attention to sales in the spring. With this development, technology became less important than distribution. Shortly thereafter McCormick found it necessary to devote much of his time to the market for his products.

And he had a definite talent for the market. In 1843 McCormick sold twenty-nine reapers to Hussey's two, and although his rival's record improved the following year he continued to trail McCormick by a wide margin. By 1849 McCormick was selling some 1,500 machines in all parts of the country, while Hussey could sell but a hundred or so, most of which were taken by eastern farmers. It was then that McCormick's competitors began to charge that he was "flooding the country with his machines." Law suits followed, but none of them were decisive. Before the end of the 1840s, McCormick had emerged as the reaper king.[17]

Throughout this period McCormick, Hussey, and the other manufacturers attempted to improve their machines, preserve patent rights, and expand markets. Patent struggles occupied much of McCormick's time, both those instituted against him and those he initiated. The fact that McCormick resorted to suits in attempts to crush rivals indicated that he still did not have the confidence that he could vanquish them in the wheat fields. In addition, the suits illustrated the crude nature of the machines of that period. Nor was this his only concern. As his business improved, McCormick found he no longer could control it as he did in the early 1840s. When he produced only a handful of machines, all the parts were made in his shop. Now he employed several parts manufacturers, many of whom were unreliable. Furthermore, the early sales agents were irregular in their operations and reporting, and now McCormick had to spend time training and often disciplining his salesmen.

McCormick was distressed with the quality of machines produced by his licensees. The Brown machines in particular were not well received. Nor were those produced by Henry Baer in Missouri and Grey and Warner in Illinois. Given the distances and transportation difficulties, it would have been uneconomical to ship reapers from his Virginia installation. If the situation were permitted to continue, the McCormick machines might earn a bad reputation in the Midwest, one that would be difficult, perhaps impossible, to overcome. In the late 1840s, McCormick began to plan for a move westward.[18]

Plant-location theory did not exist in the 1840s; most manufacturers then and earlier had little to guide them except common sense and a feel for the market. Earlier, factory masters had selected sites on the basis of their homes, the availability of power, raw materials, and the labor supply. McCormick was one of the first American industrialists to face the necessity of making a major move in order to be closer to his markets.

And the customers for reapers were moving westward. Apparently McCormick, alone among the major reaper manufacturers, thought the move would be swift. Blight and harsh summers in the late 1840s conspired to push farmers westward, while the lure of new, cheap land pulled them there as well. The geography of the Mississippi basin and lower land prices there made large farms more easily obtainable than they were in the settled East, and the larger the farm, the greater the need for machines. In the late 1840s, upper New York was the nation's wheat center, and there were relatively few large farms there. Still, a more prudent man might have decided to move his plant there not only to serve an already existing market but to ship machines westward via the Erie Canal and Great Lakes. In 1848, western Pennsylvania and Ohio were among the four leading wheat areas, and a McCormick move to either of these two places would have been understandable. But to go further would have taken great foresight and a willingness to stake all on the belief that America would continue to expand rapidly. Apparently McCormick did have this vision and was determined to make such a move.

America had recovered from the depression by then, and the move westward not only continued, but accelerated. A new depression in the early 1850s, after McCormick had moved to his new plant, might have destroyed his company and made Hussey or someone else the reaper king of America, in which case McCormick would appear today as a footnote in some history books. There was no depression, but prosperity instead. By 1859 the leading wheat states in America were Illinois, Indiana, Ohio, and Wisconsin, in that order. During the decade Illinois' wheat production more than doubled, while that of Indiana almost tripled.

McCormick had visited the Midwest on sales and promotion campaigns, and as a farmer recognized the worth of the soils of the region. Clearly the producer of a bulky object such as a reaper, one that would require servicing and close relations with purchasers, would have to be closer to the market than Virginia or even western New York or Pennsylvania. Already McCormick had a small army of agents in the Midwest. Now he would move his factory closer to these agents.

Where should the factory be located? Cincinnati, the Queen of the Ohio, seemed a logical choice. It was a factory as well as commercial center, one with an ample supply of labor and, for the time, centrally located so as to serve both the old and new wheat areas. St. Louis, which dominated the Missouri-Mississippi, and in 1848 was the prime candidate for the title of "the New York of the Interior," was another. St. Louis had a growing capital market, experienced commercial and financial leaders, and close ties with New Orleans, then bidding to become the nation's leading export port, hoping soon to surpass New York. If McCormick had ambitions to sell his reapers abroad—and he did—St. Louis would prove an admirable site. So would Cleveland, closer to the upper Midwest, with a fine harbor. There were a dozen other cities McCormick might have considered and selected, with good reason. In the end he chose Chicago.

Chicago was a small town in 1848, with a population of around 18,000. It was in the midst of smallpox and cholera epidemics, and although vigorous, seemed scarcely a rival for the other cities of the Midwest. Chicago was a lumber and grain town, and little more. Wheat farmers and lumbermen of the upper Midwest looked upon it as a natural market for their goods, which would be processed in Chicago's mills, breweries, and lumberyards. Then the wheat, flour, and lumber were exported, by lake, canal, and river, to New York. Chicago shipped more wheat in 1848 than any other midwestern city, and it may have been this that attracted McCormick. It also had a fine harbor, while the Chicago River, though hardly comparable to the Hudson, not to say the Missouri, Ohio, or Mississippi, provided it with access to the interior. Chicago's leaders were dredging the river at that time, and in 1848 the Illinois and Michigan Canal was opened, connecting the city with the Illinois River, and thus with the Mississippi. To Chicagoans it seemed the nation had a watery spine, with their city at one end and New Orleans at the other. They hoped to master the river, and so the entire Midwest.

This vision might have captivated McCormick. He visited Chicago as early as 1844, and had even formed a partnership with C. M. Gray of the city to produce and sell reapers for the 1848 harvest. Several of McCormick's key agents operated out of Chicago, and his contacts there were better than in any other city in the Midwest. Then, too, he struck an acquaintance with William B. Ogden, the city's leading booster and wealthiest businessman, and Stephen Douglas, who was elected to the Senate in 1847, and spoke glowingly of new railroads as well as a financial center. These friendships may have been a decisive element in the move, for Ogden became McCormick's partner and Douglas his attorney. These two also helped bring federal funds to Chicago and were instrumental in the formation of the Board of Trade in 1848, another sign of commercial and financial growth.

Finally, McCormick was a devout Presbyterian, and Chicago had a strong and active Presbyterian community. He must have found this attractive.

Like the other midwestern cities McCormick might have selected, Chicago had its share of assets and liabilities. Most of the liabilities were real enough in 1848, while some of the assets were based on hope and dreams. The Chicago move seems logical and sensible today; at the time it was a bold stroke.

In 1847 McCormick prepared for the move, cleaned up his accounts by ending the partnership with Gray, and spent much time on his endless court cases regarding the reaper. Work was also begun on the new factory. In October 1848, McCormick, Ogden, and William E. Jones, another Chicago tycoon, announced the formation of McCormick, Ogden and Company. Ogden had already loaned McCormick money for the new factory, the final inducement in making the move. Now he would provide the capital for the installation's operations. McCormick planned to produce 1,500 reapers at the Chicago plant for the 1849 harvest. Each machine cost him less than $65, including agent's fee and allowance for bad debts. The reaper would be sold for $115, or if purchased on credit, $120. The sales campaign was a success. McCormick was a wealthy man.[19]

Late in 1849 McCormick bought out his partners and took complete control of his business. Never again would he have to worry about money. While Hussey and other rivals spent much of their time in obtaining backing, meeting bills, and trying to finance receivables, McCormick was free to operate his business. His brothers William and Leander were taken into the firm, and they handled office duties and production. Each year the reaper was changed, improved, and developed, and Cyrus McCormick was involved in what today would be called research and development. Some of his time was spent in litigation, but by now he had a small army of lawyers to do most of this work. Increasingly, McCormick concentrated on sales and promotion, and these became the key to his operations. It is by no means certain that the McCormick reaper was the best to be had. John H. Manny's machine was considered superior by many farmers, and Manny sold almost a thousand of them in 1854 and over two thousand the following year. Seymour and Morgan had a machine that was simple, inexpensive, and most attractive. Atkins-Wright, Wood and Company, Palmer and Williams, and several other reaper manufacturers offered features not to be found on the McCormick machines. Later on Wood and Company would develop an efficient wire binder and so initiate a major breakthrough in the industry, one McCormick was forced to concede and imitate. But none of these companies could match McCormick in sales and promotion.

McCormick's advertisements continued, more in number and larger than any that had previously been seen in the nation. These contained testimonials from satisfied customers, assurances of the machine's worth, and often detailed cost analysis as well. On occasion the advertisements would take note of events that might make reaper ownership worthwhile. For example, the gold rush began shortly after McCormick made his move to Chicago. Farm labor was attracted to California, and at the same time the influx of gold into the monetary stream helped bring prosperity. Farmers needed mechanical help at harvest time, and now they could afford it. McCormick took note of both, offering what amounted to short lessons in economics in some of the advertisements. Such notices helped make the 1849 sales effort successful.

McCormick continued to give assurances as to the worth of his machines. The written guarantee and what for the time was a good service organization helped here. The farmer paid $30 down on his machine, with the rest of the price due within six months on condition that the reaper cut one and a half acres an hour. If the machine failed to live up to expectations, the $30 would be refunded. Too, McCormick operated on a fixed price. Other manufacturers would offer discounts, haggle with farmers, and in the end come down to a lower price in order to clinch the sale. McCormick's prices varied from year to year, usually reflecting increases in costs and improvements but always based on what he thought the market would bear. And seldom would a McCormick agent deviate from this price; if he did, it would come out of his commission.

McCormick recruited an able agent corps, provided them with literature, brought them to the factory to see how the machines were produced, and even instituted an *ad hoc* trainee program. He paid these men smaller commissions than did his rivals, but the McCormick agents more than made up the difference by volume sales. Also, he made his agents' tasks easier by his deep concern for goodwill. McCormick rarely sued a farmer for nonpayment, preferring instead to carry him until the next harvest. Agents' calls for spare parts were answered promptly, and McCormick would schedule more field trials—really company-paid picnics—than would his rivals. Bonuses, sales incentives, and higher commissions as rewards for success became common parts of the McCormick sales efforts. Some of the agents, A. D. Hager and D. R. Burt among them, developed into prototypes for the future. They knew their machines, cultivated the customers, and offered a good line of patter. In 1854, for example, Burt wrote of his experience in Iowa:

> I found in the neighborhood supplied from Cassville quite early in the season one of Manny's agents with a fancyfully painted machine cutting the old prairie grass to the no small delight of the witnesses, making sweeping and bold declarations about what his machine could do and how it could beat yours, etc., etc. Well, he had the start of me, I must head him some how. I began by breaking down on his fancy machine pointed out every objection that I could see and all that I had learned last year . . . gave the statements of those that had seen the one work in the grass . . . all of which I could prove. And then stated to all my opinion of what the result would be should they purchase from Manny. You pay one half money and give your note for the balance, are prosecuted for the last note and the cheapest way to get out of the scrape is to pay the note, keep the poor machine and in short time purchase one from McCormick. . . . Now gentlemen I am an old settler, have shared all the hardships of this new country with you, have taken it Rough and Smooth . . . have often been imposed upon in the way I almost know you would be by purchasing the machine offered you today. I would say to all, try your machine before you [pay] one half or any except the freight. I can offer you one on such terms, warrant it against this machine or any other you can produce, and if after a fair trial . . . any other proves superior and you prefer it to mine, keep [it]. I will take mine back, say not a word, refund the freight, all is right again. No, Gentlemen this man dare not do this. The Result you have seen. He sold not one. I sold 20. About the same circumstances occurred in Lafayette Cy.[20]

After such a sales pitch, the agent often would point out that McCormick would not press for payment if the crops were poor that year. "It is better that I should wait for the money, than that you should wait for the machine that you need." Of course, such an approach—low down payment, deferred payments in bad years—meant McCormick would have to carry a large number of accounts receivable. In some cases he lost large sums of money, such as when farmers were driven from the land due to continued drought. For the most part, however, the tactics paid off.[21] The 1850s proved a banner decade for McCormick, his first great period of prosperity.

In this period McCormick traveled to England to show his machines at the Crystal Palace Exposition of 1851, and then went on to travel the Continent, attempting to sell them on a worldwide basis. These early efforts were not very successful, and even later on, European models and those exported by some of McCormick's American rivals did better overseas than did the McCormick reapers. In America, however, McCormick easily led the field. His major problem in this regard was Manny, with whom he engaged in a running series of court battles involving patents. Manny was represented by a battery of lawyers, including Abraham Lincoln and Edwin Stanton, while Stephen Douglas spoke for McCormick. Manny won the case, in part the result of a brilliant defense and attack mounted by Stanton. McCormick was so impressed he engaged Stanton in future legal frays, while Lincoln received his first major fee from the case, money he would later use to help finance his senatorial struggle against Douglas. As for McCormick, he recovered from the suit and fought off attempts on the part of his rivals to strip him of all his patents by 1861.

The McCormick interests prospered in the field. During good times his agents outsold their competitors. During bad—such as the panic year of 1857 and the immediate aftermath—McCormick's willingness to extend credit won him additional sales, even when some of his competitors were forced to the wall. By 1860, McCormick was planning new machines and working to extend his markets in America and abroad.

McCormick was also interested in politics. A lifelong Democrat and a Virginian, he supported James Buchanan in the 1856 presidential election, and during the next four years spoke out in favor of reconciliation of sectional differences. Later on, his critics argued he had done so in the belief that a civil war would destroy his company. There is some justification for this claim. When the war broke out in 1861, McCormick ordered his treasurer to convert all the company's liquid assets into gold. He wrote of fears that the British would come to the aid of the Confederates, that Lee's army would invade the Midwest and in the process destroy machines not yet paid for as well as his prime market. Too, in 1861 there were reports of hardships on the wheat farms, and this did cripple sales for a few months. McCormick might well have anticipated this in the late 1850s when he worked for peace.

On the other hand, most of his machines were sold in the North, not

Statistics for C. H. McCormick & Co., 1849–1858

Year	Machines Made	Costs	Number on Hand	Number in Agents' Hands	Notes held by Company	Gross Sales
1849	1,490	$72,149	0	0	$2,590	$172,505
1850	1,603	?	5	0	2,976	89,017
1851	1,004	36,290	5	0	5,114	105,000
1852	1,011	38,702	17	0	3,753	112,531
1853	1,108	62,573	22	0	5,841	124,974
1854	1,558	86,737	9	0	12,855	209,374
1855	2,534	138,344	10	0	34,558	363,484
1856	4,095	194,398	56	100	125,415	574,011
1857	4,091	199,892	154	39	315,690	541,346

Source: Hutchinson, McCormick: Seed-Time, p. 369.

the South, and so secession did not mean the loss of a major market. McCormick's experiences during the gold rush taught him that any major loss of labor from the wheat fields would be translated into reaper sales, and this did occur when Lincoln began to form the Union armies. A case can be made for either interpretation of his motives, but the most likely explanation is that by ancestry and upbringing (his brothers' wives had relatives in the Confederate armies) he was a natural ally of the peace forces. There is an internal consistency in his politics throughout the 1850s and 1860s.

As it happened, McCormick and other farm implement manufacturers profited greatly during the war. The demand for harvesting equipment was such that the Chicago factory oversold its production by 500 units in 1862, causing grave distress in the Midwest that year. During the five years prior to the war, annual reaper sales (for all manufacturers) were approximately 16,000. In 1862, 33,000 reapers and mowers were sold in the United States. The following year's figure was 40,000, while 80,000 were sold in 1864 and a like number in 1865. At the start of the war, there were some 125,000 harvesting machines in operation in the United States; this number doubled by 1864, when three quarters of all farms of over 100 acres had a machine.[22] Although hundreds of thousands of men had been drawn away from the fields for the war, wheat production soared. Slightly over 4 million bushels were exported in 1860; in 1862, the figure was well over 37 million bushels, much of which went to England, the Confederacy's putative ally. Later on it would be said that in the Civil War the reaper defeated the cotton gin—that, among other things, England's need for grain was greater than her need for cotton, and so she remained neutral. This is an exaggeration, but it is true that without the reaper, the North could not have produced as much wheat as it did while at the same time taking men from the soil.

McCormick spent much of the war in England, attempting to convince the Europeans of the worth of his new machines. He competed not only with the European products but those produced by Manny, Wright, and

Seymour and Morgan, among others. McCormick selected several foreign firms as licensees, but few of these arrangements worked out well. Walter Wood, an American, had better fortune with his combination reaper-mower, which easily outsold the McCormick models, and not even the improved self-rake reaper, which McCormick introduced in 1862, could defeat it. Not until the late 1860s did McCormick's work bear fruit, and then not for long. Trouble with licensees, difficulties in adapting his machines to European fields, and a general inability to adjust to the continental market caused the world vision to fade for the time being. It would appear that McCormick, so well attuned to the needs and desires of American farmers, could not comprehend their European counterparts. The age of international manufacturing had begun, but McCormick was not a major participant in it at that time.

McCormick's business activities brought him new fortunes in the postwar years. Now he expanded into real estate and railroads, natural activities for a man who wanted to enlarge the market for agricultural machinery. He continued to dabble in Democratic politics and took stands on public issues, especially those concerning farmers. In fact, he became an unofficial spokesman for agrarian interests within the party's inner circles where, as one of the few Democratic millionaires and contributors, he was given a voice. Fortunately, he did not live to see the development of agrarian radicalism, which McCormick might not have been able to square with his innate conservative nature. Too, he became engrossed with the Presbyterian Church's development in the upper Midwest, and as he grew older this took up more of his time.

Now honors came from American organizations and foreign nations. McCormick was hailed as an inventive genius. But the praise was for what he had done prior to the 1860s, and not for what he was doing in these last years of his life. After 1860, McCormick did little to develop the industry, except of course for contributing greatly to reaper production during the war. Indeed, one might argue that by continuing his litigations he may have done more to hinder than help technological innovation. In this period McCormick was more a follower than a leader. Others developed the binders and combines, and eventually the harvester; McCormick was in the rear. Yet all the successful companies that emerged or grew in the postwar period imitated McCormick's sales techniques and distribution and finance methods.

Cyrus McCormick's place in the American business tradition is secure. But what is it? A recent text called him "the first really big businessman in American manufacturing" and "an ideal example of an innovator."[23] This is true, and more. Despite his difficulties in the foreign market, McCormick was one of the first American manufacturers to make this country's goods respectable in Europe. The McCormick factory practiced the division of labor on a scale unknown before its coming. McCormick was instrumental in the growth of Chicago; even though he was never the city's leader, he was long considered its "first citizen." When he died in 1884, at the age of seventy-five, all the city's churches held memorial services, and the Board of Trade closed down in tribute. McCormick had led the way in reconstructing the city after the great fire of 1871, and had lent important aid to distressed businessmen during the panic of 1873. He could do this not only because he was wealthy but because his company was one of the best managed in the nation, a model to others, not only in farm machinery, but almost all industries.

Still, McCormick's major contribution was not in manufacturing, invention, or finance. As has been indicated, others could and did develop farm machines at least as good as his, and in some cases, better. Manny and Hussey are all but forgotten today, probably because they did not create enterprises that survived. But John Deere, Jerome I. Case, John Oliver, and others who founded firms that remain in business to this day, are not considered McCormick's equal. Although these men all were innovators in invention, they were obliged to imitate McCormick in sales and distribution. As one scholar put it, he was a "merchandiser par excellence."

> He formed his own sales organization, guaranteed the reaper, and offered to sell it on installments. Exclusive agents were selected at key points throughout the marketing area—this was, in effect, the dealer system later employed by the automobile industry—and control of agents was just as tight under McCormick as it was to be under General Motors. He challenged builders of other reapers to field tests. It was like the automobile speed races of later years. . . .[24]

McCormick convinced farmers that he was one of them. Speaking in a language they understood, he showed farmers that the reaper was a sound investment and made them believe they would benefit by purchasing it. He knew his market and how to exploit it. As a salesman-distributor, he had no peer.

Notes

1. By 1870, the United States was responsible for almost a quarter of the world's manufacturing output. Still, such production was a small part of the world's economy, and smaller yet for America. More than nine out of ten Americans still found employment on the soil at that time. Douglass C. North, *The Economic Growth of the United States, 1790–1860* (New York, 1966) pp. v ff.

2. "Sketch of My Life," by C. H. McCormick, August 4, 1876, as quoted in William T. Hutchinson, *Cyrus Hall McCormick: Seed-Time, 1809–1856* (New York, 1930), p. 34.

3. G. E. Fussell, *The Farmer's Tools, 1500–1900: The History of British Farm Implements, Tools and Machinery Before the Tractor Came* (London, 1952), pp. 115–18.

4. Bennet Woodcroft, comp., *Specifications of English Patents for Reaping Machines* (London, 1853), pp. 16–17.

5. Hutchinson, *McCormick: Seed-Time*, p. 69.
6. Alexis de Tocqueville, *Democracy in America* (New York, 1969 ed.), p. 552.
7. Herbert N. Casson, *Cyrus Hall McCormick: His Life and Work* (Chicago, 1909), pp. 37–41.
8. Hutchinson, *McCormick: Seed-Time*, p. 89.
9. Percy W. Bidwell and John I. Falconer, *History of Agriculture in the Northern United States, 1620–1860* (New York, 1941), pp. 204–6.
10. *Ibid.*, p. 281.
11. *Ibid.*, p. 498.
12. Hutchinson, *McCormick: Seed-Time*, p. 153.
13. Samuel Rezneck, *Business Depressions and Financial Panics* (Westport, 1968), pp. 73–100, is the best description of the social and economic problems of the depression.
14. Hutchinson, *McCormick: Seed-Time*, pp. 175–79; Leo Rogin, *The Introduction of Farm Machinery in its Relation to the Productivity of Labor in the Agriculture of the United States During the Nineteenth Century* (Berkeley, 1931), p. 133.
15. *Ibid.*, pp. 172–73, 187.
16. *Ibid.*, p. 192.
17. *Ibid.*, pp. 194, 265–66.
18. Casson, *McCormick*, pp. 66–67.
19. Hutchinson, *McCormick: Seed-Time*, p. 317.
20. *Ibid.*, pp. 361–62.
21. Casson, *McCormick*, pp. 79–87.
22. Paul W. Gates, *Agriculture and the Civil War* (New York, 1965), pp. 227–233.
23. Herman E. Krooss and Charles Gilbert, *American Business History* (Englewood Cliffs, 1972), p. 100.
24. Ben B. Seligman, *The Potentates: Business and Businessmen in American History* (New York, 1971), p. 101.

Ideas for Discussion

1. What roles do the "spirit of capitalism" and "competition" play in McCormick's career?
2. What is your understanding of McCormick's place in American business history?

Topics for Writing

1. Prepare a detailed description of how the McCormick reaper operates.
2. Examine histories of farming or business for mention of McCormick. What is his reputation? Answer in a memo to your instructor.
3. Sobel's book collects portraits of important American business people. Read several of these or separate biographies of leading figures such as Andrew Carnegie or Andrew Mellon for a report on the "characteristics" of the American entrepreneur. Write a comparative essay on this topic.

4. Prepare a report on a contemporary industrial inventor.
5. Write a sales letter, a press release, and a press kit for the reaper based on the materials given in this essay and additional research you may have to conduct.
6. Write a sales proposal to an equipment dealer suggesting the company market the reaper.

The Acquisitive Spirit

The fifteen readings that constitute Part II of *Writing about Business and Industry* offer ideas on the global market, business systems, the U.S. auto industry, and ethics. The section opens with a look at how gestures are significant in business settings, then takes a close look at precapitalist Florence, which depended heavily on a global market for its cloth trade. Daniel Defoe next offers an eighteenth century guide to writing business letters. Then, "The Occupational Structure of India" is examined as it contrasts to the growth of European and American manufacturing from the 1880s to the 1950s. Next, "Foreign Assembly in Haiti" describes how this island nation struggles to survive on manufacturing contracts from the United States and elsewhere.

The *Harvard Business Review*'s essay by Hamel and Prahalad on "Strategic Intent" is a well-known discussion of the steps managers must take to become leaders in the global marketplace. These authors outline a strategic management system derived from Japanese and European models with a special focus on how Japan's Komatsu corporation has avoided takeover by Caterpillar for nearly twenty years. "Pouring Ideas into Tin Cans" is a 1936 essay on Oscar Huffman's Continental Can Company. This half-century-old article is still a fresh example of sound strategic planning and industrial foresight. The success of businessmen like Huffman contributed to the division of labor into "blue"- and "white"-collar groups, which Stuart M. Blumin shows as a developing trend in four U.S. cities from 1792 to 1845. He demonstrates how a "new middle class" emerged in Philadelphia, New York, Charleston, and Hartford in response to occupational stratification. Next, Beatrice Webb, a Fabian author, argues for legislation in factories to make them safe, sanitary, and secure for all workers, but especially for women in the textile trade, who were often considered "amateur" laborers next to men.

The three selections by Alfred Sloan, Paul Lawrence and Davis Dyer, and Andrea Gabor focus on the U.S. auto industry. The selection from Sloan's *My Years with General Motors* (1964) concentrates on his method of management by committees when he assumed GM's leadership in 1924; Lawrence and Dyer critique GM and Sloan as part of the industrywide problems with top management in a chapter from their book *Renewing American Industry*. Andrea Gabor offers a profile of W. Edwards Deming's efforts to redirect the perspective of American business from profit-motive to consumer-satisfaction in her summary of Deming's "Fourteen Points."

Jeffrey Pfeffer in "Organizations as Physical Structures" encourages both managers and nonmanagers to consider how the physical layout of an office affects office behavior and communication. Bowen McCoy approaches the subjects of individual and group ethics through his anecdotal story of mountain climbing in Nepal. The concluding essay, "The New Product Development Map," profiles how Stratovac Vacuums studied its market, competition, and consumers to develop its Challenger 6000 model. Here is an example of market forces determining the type and quality expected in a new product. Perhaps, this essay also may be said to deal with ethical issues and issues of diversity, as research and development personnel must be careful to be fair to their competitors and sensitive to consumer needs.

The global market, business and industrial development, and corporate ethics are key areas of concern to all decision makers and employees. The essays here discuss these topics from a variety of perspectives, mingling precedent with practice to engage the reader in the current debate.

Roger E. Axtell
Counting Globally

Roger E. Axtell has put his experience as a world traveler and former marketing vice-president at the Parker Pen Company to use in four books on multicultural aspects of business and travel. At the same time he has pursued a successful career as a lecturer. This selection is from his 1991 Gestures: The Do's and Taboos of Body Language Around the World, *an interesting and practical guide to verbal and non-verbal communication which pays particular attention to differences in business and social bahavior among various world cultures.*

IF YOU LIKE *counting* with your fingers, be cautious as you travel abroad because you might just add up to total confusion. In Germany, for example, to signal "one" a person uses the upright thumb instead of the forefinger. Therefore, in a German *gasthaus* beer garden, to order "one beer" you would flash the "Thumbs-Up" gesture. Many Westerners use the *forefinger* to signal "one," more or less ignoring the thumb. But, do that in Germany—casually hold up the forefinger and forget the thumb—and it could be seen as meaning "two," with the result that you'd be drinking two beers instead of one.

In Japan, they use the fingers to count visually just as most cultures do. However, in Japan counting *begins* with the index finger (not the thumb) to designate "one"; then the index and middle fingers combined equal "two"; the combination of index, middle and ring fingers is "three," and adding the little finger equals "four." Then the Japanese show the upright thumb alone to mean "five." So, if you order "One beer" with your thumb in Japan you may well receive five beers.

When one thinks about it, the thumb is a wonderfully versatile and utilitarian digit. Artists hold it in front of their eyes, peering over it to gain a sense of perspective and depth. We like to put people "under our thumb," and film critics and editorial writers frequently use "thumbs up" or "thumbs down" to indicate approval or rejection, respectively. Finally, we "thumb our noses," bite our thumb to hurl an insult à la Shakespeare, suck our thumbs as infants, and if that is not sufficient, there must be at least a million "rules of thumb."

So the thumb becomes a valuable part of our daily worldwide communication, whether in school to ridicule a schoolmate with the "nose thumb" or in space to reassure a spacemate with the "thumbs-up" gesture.

Just be careful where you point your thumb in places like Nigeria and Australia.

Ideas for Discussion

1. Have you ever considered the significance and possible meaning of gestures you commonly make, say with your eyebrows, chin, foot or ears?
2. If possible, interview students from other parts of your state, your country, or other countries and inquire about the importance of non-verbal communication in their region.

Topics for Writing

1. Compose a word table depicting the information the selection from *Gestures* presents.
2. Expand discussion question #2 into a short report on your findings.
3. Develop a communication packet which includes information on appropriate gestures, conversations between the sexes, and suitable conversation topics for a colleague or group of colleagues preparing to do business overseas.
4. As an oral presentation, locate more information on controversial gestures and give your findings in a demonstration to your class, assuming the role of communication professional asked to present such information at a meeting. Use whatever handouts and visual aids to make the presentation meaningful to the audience.

Richard Goldthwaite
The Wherewithal to Spend: The Economic Background of Renaissance Florence

Richard Goldthwaite is a Renaissance historian teaching at Johns Hopkins University in Baltimore, Maryland. He is a Columbia University graduate whose previous book is Private Wealth in Renaissance Florence. A Study of Four Families *(1968). In this chapter from* The Building of Renaissance Florence: An Economic and Social History *(1980) Goldthwaite studies the role of textiles and consumerism in Florence from the twelfth to the seventeenth centuries.*

THE STUDY OF precapitalist economic systems has not yet clarified how demand fits into its scheme of things. Economic historians seldom talk about it, and it has hardly any identity in their analysis. Demand arises from taste and needs that give direction to motivation, and such subjects generally fall outside the realm of traditional economic history. Whatever its roots, however, demand was conditioned by what the economic system would permit. The decision to spend large amounts of money for something like the building activity that is the subject of this book presupposes the availability of money to spend. The basic economic questions about the building of Renaissance Florence, therefore, are: What was the level of wealth in the city? How did the city's social structure determine the way the wealth was spent? Was there some change in the level, structure, or nature of wealth that may help explain the relatively sudden release of money at the beginning of the fifteenth century that enabled the conspicuous consumption we associate with the Renaissance to take place?

The fundamental proposition advanced here is that by the fifteenth century extraordinary amounts of wealth were accumulating in the hands of a relatively large number of Florentines. The fall in population by as much as one-half to two-thirds that Florence, like other European cities, suffered in the second half of the fourteenth century in itself accounts for higher per capita wealth of the survivors. In fact, the city was smaller in population than it had been for almost two centuries, and those who remained were enjoying the fruits of the considerable economic development that had taken place in the meantime. Moreover, the considerable expansion of the Florentine territorial state precisely in these years, over the century following the Black Death, brought greater wealth into the capital city, which clearly dominated the regional economy. More importantly—and this is the central theme in the ensuing discussion—during these same years the economy successfully adjusted its performance to a changing situation in international markets in a way that stimulated growth in the textile industry and strengthened the operations abroad of the commercial and financial sectors. The result was a highly favorable balance of payments. During the late fourteenth and early fifteenth centuries much of this wealth was absorbed by the military costs of the city's territorial expansion and its emergence as a major Italian power; but after the first third of the fifteenth century, with the stabilization of the political order abroad and the consequent lightening of the tax burden at home, the profits that were flowing into the city from international commercial and financial operations became available for consumption spending.

In short, more money was spent on luxury goods in the Renaissance because more money was available, and the spirit with which Florentines began to consume in the fifteenth century and the culture that spending generated reveal the Florentines' optimism about the economic situation. Moreover, that spending, by calling into existence new forms of production, in itself brought about some major improvements in the performance of the economy during this period. No sector better illustrates that proposition than construction, where so much money was spent.

Florence and the European Economy

The economic system in which Florence developed as one of the great centers of early European capitalism had its origins in the so-called commercial revolution of the eleventh century, when merchants from all over northern Italy ventured forth to take advantage of the opportunities for trade in the expanding European economy. The basis of that trade was the relation between an undeveloped area and a developed one, between northern Europe, where markets for luxury goods opened up once the political situation began to stabilize, and the older markets in the Levant, where many of those goods were to be found. Italy was located halfway between these market areas and lay in the midst of the sea that facilitated transport, so the Italians were destined to be the middlemen—the traders and shippers. Everywhere in northern Italy, in small towns and large ones, entrepreneurs collected what capital they could and went forth to exploit market opportunities; and in the course of the next two centuries they built up a network of commercial relations that spanned all of Europe, from England to Egypt, and that was the lifeline of the developing European economy.

What set Florence on its own course in making a way in the larger economic system was the local development of a major industrial sector whose production was geared to markets abroad. Unlike the other Italian cities it is often compared to—the maritime ports of Genoa and Venice—Florence, by the time of Dante, was a large industrial city; and yet, unlike the great wool-producing cities in the Low Countries with which it is also compared, it fully participated in the international commercial system of the Italians. In other words, Florence developed both an industrial and a commercial sector, giving it one of the strongest economies in medieval Europe.

Geography favored the development of the cloth industry. Located on one of Italy's largest rivers just at the point where, after taking in the watershed of an exceptionally long tract of the Apennines, it comes out of the hills into a flood plain, Florence had, on the one hand, an abundance of rapidly flowing water for the cleaning of wool and, on the other, easy access to the sea at the port of Pisa,, whence the entire Italian commercial network abroad, from northern Europe to the Levant, could be exploited for both the supply of raw materials and market outlets for production. The steady rise of population in Europe from the eleventh century onwards generated demand for clothing that led to the development of local industries everywhere, but few cities anywhere in Italy—and none of the neighboring hill-towns of Tuscany—had the same potential for raising the level of local production to the point that the industry became oriented primarily toward export markets.

Originally the industry took its wool from the hills of the hinterland and sold its products in the rapidly growing urban markets nearby; but as production increased, merchants went farther afield for raw materials, and as quality improved, they expanded into new markets. By the end of the thirteenth century they were bringing wool from England and also cloths from the Low Countries and elsewhere in northern Europe for finishing, dyeing, and then reexporting. The success of this industry in expanding its markets lay in the improvement of the quality of its product to a point that it could compete with northern European cloths, especially those from the Flemish cities. These northern cities had been the leaders in manufacturing luxury cloths that were sold all over Europe, including Italy; but since the marketing of much of this cloth was in the hands of Italian merchants, it was not difficult for aggressive Florentines to push their own products once these were competitive. Moreover, the heavy urban concentration of the Italian population assured Florence a large market potential close at hand. By 1300 the city had far outstripped all of its neighbors—and indeed most other European towns—in size and wealth as an industrial center.

The presence of a strong industry importing raw materials from one part of Europe and exporting its products to another part strengthened the economy's commercial sector by making available that much more capital for investment in international trade. Florentines were to be found virtually everywhere throughout the vast international commercial network of the Italians, and in some of the major markets they were the single most important group of merchants. Their gold florin, first issued in 1252, became a standard international currency for all of Europe, and the use of their commercial network for the execution of foreign exchange and international transfer of credits led to their preeminence in international banking and finance. Since Florentines established colonies in virtually every major center of trade in Europe, the financial network built up by them was as vast as the commercial system itself. The service they could perform in moving funds and extending credit had by the fourteenth century brought them into the highest sphere of papal and princely finance. They were the ubiquitous agents of transfer and exchange in the international ebb and flow of payments. In the traditional historiography the names of their leading merchantile and banking companies—Frescobaldi, Bardi, Peruzzi—symbolize the high point of all Italian banking in the Middle Ages.

Of the vigor of the leading sectors of the Florentine economy during the period of the so-called commercial revolution there can be little doubt. The full weight of the historiographical tradition is behind the proposition that Florence rapidly moved into the vanguard of this expansion, becoming one of the wealthiest cities in Europe, and the well-known details need not be rehearsed here. Yet in the most fundamental respect the economy was not successful; it could not handle the population growth that in the fourteenth century reached the upward limits of what could be supported. The situation must have been desperate by the time the city was hit by the Black Death. Florence was as devastated by that event as any place in Europe, losing perhaps one-half of its population (depending on what estimates one accepts) at the middle of the century and perhaps one-third again by the beginning of the fifteenth century (table I). A number of historiographical problems have arisen over events of the fourteenth century and the effect of the general crisis that the entire European economic system is thought to have suffered in the late Middle Ages primarily as a result of the sequence of plagues drastically reducing the population. Demographic disaster, however, did not altogether cloud the local economic scene. Over the course of the fourteenth century the Florentine economy was reasonably successful in making adjustments to the changing situation, and in many ways those adjustments add up to a strengthening of its economic system with respect to both the performance of the leading sectors abroad and the well-being of the population at home.

The major development in later fourteenth-century Europe that assured the continuing success of the Florentine economy was a rise in demand for banking services and for luxury goods of all kinds—above all, cloth. In part luxury consumption was a consequence of the more concentrated wealth in the hands of those who had survived the demographic disasters—and much has been made of their greater propensity for spending as a result of the psychological shock they suffered during those events. In part, too, this demand was generated by new needs and taste arising out of the consolidation of power by a number of princes all across Europe that resulted in the elaboration of bureaucratic government, the building-up of the military

Table 1
Population of Florence, 1172–1632
Estimates of Population

Year	Russell	Fiumi	de La Roncière	Herlihy/ Klapisch	Beloch
1172	10,000				
1200	15–20,000	50,000			
1260		75,000			
1280		85,000	100,000		
1300	96,000	95,000	110,000		
1338		90,000	100,000	120,000	
1347		76,000	90,000		
1349			32,000		
1352			41,000	42,000	
1362			70,000		
1364			54,000		
1373			60,000		
1375			53,000	60,000	
1379			56,000		
1380		54,747		54,747	
1400				60,000	
1427				37,144	
1441				37,036	
1458				37,369	
1469				40,332	
1480				41,590	
1520					70,000
1551					59,557
1562					59,216
1632					66,056

Sources: Josiah Cox Russell, *Medieval Regions and Their Cities* (Newton Abbot, 1972), p. 42; Enrico Fiumi, "Fioritura e decadenza dell'economia fiorentina," *ASI*, 116 (1958), 465–66; Charles M. de La Reconière, *Florence, centre économique régional au XIVe siècle* (Aix-en-Provence, 1976), pp. 693–96; David Herlihy and Christiane Klapisch-Zuber, *Les Toscans et leurs familles: une étude du catasto florentin de 1427* (Paris, 1978), pp. 173–88; Karl Julius Beloch, *Bevölkerungsgeschichte Italiens*, II (Berlin, 1939), 148.

complement of power, and the growth of the sedentary court with its highly ceremonial life-style. These events were particularly notable in Italy, where the heretofore fluid political situation coagulated into a more stable multistate system, for the most part in the hands of princes of one kind or another. Taking up permanent residence in cities, these men sought to consolidate their position and establish their legitimacy by engaging in the kind of consumption we associate with the spendor of the Italian Renaissance. These urban courts in Italy and their counterparts elsewhere in Europe were the markets that stimulated the production of, and commerce in, luxury goods; the resulting intensification of activity throughout the commercial system challenged the Florentines to retool their home industry for the production of the kind of cloths required by the new demand and to improve their banking services for investment in government finance. The zeal with which they went about this is evidenced by their appearance wherever business opportunities opened up—and by their success in dominating whatever market they operated in.

No level of international finance was any higher than the papacy, whose needs for credit transfers throughout its far-flung organization accounted for the rise of Tuscan banking in the first place. And no one profited more than the Florentines from the soaring fortunes of the papacy as a secular power at the the end of the Middle Ages—from the bureaucracy it built up at Avignon to tighten its hold on the international network of ecclesiastical finances, from the sumptuous court it set up once back in Rome, and from the large territorial state it carved out for itself in central Italy.

Avignon was an important base for Florentine commercial operations as a central place for the collection of wool throughout southern France for export to Italy and the Levant, an independent activity in the area that gave Florentines an advantage over the Genoese and Venetians in the competition for papal business. By plugging into the papal financial network at its center, major firms were able to build up an extensive system of branch operations all over Europe. The strongest of these firms at the end of the fourteenth century, the Alberti, with extensive control over papal finances throughout Europe, established its position in the papal network so securely that it was able to survive the break in diplomatic relations between the papacy and Florence occasioned by the War of the Eight Saints (1375 to 1378); nor was the firm's business much hurt by reverses suffered by the family in the city's internal factional strife that resulted eventually in the family's exile for over a quarter of a century, from 1401 to 1428.

After the Great Schism divided the church in 1378, Florentine firms—Alberti, Spini, Ricci, Medici—took their chances with the Roman claimants, eventually planting their roots in Rome; and Florentine commerce and banking reaped the harvest once the Council of Constance (meeting from 1414 to 1417) unified the church and the papacy was reestablished in its historic capital. The close ties both Martin V and Eugene IV had with Florence are to be seen as partly a result of Florentine banking interests, above all those of the Medici, who made their fortune in papal business. With the growth of the papal capital as a market for Florentine cloths, Florentines became more conspicuous, dominating virtually every aspect of the rapidly expanding luxury trade. Moreover, the capital accumulated from these commercial profits was available for investment in the papacy itself, now more than ever in need of funds to pursue its policy of state-building in central Italy. It was thus that Florentines were able to dig deeper and deeper into papal finances as treasurers and tax collectors, at times virtually taking over the fiscal administration of the papal state and claiming possession of the papal tiara itself as security. Rome remained one of the most important centers for Florentine banking down through the two Medici pontificates in the sixteenth century. For all their involvement with papal finances, however, Florentines held a position in the papal state that was built on the solid economic basis of the international commercial system they operated; and their business as merchants did not seriously suffer during those moments when any one of them lost papal favor in financial affairs.

A second Italian capital whose growth at the end of the Middle Ages opened important new opportunities to merchants and bankers was Naples. Under the Angevins, Florentines had had a commanding position in foreign trade, and the new regime established in 1435 with the conquest of the kingdom by the Aragonese fully appreciated their presence for the well-being of the realm. In fact, the banking services of the Florentines were essential to the fiscal health of the kingdom; and when Alfonso the Magnanimous was compelled to outlaw them in 1447 in line with his opposition to the Medici, he found that he could not so easily dispense with them. Florentines remained active despite the ban, and the king granted safe conduct to certain of them, in violation of his own policy. Florentines continued to receive special privileges from Alfonso's successor, Ferrante, who was anxious to develop commerce and banking in his capital, and for the rest of the century they dominated the markets in Naples and other cities in the kingdom, especially in Puglia. Although this aspect of the economic history of the kingdom in the fifteenth century is poorly documented, many Florentine fortunes were known to have been made there, including those of two of the most prominent palace-builders in the Renaissance—Filippo di Matteo Strozzi and Giuliano di Leonardo Gondi, both of whom had close personal ties with the king himself.

Both Rome and Naples were capitals of states with wealth out of proportion to their territorial extent in Italy—Rome was the financial capital of the official organization of European Christendom, now finally strengthened by territorial independence and a solid bureaucratic structure; and Naples, after its conquest by Alfonso I, was the capital of a state spreading across the Tyrrhenian Sea to Aragon, Valencia, and Catalonia, with its flourishing port of Barcelona. Both cities experienced rapid growth, especially Naples, whose population reached 200,000 by the mid-sixteenth century, making it one of the largest cities in Europe. Great quantities of wealth flowed into both places, much of it brought by aristocrats who were induced by the attraction of the new courts to take up urban residence. These courts, among the most sumptuous in Europe, were major new outlets for the quality products being manufactured in Florence. It has been estimated that in the fifteenth century one-half of the luxury cloth supplied to Rome came from Florence and that Roman consumption alone accounted for as much as one-tenth of the output of the Florentine industry.

A third major market for Florentine cloths that opened up later in the fifteenth century was the Turkish Empire. With the Mongol and Turkoman invasions and the disintegration of the Mamluk Empire in Syria and Egypt in the late fourteenth century, demand for local luxury crafts collapsed, and the region went into a general economic decline. In this situation Italian luxury cloths (like Chinese ceramics) enjoyed a competitive advantage over local products. These markets grew with the expansion of the Ottoman Empire, which culminated in the restoration of Constantinople, after 1453, as one of the great entrepôts of the Western world. The city's long decline under the last Byzantine rulers to a size of no more than 100,000 was now reversed, and by the third quarter of the next century it had a population of almost three-quarters of a million. The sumptuous court of the sultans and their elaborate bureaucracy endowed the place with much of its former splendor and assured its importance as an emporium for the luxury trade. Cloths were among the leading items in this, as in all, luxury markets; and because the native cloth industry was incapable of meeting the demand, Italian producers profited most of all from the new situation. Florentine merchants, formerly active in Constantinople, quickly came to terms with the new rulers, entering into trade agreements with the Sultan within a year of the conquest. Almost fifteen years later Benedetto Dei listed fifty-one Florentines active in the area, and by the end of the fifteenth century they had organized themselves into a formal trading community in Pera. Even the Venetians felt threatened by the strong presence of Florentines in what had traditionally been their territory. In addition, many merchants from the Balkan region of the Ottoman Empire showed up in Florence to purchase cloths. Most of this trade was in woolens, but Florentines were also able to push some silk cloth in this area that supplied raw silk for their home industry—a fact that marks the clear industrial advantage Italy had over the Near East.

Scattered all across the Western world, and yet linked through the technical facilities for dealing with one another, the Florentines had what was in effect Europe's only international banking system, and more than any other money their gold florin was the international standard of value. Well before the end of the fourteenth century Florentine capital and business know-how had asserted its strength to the point that even the most vigorous competitors could not resist the penetration by Florentines into their home areas. The need for their capital and banking services thus led the Venetians in 1382 to reverse a longstanding policy of protectionism and come to terms with Florentines by letting them finally set up shop in Venice and invest in Venetian maritime trade, and Florentine capital contributed significantly to the growth of the textile industry in Ragusa in the fifteenth century. In Barcelona, the great center of maritime commerce in the western Mediterranean, local merchants persisted more vigorously in their hostility to the growing presence of these foreign competitors at the end of the fourteenth century, but the Florentines eventually moved in under the protection of the king, who recognized the superiority of their financial services despite the continuing protests from his own subjects.

It was almost entirely under Florentine auspices that a major center for international clearance and speculation in exchange opened first in Geneva and then in Lyons. When the prince-bishop of Geneva announced a liberalization of his usury policy to encourage business at the fairs there, Florentine companies were quick in setting up branches to take advantage of conditions favorable to their traffic in bills of exchange (1387). With the Hundred Years War raging in France, Geneva replaced Paris as the trading-place for the entire area—but only until the situation in France finally settled down, when Louis XI, anxious to promote Lyons as a rival place of business, scheduled fairs there to compete with those in Geneva. In 1463 he granted all kinds of concessions to merchants to induce them to bring their business into his kingdom, and before the year was out the first Florentine bank opened its doors. By 1470 the Florentine colony had organized itself into an official corporation *(nazione)*. Almost five-sixths of the 169 firms that can be identified as operating in Lyons down to the end of the century were Florentine. The emigration of Florentines spelled decline for Geneva and assured the preeminence of Lyons as the principal trading and banking center north of the Alps. In the sixteenth century many Florentines in Lyons took the traditional path of their trade into the highest realm of royal finance and ended up as French noblemen at the court in Paris.

Florentine firms were found all across the continent beyond the Alps. Only in northern Germany was their penetration somewhat blunted by the organized resistance of the Hanseatic cities. In Poland Florentines were the leading Italian merchants, especially active in the commerce of luxury cloths. In Hungary they put their commercial wealth and financial skills at the disposal of the monarchy, receiving, in return for political loans, control over mines, customs, and other tax revenues; and in the early sixteenth century they had a monopoly in the purchase of all precious metals mined in Transylvania. In England, although no longer involved so much in royal finances as were their predecessors in the early fourteenth century, Florentines were still the most prominent bankers in effecting international payments, especially to the papacy and the Low Countries. The leading house there during the first third of the fifteenth century was the Alberti, the members now conducting their business in exile from their native city. From their banking place in Bruges, Florentines were essential to the flow of international payments on which the cloth industry of northwestern Europe had long been so dependent, this commercial business being more important than government finance for most of them. In the luxury trade that began to pick up in this area so quickly in the fifteenth century, however, Florentines, though active, were probably less important than their competitors from the maritime republics of Genoa and Venice.

Florentines did not fail to take advantage of the opportunities that opened up in the Atlantic trade with the expansion of Portugal and Spain. From the time that they began to buy wool in Spain, first in the eastern kingdoms of Aragon and then in Castile, they roamed all over the Iberian peninsula looking for business opportunities. Although there is as yet no study that takes an overall view of their activity in the area, scattered evidence about particular merchants indicates something about the networks they built to the Atlantic islands, to Africa, and eventually to the New World. They dealt in a great variety of items, including, besides the traditional luxury products, African slaves, sugar from Madeira, and more common goods like grain, fish, and leather from Portugal. Among the Italians doing business in the peninsula the Florentines seem to have had a notable presence in Portugal. One of the chief companies there, the Cambini, had its own ships sailing the Atlantic coast and covered the voyages of others with insurance. It provided Henry the Navigator with his copy of the Florentine edition of Ptolemy, and when his brother the cardinal died in Florence the home office became the executor of his estate, in charge of building his tomb in San Miniato. The Marchionni, another merchant family that over several generations played a major role in Portuguese colonial expansion, had companies dealing in coral and in the slave trade between Africa and Brazil. Girolamo Sernigi, of another long-established family of merchant-bankers in Lisbon, wrote letters back to Florence reporting on Vasco da Gama's arrival in India and assessing the commercial possibilities arising out of that event, and one of the first factors in Goa was yet another Florentine, Francesco Corbinelli, who for a dozen years handled much of the business of the royal treasury there.

Although there is no way to assess it quantitatively, all this activity, encircling virtually the whole of Europe, from Constantinople to the Atlantic islands, from England to Poland, adds up to a much more dynamic commercial and banking sector than Florence had had before the Black Death. Whatever the general economic situation was in Europe at the end of the Middle Ages, trade intensified, the range of commodities traded enlarged, and luxury markets everywhere boomed—and Florentines cashed in on these opportunities. Moreover, Florence's trade was rooted in a thriving home industry supplying a product that was a staple in virtually every luxury market. The need for raw materials—wool from England, Spain, and Italy; silk from the Near East and Italy—and the demand for their finished cloths sent Florentine merchants everywhere throughout the Mediterranean and northern Europe, building up a network through which they could channel any other commodity for which there was a market, be it a rare spice or ordinary salt, and improving their banking services for the extension of credit, execution of international payments, and administration of government finance. For all its geographic breadth, however, the system was oriented to the Mediterranean, where major new centers of financial activity opened up in the revitalized papal state, first in Avignon then in Rome, and in the kingdom of Aragon once it had expanded into Italy. Moreover, its success at pushing its own products in the eastern Mediterranean helped Florence's balance of payments with that part of the world—a problem that perennially plagued the European economy as a whole well into modern times. The core area of banking and commerce was thus largely conterminus with the area to which its home industry was oriented for both its raw materials and its markets. This system, more concentrated in a confined geographical area and more interrelated in its infrastructure, gave the city a much sounder economic foundation than it had ever had before.

Performance of the Economy

If at the end of the Middle Ages the times were favorable for the development of the leading sectors of the Florentine economy, those activities could achieve a high level of performance only by making adjustments to meet the changing conditions abroad. Over the fourteenth century these adjustments added up to a major transformation of the cloth industry and considerable improvement in some of the ways merchants and bankers conducted their business.

At the center of the economy was, as always, the cloth industry. It improved the quality of its products, expanded its markets abroad, and eventually diversified its production with the addition of silks. The initial phase of this expansion owed much of its success to the ability of Florentines to capture the Mediterranean markets for northern cloths, taking advantage of a favorable market situation at a time, the early fourteenth century, when the Flemish cities were beset by serious political and economic problems. It is striking, in fact, how, in upgrading the quality of their woolens to meet the competition, Florentines adopted a nomenclature for their products that derives from conscious imitation of the products against which they were competing. Eventually their production was so exclusively aimed at luxury markets abroad that cheaper cloths for local consumption were being imported, and complaints were heard about the difficulty of finding Florentine products in the local market.

Wool cloth fell roughly into two categories. The most luxurious (panni di San Martino), made from wool imported from England, came to enjoy a virtual monopoly in the traditional luxury markets of Italy and long remained a staple of the industry. The reorientation of the industry, however, resulted from the development of a strong second-line product (panni di Garbo), made from wool found closer at home, in the Mediterranean. This wool originally came from north Africa, then from Provence and Catalonia, and when political and economic problems in these areas interrupted supply at the beginning of the fifteenth century, a new source was found in Italy itself, in the Abruzzi and Lazio. Toward 1500 yet another supply channel was opened with the importation of Castilian wool, and in the sixteenth century Spain became a major source of raw material for the industry. Production had its ups and downs as it adjusted to these geographical shifts in supply, and political instability and other problems in these various areas often resulted in moments of depression in Florence; but in the long run the industry prospered. Markets for these cloths, as we have seen, lay almost entirely within the Mediterranean area.

This industrial sector was further strengthened by expansion into the production of silk, the most luxurious of cloths. Although silk had long been the leading industry in nearby Lucca, there is little evidence for an industry in Florence much before the end of the fourteenth century. The 1335 statutes of the guild of Por Santa Maria, made up mostly of retail sellers of clothing including silk, do not suggest that much manufacturing was going on. Silk workers exiled from Lucca began showing up in Florence about that time, however; and it is generally thought that the industry got underway as a result of their enterprise. In any case, guild records in the second half of the century identify silk workers as a distinct category (*membrum*) of the membership and reveal a growing concern with production.

By 1429, when the statutes were revised, silk manufacturers clearly dominated the guild, which afterward was commonly identified as the silk guild. By this time, too, the commune was making efforts to promote the industry, holding out the advantage of tax relief for reelers and spinners who took up residence in the city and encouraging the planting of mulberry trees in the countryside. Most of the raw silk for the industry, however, was imported from the Caspian Sea area, from Spain, and from elsewhere in Italy (the Romagna, the Marches, the Abruzzi, and Calabria). It was not until later, when the grand-ducal government more energetically pursued a policy of getting mulberry trees planted, that the local source of supply became significant.

As with the wool industry, growth in the production of silks cannot be quantitatively assessed with precision. Benedetto Dei counted 83 shops around 1470, a figure that may be seriously questioned but yet may have some significance for the relative importance of the industry if compared to the 280 wool shops counted in the same survey. In 1527 the Venetian ambassador reported the value of the output of the industry as two-thirds that of wool production. The quality was renowned throughout Europe, even in the north, where Florentine wool did not sell, although the best markets were largely in Italy itself and the Mediterranean, including the Turkish Empire.

In summary, the performance of the textile sector over the period of the Renaissance was very strong. Even in view of the sharp decline of the working force in the half-century following the Black Death, the few production figures we have for wool indicate that there was no decline in per capita output in the industry. Furthermore, the quality of that output increased, and the sector was further strengthened in the fifteenth century by an increasingly vigorous silk industry. With virtually all of this production directed to foreign luxury markets, the city was assured a secure source of wealth from abroad.

Earlier in the fourteenth century dependence on English wool meant a disjunction between the source of raw materials for the home industry and the markets where Florentines sold their wares, with the result that there was too much of a demand for money in a backward area on the edge of the international luxury market; but later, as Florentines began to take more wool from places closer at hand, from Spain and from Italy itself, the cloth industry, the mainstay of the city's economy, became oriented toward the core of the European commercial system around the Mediterranean for the supply of its raw materials, just as it had always been for its markets. This reorientation was strengthened by the contemporaneous diversification of the industry into silk, which was found only in the Mediterranean area.

There was, in other words, considerably more of an overlap of the markets where Florentines bought and sold. Moreover, this recession of England from the Florentine industrial-commercial system occurred at a time when both the English payments to the papacy and military expenditures on the Continent began to fall off, so that England had less need of Italian bankers to handle international payments. By the fifteenth century supply channels and sales outlets were tied closely together in the old and well-developed trade network at the Mediterranean core of the European commercial system, and credit and goods flowed throughout the area without the likelihood of the kind of glut that dependence on a single remote and peripheral area had created earlier.

It was an altogether healthier economic situation as a result. Nothing like the banking failures of the 1340s, with their disastrous consequences throughout the city, ever occurred again. Still at a stage of development where foreign trade involved a high degree of speculation, the Florentine economy of the Renaissance saw its share of bankruptcies; an entire archive of impounded company accounts survives as testimony to the business that came before the special section of the merchants' court (the Mercanzia) set up to handle the resulting legal problems. The biggest crisis of the fifteenth century, in 1464 and 1465, however, produced only ripples that left most companies in the banking community untouched. It was presumably caused by an entirely local situation in the Levant, at the edge of the European banking system, and had little to do with the kind of problem inherent in the very structure of the system that lay behind the earlier banking failures.

No small part of the continuing success of banking and commerce was due to the business technology that made the mechanism of the system work. The roots of business practice were in the earlier period; and the entire subject of this history of business techniques is riddled with controversy about when and where the first instance of this or that device occurred. The isolated example in the earlier period, however, becomes common practice later on. Business practice improved steadily over the fourteenth century with the refinement of bookkeeping (and recordkeeping generally), credit instruments, insurance, business organization, and other techniques; and the development of elementary schools of commercial arithmetic taught men the fundamentals of how to deal with these practices. Virtually everyone in Florence who was in business, whether artisan or international merchant, used these techniques. The following discussions include descriptions of how they were exemplified in the way the kilnmen Da Terrarossa organized the business side of their enterprise; in the way the stonemason-foundryman Maso di Bartolomeo kept his accounts; in the way ordinary construction workers handled written orders of payment, used giro operations, "thought" in terms of a money of account, and, in general, trusted to a system of private recordkeeping for so many of their transactions.

This technology made the mechanisms of the business system function more efficiently at home and abroad; the merchants' court developed in the fourteenth century to handle business affairs with increasing effectiveness engendered confidence in that system; and a personal rapport among operators of all classes—*fiducia* is the term used by business historians—which was hardly disturbed even by political exile from their native city, held everything together. The vast international operations of the well-known merchant of Prato, Francesco di Marco Datini (who died in 1410), marks the singular success, not of a lone merchant-adventurer or an exalted royal favorite, but of a sedentary and diversified investor who—as an "establishment man"—worked efficiently through the highly sophisticated system of international commerce and banking Florentines had built up among themselves by the end of the fourteenth century.

The Florentine economy was additionally strengthened in the fifteenth century by greater control over the transportation facilities on which trade depended. The importance of shipping in an economy like Florence's hardly needs comment: the records of just one merchant, Francesco di Marco Datini, mention the names of 3,000 different ships in a fourteen-year period, from 1391 to 1405. Unlike its great rivals, Genoa and Venice, Florence built up its commercial system in the thirteenth and fourteenth centuries as an inland city without its own port and without a fleet. With the decline of Pisa after the battle of Meloria in 1284, Florence was in a better position to obtain port privileges; but since the relations between the two cities were not always friendly, Florentines by no means depended entirely on this one port as a commercial outlet. Much of its shipping was in the hands of the Genoese, who often used several other Tuscan ports as points of entry and exit; and much of the eastern trade was directed overland to Ancona and put in the hands of Venetians. Increasingly toward the end of the fourteenth century, however, we hear of Florentine merchants owning their own ships—the Alberti, for example, had ships serving England and Catalonia, and in the fifteenth century the Cambini had ships sailing the Atlantic coast in the expanding Portuguese Empire. The conquest of Pisa in 1406 and the purchases (from Genoa) of Porto Pisano and Livorno in 1421 gave Florence its own outlet, and many firms immediately established offices there. The next step, the setting up of a state-controlled galley system in imitation of that of the Venetians, was taken immediately, and although this system never established Florence as a maritime power (and in fact was abandoned in 1480), it contributed substantially, at least for a time, to the improvement of its control over an essential facility in its commercial system.

The investment habits of Florentines in the fifteenth century, if not further proof of the vitality of the economy, at least indicated that their confidence in that economy was not flagging. The vast expenditure for private building is not to be seen as a withdrawal of capital from more productive enterprise by a propertied class undergoing a change of heart about their business traditions for either cultural or economic reasons. Some of the men who spent most lavishly on building projects were successful merchants whose investment portfolios reveal no intention of closing shop. For example, Filippo Strozzi, who had enough cash on hand to pay the entire cost of his new palace with plenty left over, kept up his businesses to the day of his death; Giuliano Gondi, who stipulated in his will that his sons were to finish building the palace he had started, also exhorts them to keep up the family business, since it had been going for such a long time; and we have Giovanni Rucellai's own words about the satisfaction he derived from his successful career as a banker.

Rich Florentines took the cloth industry for granted as an investment, most of them, even merchant bankers for whom such a business was a small operation in relation to their international trading and banking companies, participating as partners in either a wool or silk shop. Giannozzo Alberti (in *Della famiglia*) strongly recommends investment in a cloth shop on the grounds that it provided the income necessary for the maintenance of a family with suitable dignity but did not require much work and bother (and, he adds, since this kind of investment had a social utility in the employment it provided, it also endowed the investor with a certain moral virtue). Giannozzo reveals here the mentality of a veritable rentier, such was his confidence as an investor in the stability of the industry. Other Florentines, moreover, worked hard as active partners in their cloth shops, and not the slightest stigma of ignobleness was associated with such employment—something that always impressed the Venetian ambassadors. The merchant-aristocrats from the lagoons marvelled that these upper-class Florentines, with their cloaks draped over their shoulders so they would have greater freedom of movement, were seen in their shops doing manual labor without any concern about what the public thought. At the beginning of the seventeenth century the Venetian ambassador noted the withdrawal of the upper classes from involvement in business affairs as younger men found court life more attractive; and such a late date for his observation on this change in life-style is evidence of how long the cloth industry maintained its central importance in Florentine life.

In the Renaissance the upper classes made no marked move to "return to the land," a theme that occurs repeatedly in the history of European urban elites at that point where they lose their nerve and seek the security and prestige of investment in landed estates. In 1427 the upper 2 to 3 percent of the city's wealthiest men had no more invested in real estate than in business. Like Giovanni Rucellai they recognized the pros and cons of land as an investment and concluded that land helped round out and balance—but should not dominate—an investment portfolio. Many of the biggest portfolios show no higher share of total investment in land than the 7 percent in Filippo Strozzi's and the 12 percent in Francesco Sassetti's, the manager of the Medici bank; and like these two successful merchant-bankers, many probably had more tied up in their town house than in income-yielding real estate, be it in the city or in the countryside. Most all Florentines of any substance, including even artisans, had property outside the city—many still do—and we can always turn up a few men who had extensive landholdings and nothing else. The Medici had vast estates already in Cosimo il Vecchio's generation, so vast that in some areas the family enjoyed what was virtually a feudal presence; and Lorenzo's interest in his cheese factory and stock farming (especially on his estate at Poggio a Caiano) anticipated the relatively enlightened agricultural policy of Cosimo I. Before the sixteenth century, however, there was no marked tendency for the upper classes to build up large compact estates and virtually no interest in introducing new methods of estate management or farming technology. Purchasing land, which yielded a much lower return for one's investment than business, was no way to build up a fortune. Moreover, no particular prestige was attached to owning land. For all the interest in building villas, temporary retirement to them did not induce anything like a rentier mentality or (what has been called) a villa psychology.

The history of just about any family whose economic fortunes can be traced over the four or five generations spanning the Renaissance will turn up more entrepreneurs of one kind or another than rentiers, and not one of these families could be classified a rentier family over the entire period. Since the practice of partible inheritance could reduce even a rich man's sons to a modest status, most men were compelled to invest both time and capital in the advanced sectors of the economy if they wanted to make their own fortunes. Indeed, the lack of legal devices to protect the integrity of estates beyond one generation, especially in view of the nascent dynastic sense that explains much of their interest in building, can be seen as a mark of the confidence Florentines continued to have in the business traditions of their economy.

The story of the performance of the Florentine economy in the Renaissance can be recapitulated in the language of one of the economist's models—the linkage theory of economic development. From the thirteenth century onward, Florence experienced an impressive export-led growth in her economy, the staple being cloth—first wool, then both wool and silk. The quality of the product was continually improved, and by the fifteenth century virtually all the production linkages around the staple were fully developed. The "nonindustrial forwarding operations" of selling abroad and finance went beyond the business of the staple to become independently one of Europe's leading commercial and banking systems; only the link to transportation remained undeveloped. Development of backward linkages was blocked only by the geographical requirements of the raw materials, although even here the situation much improved in the fourteenth century as supply channels were open to sources closer at hand in the Mediterranean; and by the sixteenth century, one of the final links in the chain of supply was being forged with the establishment of the silkworm culture in the area of Pescia.

Ideas for Discussion

1. Why was the fourteenth century crucial to the development of the Florentine economy?
2. What impact did the cloth trade have on construction of buildings and towns, mercantilism and banking?
3. How would you characterize the business system developed in Florence in this period?

Topics for Writing

1. Create a map of Florence's global textile market.
2. Prepare a report on the impact of the Black Death on European economies.
3. Write an essay on how Florentine cloth firms responded to the spirit of competition.

Daniel Defoe
The Tradesman's Writing Letters

Daniel Defoe is best known as the author of such popular and classic novels as Robinson Crusoe *and* Moll Flanders, *but he also had an extensive career as a political journalist, magazine editor, and author of nonfiction tracts and books in eighteenth-century London. Born in 1660, he published over 250 individual works in his life. Defoe died in 1731. The guidelines on writing business letters reprinted here are excerpted from his 1726,* The Complete English Tradesman. *In this book Defoe employs both his own experiences in a variety of businesses (successful and unsuccessful) as well as his knowledge of current, acceptable practices in the English commercial community. As is usually the case in pre-twentieth-century writing about business and industry, the language of this selection is gender-based, using the male pronoun and "tradesman" to encompass all business people. In the eighteenth century, men dominated commercial acitivity, but women were a vital part of the business world as well as they are now.*

As PLAINNESS, and a free unconstrained way of speaking, is the beauty and excellence of speech, so an easy free concise way of writing is the best style for a tradesman. He that affects a rumbling and bombast style, and fills his letters with long harangues, compliments, and flourishes, should turn poet instead of tradesman, and set up for a wit, not a shopkeeper. Hark how such a young tradesman writes, out of the country, to his wholesaleman in London, upon his first setting up.

>'SIR—The destinies having so appointed it, and my dark stars concurring, that I, who by nature was framed for better things, should be put out to a trade, and the gods having been so propitious to me in the time of my servitude, that at length the days are expired, and I am launched forth into the great ocean of business, I thought fit to acquaint you, that last month I received my fortune, which, by my father's will, had been my due two years past, at which time I arrived to man's estate, and became major, whereupon I have taken a house in one of the principal streets of the town of———, where I am entered upon my business, and hereby let you know that I shall have occasion for the goods hereafter mentioned, which you may send to me by the carrier.'

This fine flourish, and which, no doubt, the young fellow dressed up with much application, and thought was very well done, put his correspondent in London into a fit of laughter, and instead of sending him the goods he wrote for, put him either first upon writing down into the country to inquire after his character, and whether he was worth dealing with, or else it obtained to be filed up among such letters as deserved no answer.

The same tradesman in London received by the post another letter, from a young shopkeeper in the country, to the purpose following:—

> 'Being obliged, Sir, by my late master's decease, to enter immediately upon his business, and consequently open my shop without coming up to London to furnish myself with such goods as at present I want, I have here sent you a small order, as underwritten. I hope you will think yourself obliged to use me well, and particularly that the goods may be good of the sorts, though I cannot be at London to look them out myself. I have enclosed a bill of exchange for £75, on Messrs A. B. and Company, payable to you, or your order, at one-and-twenty days' sight; be pleased to get it accepted, and if the goods amount to more than that sum, I shall, when I have your bill of parcels, send you the remainder. I repeat my desire, that you will send me the goods well sorted, and well chosen, and as cheap as possible, that I may be encouraged to a further correspondence. I am, your humble servant,
>
> C.K.'

This was writing like a man that understood what he was doing; and his correspondent in London would presently say—'This young man writes like a man of business; pray let us take care to use him well, for in all probability he will be a very good chapman.'

The sum of the matter is this: a tradesman's letters should be plain, concise, and to the purpose; no quaint expressions, no book-phrases, no flourishes, and yet they must be full and sufficient to express what he means, so as not to be doubtful, much less unintelligible. I can by no means approve of studied abbreviations, and leaving out the needful copulatives of speech in trading letters; they are to an extreme affected; no beauty to the style, but, on the contrary, a deformity of the grossest nature. They are affected to the last degree, and with this aggravation, that it is an affectation of the grossest nature; for, in a word, it is affecting to be thought a man of more than ordinary sense by writing extraordinary nonsense; and affecting to be a man of business, by giving orders and expressing your meaning in terms which a man of business may not think himself bound by. For example, a tradesman at Hull writes to his correspondent at London the following letter:—

> 'SIR, yours received, have at present little to reply. Last post you had bills of loading, with invoice of what had loaden for your account in Hamburgh factor bound for said port. What have farther orders for, shall be dispatched with expedition. Markets slacken much on this side; cannot sell

the iron for more than 37s. Wish had your orders if shall part with it at that rate. No ships since the 11th. London fleet may be in the roads before the late storm, so hope they are safe: if have not insured, please omit the same till hear farther; the weather proving good, hope the danger is over.

My last transmitted three bills exchange, import £315; please signify if are come to hand, and accepted, and give credit in account current to your humble servant.'

I pretend to say there is nothing in all this letter, though appearing to have the face of a considerable dealer, but what may be taken any way, *pro* or *con*. The Hamburgh factor may be a ship, or a horse—be bound to Hamburgh or London. What shall be dispatched may be one thing, or any thing, or every thing, in a former letter. No ships since the 11th, may be no ships come in, or no ships gone out. The London fleet being in the roads, it may be the London fleet from Hull to London, or from London to Hull, both being often at sea together. The roads may be Yarmouth roads, or Grimsby, or, indeed, any where.

By such a way of writing, no orders can be binding to him that gives them, or to him they are given to. A merchant writes to his factor at Lisbon:—

'Please to send, per first ship, 150 chests best Seville, and 200 pipes best Lisbon white. May value yourself per exchange £1250 sterling, for the account of above orders. Suppose you can send the sloop to Seville for the ordered chests, &c. I am.'

Here is the order to send a cargo, with a *please to send*; so the factor may let it alone if he does not please.[1] The order is 150 chests Seville; it is supposed he means oranges, but it may be 150 chests orange-trees as well, or chests of oil, or any thing. Lisbon white, may be wine or any thing else, though it is supposed to be wine. He may draw £1250, but he may refuse to accept it if he pleases, for any thing such an order as that obliges him.

On the contrary, orders ought to be plain and explicit; and he ought to have assured him, that on his drawing on him, his bills should be honoured—that is, accepted and paid.

I know this affectation of style is accounted very grand, looks modish, and has a kind of majestic greatness in it; but the best merchants in the world are come off from it, and now choose to write plain and intelligibly: much less should country tradesmen, citizens, and shopkeepers, whose business is plainness and mere trade, make use of it.

I have mentioned this in the beginning of this work, because, indeed, it is the beginning of a tradesman's business. When a tradesman takes an apprentice, the first thing he does for him, after he takes him from behind his counter, after he lets him into his counting-house and his books, and

1 [The practice of trade now sanctions courteous expressions of this kind.]

after trusting him with his more private business—I say, the first thing is to let him write letters to his dealers, and correspond with his friends; and this he does in his master's name, subscribing his letters thus:—

I am, for my master, A. B. and Company, your
   humble servant,   C. D.

        And beginning thus:—
Sir, I am ordered by my master A. B. to advise you that

        Or thus:—
Sir, By my master's order, I am to signify to you that———

Orders for goods ought to be very explicit and particular, that the dealer may not mistake, especially if it be orders from a tradesman to a manufacturer to make goods, or to buy goods, either of such a quality, or to such a pattern; in which, if the goods are made to the colours, and of a marketable goodness, and within the time limited, the person ordering them cannot refuse to receive them, and make himself debtor to the maker. On the contrary, if the goods are not of a marketable goodness, or not to the patterns, or are not sent within the time, the maker ought not to expect they should be received. For example—

The tradesman, or warehouseman, or what else we may call him, writes to his correspondent at Devizes, in Wiltshire, thus:—

'SIR—The goods you sent me last week are not at all for my purpose, being of a sort which I am at present full of: however, if you are willing they should lie here, I will take all opportunities to sell them for your account; otherwise, on your first orders, they shall be delivered to whoever you shall direct: and as you had no orders from me for such sorts of goods, you cannot take this ill. But I have here enclosed sent you five patterns as under, marked 1 to 5; if you think fit to make me fifty pieces of druggets of the same weight and goodness with the fifty pieces, No. A.B., which I had from you last October, and mixed as exactly as you can to the enclosed patterns, ten to each pattern, and can have the same to be delivered here any time in February next, I shall take them at the same price which I gave you for the last; and one month after the delivery you may draw upon me for the money, which shall be paid to your content. Your friend and servant.

P.S. Let me have your return per next post, intimating that you can or cannot answer this order, that I may govern myself accordingly. To Mr H. G., clothier, Devizes.'

The clothier, accordingly, gives him an answer the next post, as follows:—

'SIR—I have the favour of yours of the 22d past, with your order for fifty fine druggets, to be made of the like weight and goodness with the two packs, No. A.B., which I made for you and sent last October, as also the five patterns enclosed, marked 1 to 5, for my direction in the mixture. I give you this trouble, according to your order, to let you know I have already put the said fifty pieces in hand; and as I am always willing to serve you to the best of my power, and am thankful for your favours, you may depend upon them within the time, that is to say, some time in February next, and that they shall be of the like fineness and substance with the other, and as near to the patterns as possible. But in regard our poor are very craving, and money at this time very scarce, I beg you will give me leave (twenty or thirty pieces of them being finished and delivered to you at any time before the remainder), to draw fifty pounds on you for present occasion; for which I shall think myself greatly obliged, and shall give you any security you please that the rest shall follow within the time.

As to the pack of goods in your hands, which were sent up without your order, I am content they remain in your hands for sale on my account, and desire you will sell them as soon as you can, for my best advantage. I am,' &c.

Here is a harmony of business, and every thing exact; the order is given plain and express; the clothier answers directly to every point; here can be no defect in the correspondence; the diligent clothier applies immediately to the work, sorts and dyes his wool, mixes his colours to the patterns, puts the wool to the spinners, sends his yarn to the weavers, has the pieces brought home, then has them to the thicking or fulling-mill, dresses them in his own workhouse, and sends them up punctually by the time; perhaps by the middle of the month. Having sent up twenty pieces five weeks before, the warehouse-keeper, to oblige him, pays his bill of £50, and a month after the rest are sent in, he draws for the rest of the money, and his bills are punctually paid. The consequence of this exact writing and answering is this—

The warehouse-keeper having the order from his merchant, is furnished in time, and obliges his customer; then says he to his servant, 'Well, this H. G. of Devizes is a clever workman, understands his business, and may be depended on: I see if I have an order to give that requires any exactness and honest usage, he is my man; he understands orders when they are sent, goes to work immediately, and answers them punctually.'

Again, the clothier at Devizes says to his head man, or perhaps his son, 'This Mr H. is a very good employer, and is worth obliging; his orders are so plain and so direct, that a man cannot mistake, and if the goods are made honestly and to his time, there's one's money; bills are cheerfully accepted, and punctually paid; I'll never disappoint him; whoever goes without goods, he shall not.'

On the contrary, when orders are darkly given, they are doubtfully observed; and when the goods come to town, the merchant dislikes them, the warehouseman shuffles them back upon the clothier, to lie for his account, pretending they are not made to his order; the clothier is discouraged, and for want of his money discredited, and all their correspondence is confusion, and ends in loss both of money and credit.

* * *

I might have made some apology for urging tradesmen to write a plain and easy style; let me add, that the tradesmen need not be offended at my condemning them, as it were, to a plain and homely style—easy, plain, and familiar language is the beauty of speech in general, and is the excellency of all writing, on whatever subject, or to whatever persons they are we write or speak. The end of speech is that men might understand one another's meaning; certainly that speech, or that way of speaking, which is most easily understood, is the best way of speaking. If any man were to ask me, which would be supposed to be a perfect style, or language, I would answer, that in which a man speaking to five hundred people, of all common and various capacities, idiots or lunatics excepted, should be understood by them all in the same manner with one another, and in the same sense which the speaker intended to be understood—this would certainly be a most perfect style.

All exotic sayings, dark and ambiguous speakings, affected words, and, abridgement, or words cut off, as they are foolish and improper in business, so, indeed, are they in any other things; hard words, and affectation of style in business, is like bombast in poetry, a kind of rumbling nonsense, and nothing of the kind can be more ridiculous.

The nicety of writing in business consists chiefly in giving every species of goods their trading names, for there are certain peculiarities in the trading language, which are to be observed as the greatest proprieties, and without which the language your letters are written in would be obscure, and the tradesmen you write to would not understand you.

Ideas for Discussion

1. Comment on how expectations for letter writing have changed since the eighteenth century.
2. Discuss the importance of plainness in commercial writing.
3. Why does Defoe concentrate on "harmony" when describing business letter writing?

Topics for Writing

1. In an oral presentation, discuss the primary features of effective business writing as described by Defoe.
2. Prepare a series of handouts for the tradesman highlighting the central points as described by Defoe.
3. Write a manual, like Defoe's, for the modern letter writer.

Dharma Kumar with Meghnad Desai
The Occupational Structure of India

Dharma Kumar with Meghnad Desai edited The Cambridge History of India *published in 1989. This excerpt compares the occupational structures in agriculture and manufacturing in India overall and in four regions: Kerala, West Bengal, Rajasthan, and East Punjab. Drawing on both census figures and Indian history, Kumar reveals the tenuous hold manufacturing had in India in the late nineteenth and early twentieth centuries.*

ECONOMIC GROWTH is associated with a relative shift in the structure of the workforce away from agriculture, towards industry and services. This happened not only to primary product importers like Great Britain, where agriculture's share has declined to less than 4 per cent, but also to specialized primary producers like Australia, Denmark and New Zealand, where agriculture's share has declined to 10 or 15 per cent. Simon Kuznets, who restricts his comparison of long-term trends in the structure of the labour force to twenty-five countries for which reasonably comparable information is available, finds India to be the only case of a virtually unchanged employment structure. This makes the Indian experience an important one in the history of economic development, worthy of careful, detailed examination.

Reasonably reliable information on the structure of the workforce is available for the twentieth century from the population censuses. Census data also exist for 1871–2, 1881 and 1891, but the 1871–2 data which relate to British India and some native states are clearly unreliable as the adult male workers exceed the adult male population by about 4.6 million; the 1891 data relate to the occupations of the entire population and cannot therefore be compared with the twentieth century estimates which generally relate to the workforce. The 1881 evidence can be used provided its limitations are realized. The 1881 estimates relate to males and to a territory which comprises Ajmer, Bengal, Berar, Bombay, Central Provinces, Coorg, Madras, North-West Provinces, Punjab, Baroda, Central India, Mysore and Travancore. A further deficiency is that the proportion of "general labour" (i.e., unspecified workers) is as high as 8.3 per cent of the workforce. We compare the 1881 estimates with similar estimates for 1911, a fairly normal year.

Between 1881 and 1911 the share of agriculture (defined to include activities allied to agriculture as well as general labour) hardly changed; it rose from 72.4 to 74.5 per cent of the workforce (see table 1). The share of manufacturing fell, but this is partly due to our having included all manufacturers-cum-sellers in 1881 under "Manufacturing"; in 1911, there was no manufacturer-seller category and such persons were classified under *either* manufacturer *or* seller, depending on which activity occupied more time. But even if manufacture and trade and commerce are pooled, there is still a decline from 15.5 to 14.6 per cent of the workforce. Within services there is a decline in "other services" from 9.8 to 7.7 per cent.

Given the quality of the evidence, one must not read too much into the very slight increase in agriculture's share, and the slight decrease in manufacture's share in the workforce especially since real output did not decline in absolute or per head terms. In fact, it probably increased: Heston estimates that between 1868–9 to 1872–3 and 1908–9 to 1912–13 real net domestic product (NDP) rose by 53 per cent, while population increased by only 18 per cent.

For the twentieth century, reasonably good evidence is available on both output and employment, though the output estimates for some sectors are partly derived from employment estimates and hence are not entirely independent. Between 1900 and 1947 real NDP per head in undivided India rose by about 12 per cent but if we exclude all services there was virtually no increase over the period. In spite of the virtual stagnancy of real output per head, there was a relative shift away from the agricultural sector to the industrial and services sectors.

In contrast, the structure of the workforce did not change much between 1901 and 1951. As table 2 indicates, the shares of the three sectors in the

Table 1
Structure of the Male Workforce 1881–1911 (percentages)

Sector	1881	1911
1 Cultivators	51.7	53.5
2 Agricultural labourers	10.7	13.4
3 General labour	8.3	2.7
4 Plantations, forestry, fishing, etc.	1.7	4.9
5 Mining & quarrying	0.1	0.2
6 Manufacturing	10.6	9.1
7 Construction	0.5	1.1
8 Trade & commerce	4.9	5.5
9 Transport, storage and communications	1.8	1.7
10 Other services	9.8	7.7
Total workers	100	100

Table 2
The Industrial Distribution of the Workforce in Undivided India, 1901–51
(percentages)

No. (1)	Activity (2)	Males/females/persons (3)	1901 (4)	1911 (5)	1921 (6)	1931 (7)	1951 (8)
1	Cultivators	M	53.2	53.5	56.1	49.8	54.4
		F	43.6	41.0	48.1	30.4	45.7
		P	50.3	49.6	53.5	44.3	52.2
2	Agricultural labourers	M	14.3	15.4	14.4	19.5	16.4
		F	30.2	32.5	28.0	43.8	34.5
		P	19.1	20.8	18.6	26.3	21.1
3	Livestock, forestry, fishing, hunting & plantations, orchards & allied activities	M	4.2	4.9	4.1	4.9	2.4
		F	2.9	3.2	3.2	3.8	2.3
		P	3.8	4.4	4.0	4.6	2.4
4	Mining & quarrying	M	0.1	0.2	0.2	0.2	0.4
		F	0.1	0.2	0.3	0.2	0.3
		P	0.1	0.2	0.2	0.2	0.4
5	Manufacturing	M	9.5	9.1	9.0	8.4	9.1
		F	11.4	10.9	8.4	8.8	7.7
		P	10.1	9.6	8.8	8.5	8.7
6	Construction	M	1.1	1.3	1.1	1.2	1.4
		F	0.8	0.8	0.8	0.9	0.9
		P	1.0	1.2	1.0	1.1	1.3
7	Trade and commerce	M	5.8	5.5	5.9	5.9	6.1
		F	3.5	5.3	5.2	4.8	2.8
		P	5.1	5.4	5.7	5.6	5.2
8	Transportation, storage & communications	M	1.5	1.7	1.3	1.4	1.9
		F	0.2	0.2	0.2	0.1	0.3
		P	1.1	1.2	1.0	1.1	1.5
9	Other services	M	10.2	8.3	7.8	8.6	7.8
		F	7.2	5.9	5.7	7.1	5.6
		P	9.3	7.6	7.2	8.2	7.2
	Total workers	M	100	100	100	100	100
		F	100	100	100	100	100
		P	100	100	100	100	100

workforce did not change much between 1901 and 1951; there was probably a slight rise in agriculture and slight declines in industry and services.

By comparing the industrial distributions of real output and employment, we can draw some useful conclusions about long-term changes. The main magnitudes are indicated in table 3. Initially, output per worker in the agricultural sector was slightly below average; by the end of the period it had fallen further below average output per worker in the economy. Output per worker in the industrial sector which was initially above the average became even more above average, while the output per worker of the services sector rose from just below average to well above average by the end of the period.

Table 4 juxtaposes the Indian and the international experience. In a number of important respects, the Indian experience conforms to the general pattern. As in most other cases, the relative product per worker in the agricultural sector declines from an initial value of less than 1 and the corresponding ratio for the industrial sector tends to increase over time, though the initial level in the Indian case is rather high compared to most other countries except the USA. However, the behaviour of the services sector in India is the converse of the international experience: it rises from below 1 to well above 1. Again, in marked contrast to most other countries, the index of intersectoral inequality rises in India from 17.5 to 36.2, elsewhere it declined from around 25 or more to about 20.

While the other countries considered underwent substantial increases in output per head, in India, the increase, if any, was modest. The lopsided pattern of development in India, the highly unequal income distribution and the rapid growth of the urban sector after 1931, unaccompanied by higher levels of income per head, show up in the curious behaviour over time of the services sector and in the widening of intersectoral inequality in product per worker.

We examine more closely changes in the employment pattern for undivided India between 1901 and 1951, concentrating on the results for males. For both cultivators and agricultural labourers there was a slight rise in relative shares. But the share of plantations, forestry, fishing etc. sharply declined in 1951, possibly reflecting heavy underenumeration. Employment in mining and quarrying rose considerably, but the sector was always rather small. Mining and quarrying was concentrated in Bihar and West Bengal, the latter accounting for a little less than half the employment in 1911. While employment in West Bengal declined slightly, Bihar was responsible for about 40 per cent of the increase in employment between 1911 and 1951.

Table 3
The Structure of National Income and the Workforce: Undivided India between 1901 and 1951

Sectors	Percentage of the workforce, 1911 (1)	Percentage of national income 1900–1 to 1904–5 (2)	Relative product per worker (2) ÷ (1) = (3)	Percentage of the workforce 1951 (4)	Percentage of national income 1942–3 to 1946–7 (5)	Relative product per worker (5) ÷ (4) = (6)
A	74.8	66.6	0.89	75.7	57.6	0.76
I	12.2	20.9	1.71	11.9	25.3	2.13
S	13.0	12.4	0.95	12.4	17.1	1.38

Note: A = agriculture including ancillary activities; I = mining, manufacturing, transport, storage and communications; S = trade and commerce and other services (but excludes income from house property).

Table 4
Some International Comparisons of Relative Product per Worker

Country	Period	Relative product per worker A	I	S	Index of intersectoral inequality
1 Great Britain	1800–11	1.16	0.63	1.28	27.6
	1907	0.96	0.87	1.27	16.6
2 Germany	1850–9	0.87	0.71	2.20	33.2
	1935–8	0.65	1.24	0.91	22.6
3 Italy	1861–70	0.88	0.81	1.89	24.8
	1963–7	0.58	1.13	1.17	21.4
4 USA	1839	0.75	1.80	1.16	32.4
	1929	0.56	1.27	0.96	21.4
5 Japan	1879–83	0.75	(I+S) 2.25		41.6
	1959–61	0.42	1.28		38.0
6 India	1911	0.89	1.71	0.95	17.5
	1951	0.76	2.13	1.38	36.2

Note: All figures, except those relating to India, are from: Kuznets, *Economic Growth of Nations*, table 45. The index of intersectoral inequality is defined by Kuznets as the sum (signs ignored) of the difference in the shares of each of the three sectors in product and labour force. The relative product per worker is the product per worker of a particular sector as a ratio of product per worker for the economy as a whole.

The share of manufacturing in the workforce remained constant for males in undivided India. We will later examine changes within the manufacturing sector in both male and female employment.

The share of construction rose slightly over the period, but through the period the share remained small. There was an increase in the share of trade and commerce: some part of this increase may have been at the expense of female employment. In transport, storage and communications there was a slight increase in the relative share over the period. Finally, in other services, there was a slight fall, though the numbers in public, educational, medical and legal services certainly increased.

For the manufacturing sector something more can be said about changes in its internal composition between 1901 and 1951. The figures here relate to males and females taken together. We estimate that while factory employment in manufacturing in undivided India rose from 0.6 to 2.9 million, employment in the rest of manufacturing, i.e., in small-scale enterprises, declined from 12.6 to 11.4 million. The growing weight of factory employment in the total tended to raise average output per worker in manufacturing and hence aggregate manufacturing output. Heston's national income estimates indicate that the real income produced by the manufacturing sector rose by 98 per cent between 1900–1/1904–5 and 1942–3/1946–7. The real output of small-scale industry according to these estimates rose by 14 per cent between 1900–1/1904–5 and 1942–3/1946–7. Even this could be an underestimate since the declines in employment (as we shall see) were often in very low productivity activities, and some modern small-scale units were emerging. Also within a traditional activity like handloom weaving considerable technological change occurred—as we point out later in this chapter.

A detailed breakdown of the manufacturing sector can only be provided for the Indian Union, as the required information was not available for Pakistan in 1951. Between 1911 and 1951 employment increased absolutely in beverages, tobacco, jute textiles, miscellaneous textiles, wood and wooden products, paper and paper products, printing and publishing, rubber, petroleum and coal, chemicals and chemical products, metals and metal products, machinery, electrical equipment and transport equipment (see table 5). The increases in beverages and tobacco reflected changed tastes, possibly partially associated with the trend towards urbanization after 1931, while the increase in miscellaneous textiles (mainly made-up textiles) reflected the trend towards stitched clothing. The increase in wood and wooden products was absolute, but not relative: while the demand for wood in construction may have grown—though losing some ground to cement, bricks and steel—the advent of sawmills may have acted as a damper upon employment growth. In the other expanding activities their relative novelty and the absence of a pre-existing traditional sector made for expansion. Yet we find that though their share in manufacturing employment doubled from 8.4 to 15.8 per cent, even in 1951 they formed a relatively small part of the manufacturing sector.

Foodstuffs, cotton textiles, silk textiles, wool textiles, leather and leather products, and non-metallic mineral products (other than petroleum and coal products) declined absolutely. In foodstuffs the decline was mainly in the milling, dehusking and processing of grain, and in the production of edible oils. In both, mechanical processing resulted in labour displacement; additionally, kerosene was progressively replacing vegetable oil as an illuminant. At the same time employment in sugar, in milk, butter and cheese production and in other food products (including cashew-nut processing) expanded. In cotton, silk, and wool textiles small-scale producers declined in number. In cotton spinning and weaving the number of small-scale producers declined from 2.4 million in 1911 to 2.2 million in 1951. However, as machine-spun yarn supplanted hand-spun yarn and as the fly-shuttle replaced the throw-shuttle in many parts of India, output per worker in handloom-worked production must have risen. While the available estimates for handloom-worked production depend greatly on a number of unverifiable assumptions, we estimate that handloom-worked output in undivided India rose from 965 million yards in 1902–3/1912–13 to not less than 1,068 million yards in 1930–1/1937–8. So this major traditional activity was marked by rising output per worker. The substantial decline in employment in leather and leather products resulted from tanneries and boot and shoe factories replacing the labour-intensive village worker. Also, the demand for more traditional leather articles like saddles, bags and jars tended to disappear. There was a decline in non-metallic mineral products due

Table 5
Changes in Manufacturing Employment at the Major Group Level: Indian Union, 1911–51 (in thousands)

	Persons				Males				Females			
	1911 (1)	1921 (2)	1931 (3)	1951 (4)	1911 (5)	1921 (6)	1931 (7)	1951 (8)	1911 (9)	1921 (10)	1931 (11)	1951 (12)
20 Foodstuffs	2,077	1,617	1,445	1,431	749	671	674	862	1,328	946	771	569
21 Beverages	20	10	35	182	12	7	29	171	8	3	6	10
22 Tobacco and tobacco products	30	35	114	572	21	23	78	361	9	12	36	212
23 Cotton textiles	3,340	2,930	3,169	3,161	2,015	1,767	2,003	2,149	1,326	1,163	1,166	1,012
24 Jute textiles	229	316	277	397	187	265	242	316	41	52	36	81
25 Wool textiles	170	115	111	108	104	72	67	60	66	43	44	48
26 Silk textiles	123	60	75	80	72	32	55	54	51	28	21	27
27 Miscellaneous textiles	713	765	717	1,174	451	499	474	908	262	267	142	26
28 Wood and wooden products	1,448	1,316	1,384	1,608	1,075	1,017	1,088	1,235	374	299	296	373
29 Paper and paper products	4	4	—	35	3	3	—	31	1	—	—	4
30 Printing and publishing	43	41	44	130	42	40	43	126	1	1	1	3
31 Leather and leather products	1,102	1,054	929	750	907	870	808	677	194	184	121	73
32 Rubber, petroleum and coal products	—	2	3	32	—	1	2	31	—	1	1	2
33 Chemical and chemical products	37	25	55	128	25	19	39	106	12	6	16	22
34/35 Non-metallic mineral products	1,128	1,124	917	960	726	761	636	709	403	363	280	251
36 Metals and metal products	660	655	639	879	584	569	588	818	76	86	51	61
37 Machinery and electrical equipment	—	—	—	170	—	—	—	167	—	—	—	3
38 Transport equipment	10	15	20	233	10	14	19	228	—	1	1	5
39 Miscellaneous manufacturing	573	568	599	644	536	529	558	592	37	39	41	52
	11,707	10,652	10,433	12,674	7,518	7,158	7,403	9,601	4,189	3,494	3,030	3,074

largely to massive declines in earthenware and earthen pottery displaced by metal, china and glassware which flooded the countryside.

While employment in the small-scale manufacturing sector declined from 12.6 million in 1901 to 11.4 million in 1951, the decline in employment in handicrafts may have been somewhat sharper, for additional employment in some of the new small-scale activities which emerged, like powerlooms, may have offset part of the decline in handicrafts employment. Also, the Indian economy in 1950 was much more urbanized and employment in modern industry (large and small) in the towns and cities is certain to have grown. In rural areas, therefore, the decline in employment in handicrafts may have been larger. However, a declining output of handicrafts cannot be assumed. Certainly this was not the case for handloom textiles, bidi, and gur production, though declines in leather and leatherware, earthenware and earthen pottery, oil-pressing, foodgrain processing and traditional food products might have occurred. But taking the entire manufacturing sector into account, its share in total employment did not decline if we rely on the figures for males. Even if we accept the estimates for males and females together, while we do get a decline from 9.6 per cent in 1911 to 8.7 per cent in 1951, this cannot be described as deindustrialization, for there was a significant relative and absolute increase in the output of the manufacturing sector. Similarly even for the period 1881–1911, the term de-industrialization cannot be applied, for the decline in manufacturing's share in employment occurred alongside a rise, relative and absolute, in manufacturing output.

For India as a whole the structure of the workforce changed little during the period. In many states of India, too, there was virtually no change in the structure of the workforce. These were mainly concentrated in central India: Uttar Pradesh, Bihar, Madhya Pradesh, Andhra Pradesh and Gujarat. In four states, there was a marked shift away from agriculture. These were Kerala, Maharashtra, Madras and West Bengal. In four other states, Rajasthan, Orissa, east and west Punjab and East Bengal, there was a marked rise in agriculture's share over the period (see tables 6 and 7).

It would be futile to attempt to explain these changes or draw conclusions about them without examining closely the economy of each of these states. Such a task cannot be attempted here. We instead pick four areas for closer scrutiny: Kerala, West Bengal, Rajasthan and east Punjab.

Kerala

Throughout the twentieth century, less than 60 per cent of Kerala's working population was engaged in agriculture, while more than 6 per cent worked in livestock, forestry, fishing, and plantations, and 12–15 per cent were in manufacturing. These features differentiate Kerala from the rest of India, and as the 1911 census superintendent for Cochin noted:

> This comparative preponderance of industrial population in these two States [Travancore and Cochin] is due not to the infertility of the soil or its unsuitability to agriculture but to certain natural advantages possessed by them which have directed a larger proportion of people than in most other parts of India from agriculture to industrial occupations. Among these may be mentioned the existence of a large extent of back waters and canals teeming with fish life and providing occupation to a large number of fishermen, fish-curers and dealers, and boat and bargemen; of valuable forests covering nearly one-half of the States and providing employment to numbers of wood-cutters, sawyers, carpenters and collectors of forest produce; and of the facilities for the cultivation of coconut palm, the raw produce of which affords scope for important and extensive industries, such as toddy drawing, jaggery making, arrack distilling, oil pressing, coir making etc. . . .

During the half-century that followed, there was a further shift away from agriculture and towards manufacturing and services. Employment of persons in plantations increased from about 22,000 to about 170,000 between 1911 and 1951, while employment of persons in manufacturing more than doubled, from 466,000 to 877,000. Within services the bulk of the increase was in transport, storage and communications and other services. Curiously enough, employment in trade and commerce declined relatively.

Within manufacturing the bulk of the increase came in sugar production, cashew-nut processing, cotton textiles (mill and non-mill), coir and coir products, made-up textiles, wood and wooden products and non-metallic mineral products. These were generally traditional items made with locally available and abundant raw materials. Coir and cashew were, in addition, major exportables. Over half the persons engaged in manufacturing were women and they were mainly working in the coir and cashew industries.

The development of tea, rubber, coffee, spices, coir and cashew owed much to the existing system of inland waterways and sea-ports which even in the past had fostered trade and commerce. The trade figures for Cochin and British Malabar indicate that the aggregate value of trade rose from Rs. 150,000 in 1870 to about Rs. 1,050,000 in the 1920s. The major exports of Kerala in the 1920s included coffee, coir, lemongrass oil, coconut oil, rubber, spices, tea and rope. In the 1930s there was a substantial expansion of exports, as a new commodity, cashew kernels from Travancore, entered the world market. Cashew exports alone were worth in 1930 about Rs. 75,000. The effects of expanding foreign trade on the domestic economy of Kerala must have been considerable.

Expanding exports must have meant an increased demand for services. This is reflected in the expansion of employment in transport, storage and communication and other services. In trade and commerce, the enormous growth in turnover did not result in a rapid growth in the numbers em-

Table 6
Comparative Employment Distribution

	Andhra Pradesh		Assam		Bihar		Gujarat		Kerala		Madhya Pradesh		Madras	
	1911	1951	1911	1951	1911	1951	1911	1951	1911	1951	1911	1951	1911	1951
Cultivators	41.3	41.3	67.6	61.6	58.5	60.3	39.8	45.3	31.5	25.6	48.9	54.0	51.0	40.9
Agricultural labourers	22.1	26.5	1.9	4.5	21.6	24.6	20.9	14.2	24.8	25.3	21.0	23.2	18.1	19.4
Plantations, forestry, fishing, livestock, hunting	6.3	2.9	18.1	12.7	3.9	0.7	4.7	3.3	12.0	6.5	5.3	2.0	4.2	2.6
Mining and quarrying	0.2	0.5	0.2	0.2	0.2	1.4	0.2	0.3	0.1	0.5	0.1	0.6	—	0.3
Manufacturing	9.2	10.0	1.6	4.0	4.8	3.5	12.3	13.5	11.7	15.8	9.3	8.0	9.4	13.6
Construction	1.2	1.2	1.1	0.6	0.7	0.5	1.2	1.2	2.2	1.9	0.6	0.7	2.6	1.8
Trade and commerce	5.3	6.2	3.2	4.9	4.1	3.9	7.3	8.2	8.7	8.4	4.3	4.0	5.8	7.8
Transport, storage and communications	1.5	2.0	1.6	1.7	1.3	1.2	1.8	2.2	2.8	4.2	0.8	1.3	1.6	2.4
Other services	8.7	8.6	4.4	6.8	5.0	3.8	11.7	11.8	11.2	12.0	9.4	6.2	7.3	11.2
Total	100	100	100	100	100	100	100	100	100	100	100	100	100	100

	Maharashtra		Mysore		Orissa		Punjab		Rajasthan		Uttar Pradesh		West Bengal	
	1911	1951	1911	1951	1911	1951	1911	1951	1911	1951	1911	1951	1911	1951
Cultivators	39.5	39.4	50.8	52.4	55.5	58.0	52.5	54.2	53.8	64.6	61.3	65.3	45.8	36.3
Agricultural labourers	25.2	22.1	16.1	15.9	20.1	19.9	6.9	12.1	6.2	6.5	12.0	8.1	17.2	17.1
Plantations, forestry, fishing, livestock, hunting	5.8	1.9	5.4	2.4	5.3	2.0	3.7	1.3	3.3	3.2	1.9	1.1	5.5	3.2
Mining and quarrying	0.2	0.3	0.7	0.7	0.1	0.2	0.1	—	0.2	0.3	—	—	1.5	1.2
Manufacturing	10.4	13.8	8.9	9.7	6.8	6.2	14.4	7.9	12.3	8.1	8.6	9.9	10.9	14.7
Construction	1.3	1.5	1.4	2.2	0.4	0.7	1.5	1.0	1.2	1.3	0.6	0.9	2.2	1.6
Trade and commerce	5.4	7.4	5.4	5.8	4.2	3.6	6.2	8.6	7.9	6.2	4.2	4.7	5.5	10.3
Transport, storage and communications	2.3	2.8	0.9	1.5	0.7	0.7	2.2	2.2	1.5	1.1	1.0	1.5	3.5	4.7
Other services	9.9	10.7	0.2	9.4	6.8	8.6	12.5	12.7	14.2	8.7	8.6	8.2	9.5	10.9
Total	100	100	100	100	100	100	100	100	100	100	100	100	100	100

Source: *Census of India*, 1961, Paper 1, 1967 (Delhi, 1967).

Table 7
The Distribution of the Male Working Force in the States of Undivided India, 1911 and 1951

	East Bengal including Sylhet		West Punjab		Undivided Bengal		Undivided Punjab	
	1911	1951	1911	1951	1911	1951	1911	1951
1 Productions of field crops	76.7	82.8	49.8	63.0	70.2	70.3	54.8	63.5
2 Plantations, forestry, fishing, hunting, livestock	3.9	2.1	8.3	3.0	4.6	2.5	5.8	2.2
3 Mining and quarrying	—	—	—	0.6	0.5	0.1	—	—
4 Manufacturing	4.1	3.4	15.6	11.4	6.2	7.9	14.9	9.9
5 Construction	1.8	1.2	3.1	0.6	1.8	1.3	2.2	0.7
6 Trade and commerce	5.0	4.1	6.4	6.3	5.2	6.6	6.3	7.3
7 Transport, storage and communications	1.9	1.7	3.3	0.9	2.5	2.9	2.7	1.4
8 Other services	3.1	3.8	11.6	8.0	5.0	6.0	11.8	9.0
9 Unspecified	3.5	0.9	1.8	6.7	3.9	5.9		
Total	100	100	100	100	100	100	100	100

Source: Retabulated Census data for India and Pakistan.

ployed. It is worth noticing that in Kerala in 1911 for every person working in educational, scientific, medical, and health services there was about half a person in public services (i.e., general administration); for the Indian Union in the same year, the corresponding proportion was 1:4. As a result there was a more rapid increase in literacy and a sharper decline in mortality in Kerala than in India as a whole. The proportion of literates to the population aged over 5 rose in Kerala from 26.7 per cent in 1911 to 53.8 per cent in 1951; in the Indian Union the corresponding percentage even in 1951 was less than 20 per cent. The death rate in Kerala in 1951 had come down to 16.1 per 1,000, while the Indian rate was 27.4 per 1,000.

In 1951 Kerala had a remarkably low degree of dependence on agriculture for a state with a still rather low per-head income. The growth of labour-intensive processing industries, partly in response to foreign demand, probably had more dramatic effects on the employment pattern and other correlates of welfare like literacy and mortality than on per capita income.

West Bengal

During the latter half of the nineteenth century the growth of tea and jute had been important factors in the growth of Calcutta. Employment in tea plantations was concentrated in Assam, but jute was grown largely in East Bengal. In West Bengal in 1911, the employment in jute processing was large, accounting for about 40 per cent of total (male and female) manufacturing employment: by 1951, its share had declined to 24 per cent. In the meanwhile, the lead within manufacturing had shifted to the iron and steel engineering industries.

Between 1920 and 1937 pig-iron production in India grew by about 400 per cent and steel by even more. Almost the entire output came from Tata's Jamshedpur plant in Bihar, and the Indian Iron and Steel Company's plant at Asansol in West Bengal. One more plant was set up in Bengal by the Steel Corporation of Bengal to meet the enormous increase in the demand for iron and steel during the Second World War.

The availability of coal and iron along with an extensive rail network favoured the development of a substantial engineering complex in the Asansol-Calcutta belt. Employment of persons in West Bengal in the production of basic metals and metal products (including machinery) quadrupled between 1911 and 1951. The expansion of iron and steel and engineering did not have significant linkages with the Bengal countryside. The effects were probably more strongly felt in neighbouring Bihar which provided the iron ore and part of the coal and steel. The percentage of persons employed in mining and quarrying rose from 0.2 to 1.4 per cent of the workforce in Bihar.

In West Bengal the expansion in manufacturing employment was due to the factory sector: employment of persons in factories grew from 300,000 to about 700,000 between 1911 and 1951; employment in small-scale production remained at about 600,000. Factory employment was highly concentrated in location: three districts, Howrah, Twenty-Four-Parganas and Hooghly accounted for 83 per cent of factory workers. Also, unlike other states, in West Bengal over half the manufacturing workforce worked in factories.

Rural Bengal did not progress during the period. Agricultural growth may have been positive, but it certainly was slow. In East Bengal, there was a marked shift in the workforce towards agriculture and away from manufacturing. The growth of the jute industry did not significantly alter living standards in rural Bengal, partly due to the monopsonistic structure of the jute market.

Bengal underwent dualistic development, not pervasive transformation. The backward linkage effects of jute processing were probably unfavourable; the backward linkages of steel and steel-using industries were spread mainly in Bihar; the new industries of the Calcutta-Asansol belt largely depended on Marwari capital; and the employment created, though it benefited local labour, was relatively small. While the indirect effects of this expansion in urban Bengal may have been significant, they generally bypassed the countryside. Positive change was concentrated in urban areas where the railways were born in the nineteenth century, the commercial and financial institutions created by the jute and tea trade, combined with cheap coal and steel to create a modern industrial structure. But growth remained narrow: it reinforced, rather than overcame, the rural-urban dichotomy.

Rajasthan

In Rajasthan there was a massive shift away from manufacturing and services towards agriculture. This resulted from the collapse of premodern manufacturing activity within unintegrated sub-economies following the economic integration brought about by the railway; at the same time, employment opportunities in agriculture expanded.

Before the railway, camels provided the main form of transport. The transport constraint dictated a pattern of local self-sufficiency and most of the items of daily use were produced locally. Each settlement had to carry out the entire spectrum of economic activity and this implied a diversified employment structure. In 1881 there were hardly 400 miles of railways in the area, but by 1931 the system had been greatly extended: there were over 2,900 miles of track and most places were within 50 miles of a railway station. This revolution in transport had far-reaching consequences, as a wide range of local manufactures faced national and international competition for the first time. Employment declined substantially in cotton textiles, leather and leather products and in the manufacture of earthenware and earthen pottery.

Information collected from the Rajputana states in 1931 on the "disappearing industries" is revealing. According to the summary of the replies received to a questionnaire: "Reza or homespun cotton weaving is still carried on in most parts of the country but except in very isolated tracts such as Jaisalmer, mill made cloth had largely supplanted it as an article for wear by men." Camel and goat hair bags and sacks" . . . were at one time in great demand for carrying grain on camels but the machine made gunny bag now so common all over the country is responsible for a falling off in this industry." On the leather industry, "Vessels of camel hide, often ornamented on the outside, are a product peculiar to Bikaner State . . . the industry is a rural one subsidiary to agriculture. The demand for these vessels is now no more than local. . . ." The industries surveyed were generally urban and specialized. Rural production catering for the village is likely to have been affected even more adversely.

At the same time, important changes occurred in Rajasthan's agriculture. In the dry tracts the area under cultivation and the area irrigated were expanding rapidly, while in the already irrigated area multiple cropping was being extended. As a result, cultivated area per head of population did not decline, and irrigation made it possible for the land to support a larger population. There was a great expansion in the production of high-value crops like wheat, pulses, oilseeds and cotton which were probably on the average more labour-intensive than the dry crops they replaced.

Did agricultural expansion outweigh the decline in manufacturing? One cannot tell, as the required output figures do not exist. Even if, on balance, per head income did rise, it was not adequate to create substantial additional demand and employment in local manufacturing and services. The transport revolution rendered redundant a whole range of local industries and services catering to the needs of isolated self-sufficient villages. In this special context economic growth, at least up to a point, implied a shift of the workforce *towards* agriculture.

East Punjab

There was marked shift away from manufacturing and services towards agriculture in east Punjab. While output data for agriculture must be treated with scepticism, it is nevertheless clear that rapid agriculture growth did not take place in east Punjab, the growth rate being less than 1 per cent per annum between 1900 and 1947. This was because the effects of irrigation developments in the late nineteenth century had tapered off in this region.

As one would expect, male employment in manufacturing in east Punjab declined from about 572,000 to about 378,000 between 1911 and 1951. This decline was mainly in foodstuffs, cotton textiles, wood and wooden products, leather and leather products and non-metallic mineral products: largely traditional activities unable to withstand competition from the modern sector. Of course the 1951 figures could reflect the effects of the Partition which considerably dislocated normal patterns of activity after 1947. But, the 1941 results, for what they are worth, do not suggest very different trends.

It may be argued that the focus of growth in the Punjab during this period was in the west and that perhaps men, materials and enterprise gravitated westward leaving behind a stagnant east. The development of the new canal colonies meant that many new population settlements arose, and crop production, especially for the market, expanded rapidly. Available estimates suggest that the annual compound rates of growth of agricultural output in the western districts could have been as high as 4 per cent per annum between 1900 and 1947.

One would therefore expect that employment in manufacturing expanded greatly in West Punjab, but this does not appear to be the case: male employment rose from 519,000 to 685,000, but its share in the male workforce declined from 15.6 to 11.4 per cent between 1911 and 1951. The undoubted increase in agricultural incomes must have increased the demand for non-agricultural goods; but, to a large extent, this must have been met by "imports" from other parts of India. Also, since the increase in agricultural output in west Punjab came from the enormous increases in the area under cultivation, employment in agriculture grew rapidly, perhaps providing superior alternatives to pastoral activities and pre-modern manufacture.

Conclusion

The Indian occupational structure showed little sign of change over the whole period 1881–1951. Agriculture's share remained at about 70 per cent, manufacturing at about 10 per cent and services at about 15–20 per cent over the period. But there were changes which lie hidden behind these aggregates. Between 1901 and 1951 factory employment expanded partly at the expense of the non-factory sector; the more modern branches grew at the cost of a number of traditional ones, and manufacturing output per head increased. While the share of transport, storage and communications rose, for the other branches of services trends are unclear. Many services associated with modernization under colonial rule expanded, in particular public, educational, medical and legal services.

Patterns of change varied widely over the sub-continent; in Kerala, Madras, Maharashtra, and West Bengal the share of agriculture declined and the shares of manufacturing and services increased; in Orissa, Rajasthan, East Bengal and the Punjab the share of agriculture rose and the shares of manufacturing and services fell; and in the other states trends are unclear.

In Kerala, economic progress centred around the development of labour-intensive processing industries and plantations often catering to foreign markets. In West Bengal, foreign demand stimulated the jute industry until about 1920. Subsequently, a conjunction of favourable circumstances led to an import-substituting group of iron and steel and engineering industries especially in the Asansol-Calcutta region. In contrast to Kerala, the developments in West Bengal largely bypassed the countryside. In Rajasthan, railway development destroyed premodern manufacturing, but this was at least partially compensated by the expansion of agriculture. In Punjab, the east declined as agriculture stagnated; in the west, agriculture boomed, but manufacturing lagged behind as 'imports' partially supplanted domestic production. These examples indicate the importance of foreign and inter-regional trade, resource availabilities, transport facilities, and expansion in cultivable areas in explaining the changes that occurred.

It is significant that the states in which there was a marked shift away from agriculture—Kerala, Madras, Maharashtra and West Bengal—had extensive coastal tracts with sea-ports and in each the ports were well connected with the hinterland. These regions had long traditions of foreign contact and commercial development, and the expansion of the railway system in the latter half of the nineteenth century widened the possibilities of development for some coastal areas.

Irrigation and railway development varied greatly in terms of extent and impact in different parts of India. In Kerala assured rainfall and an adequate water transport system facilitated the process of export expansion without substantial investments in overheads. On the other hand, in Maharashtra and West Bengal, and to a lesser extent in Madras, trade expansion was contingent upon railway development. It is well known that the structure of railway rates favoured the movement of finished goods from the ports into the hinterland and of raw materials from the hinterland to the ports. What is often forgotten is that it also favoured industries set up in the ports which could get raw materials from the hinterland and 'export' finished products to the hinterland. The cotton textile industry illustrates the operation of this process over time. In 1905–6, imports provided 59 per cent of domestic cloth supply, by 1936–7, the figure had dropped to 7 per cent. The huge decline in employment in cotton textiles in Rajasthan must therefore be attributed to the flood of Indian, not imported, textiles from Western India taking advantage of the system of railway rates. The same system made it worth Rajasthan's while to produce raw cotton for 'export' to the rest of India. The eastern Indian rail system in conjunction with the other facilities already present in Calcutta for the jute and tea trades, helped develop steel and steel-using industries in the Asansol-Calcutta region.

In Rajasthan and Punjab irrigation had important effects. In Rajasthan it moderated the effects of the breakdown of pre-modern manufacturing by making agriculture a viable proposition. In the Punjab, it transformed the west, but probably drained away resources from the east. At the same time, it made west Punjab specialize in agriculture, and depend more on 'imports' of manufactures at some cost to her pre-modern manufacturing.

Many questions lie unanswered, hidden behind the aggregates we have examined. Why did growth not spread rapidly from the coast to the hinterland?; did government policy help or hinder the process?; who were the gainers and who were the losers from the process of lopsided development?; why did agriculture grow so slowly?; why did handlooms decline in Rajasthan and expand in Madras? These are questions which will occupy Indian economic historians for a long time to come.

Ideas for Discussion

1. What functions do the several tables serve in this passage? What do they reflect about the intended audience for the piece?
2. How do the four regions of India compare to the overall profile of agricultural and manufacturing interests?
3. Why is the Indian experience important to economists?

Topics for Writing

1. Prepare a summary of "Occupational Structures of India."
2. Prepare a report on the current economic condition in one of the four regions the authors discuss.
3. Research the role of technology in modern India.
4. Write a letter soliciting a report on construction industry prospects in India.
5. Write a proposal for doing business in one of the regions of India mentioned in this essay.

Joseph Grunwald, Leslie Delatour, and Karl Voltaire
Foreign Assembly in Haiti

The authors are part of the Washington, D.C.–based Brookings Institution created in 1927 from three separate institutions: The Institute for Government Research (founded in 1916); The Institute of Economics (founded in 1922); and the Robert S. Brookings Graduate School of Economics and Government (founded in 1924). Publications such as The Global Factory, *edited by Joseph Grunwald and Kenneth Flamm, reflect the Brookings Institution's ongoing mission of making economic information available to policymakers, businesses, and the government. "Foreign Assembly in Haiti" is reprinted from* The Global Factory.

HAITI is among the poorest countries of the world. The country with the next lowest income level in the Americas, Honduras, has more than twice the GNP per capita of Haiti, which was estimated by the Inter-American Development Bank in 1982 at $311. The Dominican Republic, with which Haiti shares the island of Hispaniola and which had a population just a few percent higher than Haiti's 5.3 million in 1983, has earnings per person nearly four times as great.

Manufacturing in Haiti is of recent origin. It is confined primarily to the processing of food, beverages, agricultural materials, and basic items such as shoes, clothing, soap, and cement. National income figures—as well as most other statistics—are still rudimentary, but the World Bank estimates the contribution of manufacturing to gross domestic product (GDP) to have reached about 18 percent by 1981. Most of the increase took place during the 1970s, after a decade of near stagnation.

The Importance of Assembly Production in Haiti

The growth of assembly industries has been an important element in the recent dynamism of the Haitian economy. Between 1970 and 1980 the value added in Haitian assembly plants increased more then twenty-three times, or at an average annual rate of 37 percent (see table 1). Even in real terms, the growth was 22 percent a year, deflated by the implicit import price deflator of the United States. In part the dramatic upsurge was aided by government incentives and by changes in the political environment.

Assembly activities have benefited from two types of incentives: new production firms are exempt from income taxes for five years, following which they are subject to partial payments in increasing yearly proportions, so that after ten years they must pay the full tax; and a franchise is granted to assembly firms that exempts them from any tariff duties on imports. The franchise is given indefinitely for production all of which is exported. This covers assembly production. For production that is sold on the domestic market, the franchise is given on a temporary basis, because the firm is expected to use local materials as inputs after the first few years.

There are no free-trade zones in Haiti that cater to export industries. While in Port-au-Prince there is a successful industrial park which is run by a government agency, it emerged as a result of the growth of assembly plants rather than contributing to their emergence.

Thus, unlike that in other countries, assembly production in Haiti is not an in-bond activity in the strict sense. Although assembly production cannot be sold domestically, any firm can obtain a duty-free import franchise whether it exports or not. More than half of all manufacturing output, including assembly, is exported, and there are many firms that produce for both the home market and the export market.

With the advent of a new government in the early 1970s, investors' perceptions of political stability improved. The new environment permitted the huge wage gap between Haiti and the United States to become an enormous attraction for assembly production. Many Haitian professionals and entrepreneurs who were abroad during the preceding regime returned and established businesses for the purpose of assembling U.S. components.

While there are no exact figures, estimates by the U.S. embassy in Haiti and by the World Bank put the number of assembly plants in 1980 at about 200, employing approximately 60,000 persons. All the assembly plants are in Port-au-Prince, the capital and by far the largest city. Assuming a dependency ratio of 4 to 1, this means that assembly operations supported about a quarter of the population of Port-au-Prince in 1980.

Assembly activities in Haiti have been concentrated in sporting goods, particularly baseballs, and textiles, including apparel (see table 2). Haiti is the world's principal exporter of baseballs and ranks among the top three in overseas assembly of textile products and stuffed toys. During the period 1969–72, more than half of Haiti's assembly exports that entered the United States under U.S. tariff item 807.00 consisted of baseballs. While still growing in absolute amounts, their importance in the total declined to less than a quarter during the mid 1970s and less than a sixth in 1983. Clothing and textile products have grown much faster, from less than a quarter during 1969–72 to well over a third of exports to the United States

Table 1
Total Value, Dutiable Value, and Duty-free Value of U.S. Imports from Haiti under Tariff Items 806.30 and 807.00, 1969-83[a]

Millions of U.S. dollars unless otherwise specified

Year	Total value of 806/807 imports (1)	Duty-free value of U.S. components (2)	Dutiable value added in Haiti (3)	Dutiable value as percent of total value (3 ÷ 1) (4)
1969	4.0	2.4	1.6	39.8
1970	6.1	4.0	2.1	34.4
1971	9.1	5.9	3.2	35.2
1972	16.1	11.0	5.1	31.7
1973	28.5	20.3	8.2	28.8
1974	56.6	43.0	13.5	23.9
1975	54.7	40.2	14.5	26.4
1976	78.1	56.6	21.5	27.5
1977	84.3	61.3	23.0	27.3
1978	104.9	76.1	28.7	27.4
1979	133.7	94.5	39.2	29.3
1980	153.8	105.3	48.5	31.5
1981	171.3	117.1	54.2	31.6
1982	181.0	126.2	54.8	30.3
1983	197.4	139.4	58.0	29.4

Sources: Special magnetic tapes from the U.S. International Trade Commission (ITC); and ITC, U.S. Tariff Items 807.00 and 806.30: Imports for Consumption, various issues.

a. U.S. imports from Haiti under tariff item 806.30 amount to less than $50,000 a year.

under tariff item 807.00 since 1973. More than 90 percent of Haiti's textile exports go to the United States under tariff item 807.00; in 1983 Haiti accounted for 12 percent of such imports by the United States. It should be noted, however, that the volume of Haitian shipments of textiles has not grown significantly in recent years, so Haiti's share of the total U.S. import market declined from 1.2 percent in 1979 to 0.8 percent in 1983. There is little doubt that the textile quotas imposed on Haiti by the United States in 1973 under the International Multi-Fiber Agreements have limited the expansion of Haitian textile exports, particularly during recent years.

The facts that about 60 percent of the exports from Haiti are not subject to any restraints and that on the average less than half the aggregate quotas are used do not mean that the bilateral agreements have no negative effects on Haitian output. First, categories not specifically under restraint are still subject to large uncertainties. At any moment the U.S. government may request consultation for items that are perceived as threatening disruption of the market. Unless shipments fall off significantly, the results of these consultations are either mutually agreed-upon limits or unilaterally imposed restraints. Second, the output of potential investors may be too large to be accommodated by the unused portion of the quota, which means that the quota will remain underused. Third, the quota limitation becomes a binding constraint whenever there is a clear comparative advantage, such as the highly labor-intensive sewing of brassieres in Haiti. Finally, textile exports grew rapidly until 1973, then slowed after the signing of the first bilateral restraint during that year.

During the mid 1970s, stuffed toys and dolls became number three in Haiti's assembly production but then declined in rank, as did other "traditional" products that were significant initially. Electrical (and electronic) machinery, equipment, and parts have rapidly grown in importance and now constitute more than 30 percent of Haitian assembly exports to the United States; its share was a little more than 10 percent in 1970. Equipment for electric circuits alone accounted for more than 10 percent of Haitian assembly exports, making it the third most important assembly product group in 1981 (see table 2). The establishment of luggage assembly at the end of the 1970s has put this product group in fourth place among Haitian assembly industries.

The available data indicate that value added in assembly plants now constitutes about a quarter of all industrial value added in Haiti, whereas it constituted less than 6 percent in 1970. Contributing about two-thirds of the country's industrial exports, assembly production now earns almost a third of Haiti's yearly foreign exchange receipts from exports.

It is clear that in the case of Haiti, the effect of assembly operations on employment, income, and the balance of payments is of considerable importance. These effects and some more subtle ones will be discussed later in the chapter. At this point, let it be noted that Haiti's weight in the world coproduction system is far greater than its size—in both population and income—in the world community. Of about fifty countries that imported substantial quantities of U.S. components for assembly in 1980, Haiti ranked ninth, with imports of more than $105 million. It held a similar position among developing countries in respect to value added in the assembled articles that were exported to the United States. While its share in third world assembly operations is still small—about one twenty-fifth of Mexico's value added and one-eighth of that of Singapore—after Mexico it is the most important assembly country in Latin America.

Characteristics of Assembly Operations in Haiti

The nature of the assembly industry is not the same in all countries. Haitian assembly operations differ from the Mexican maquiladoras' in most respects. The differences are explained not only by the small size or the level of development of the country but also by political and social factors. Data in this section were obtained from two sample surveys undertaken as part of the research for this chapter—one a survey of enterprises, the other, of workers.

Table 2

The Ten Most Important U.S. Imports from Haiti under Tariff Item 807.00, by ITC Product Group, as Percent of Total U.S. 807.00 Imports from Haiti, Ranked among U.S. 807 Imports from Haiti and among U.S. 807 Imports from all Countries, 1970, 1978, and 1981[a]

Product group	1970			1978			1981		
	Percent of total value of U.S. 807 imports from Haiti	Rank among U.S. 807 imports from Haiti	Rank among U.S. 807 imports from all countries	Percent of total value of U.S. 807 imports from Haiti	Rank among U.S. 807 imports from Haiti	Rank among U.S. imports from all countries	Percent of total value of U.S. 807 imports from Haiti	Rank among U.S. 807 imports from Haiti
Baseballs and softballs	52.2	1	1	23.3	2	1	18.2	2
Textile products	24.6	2	6	37.8	1	3	35.6	1
Equipment for electric circuits	1.9	6	9	2.9	7	4	10.3	3
Television receivers and parts	2.9	4	8	—	—	—	—	—
Fur and leather products	2.4	5	—	—	—	—	—	—
Office machines	—	—	—	5.3	4	11	—	—
Toys, dolls, and models	3.6	3	5	7.2	3	2	4.3	5
Resistors and parts	1.3	10	—	—	—	—	—	—
Electric motors, generators, and so on	1.6	8	—	1.9	9	3	3.1	8
Tape recorders and players	1.6	9	—	—	—	—	—	—
Valves and parts	1.8	7	—	—	—	—	—	—
Miscellaneous other machinery	—	—	—	—	—	—	1.8	9
Capacitors	—	—	—	3.4	6	3	3.9	6
Miscellaneous electrical products and parts	—	—	—	4.1	5	—	3.8	7

Luggage, handbags, and the like	—	—	—	—	—	4
Gloves	—	—	1.7	10	4.8	10
Miscellaneous other articles	—	—	2.4	8	1.7	—
Addendum						
Total 807.00 value (millions of dollars)	6.1	—	104.9	—	171.1	—

Source: Special magnetic tapes from the ITC.

a. U.S. imports from Haiti under tariff item 806.30 were negligible.

Ownership and Structure

Unlike Mexican assembly production, which is done largely through subsidiaries of U.S. companies, Haitian assembly activities rest to a great extent on Haitian entrepreneurship. Almost 40 percent of the firms in the sample are Haitian-owned, a third are foreign-owned, and 30 percent are joint ventures. Most of the Haitian-owned firms are concentrated in textile assembly, where almost two-thirds of the firms are locally owned. In electronics and the miscellaneous category, joint ventures predominate; in baseballs, foreign companies. Even in foreign firms, most of the managers are Haitian. Of forty-nine firms that responded to the question, about three quarters had local managers.

It should be noted that several firms, especially in the baseball category, that had been wholly owned by Haitians were sold to foreign interests before the sample was taken. The oldest assembly enterprises are Haitian, and their average age is ten years. The youngest are foreign-owned, with an average age of seven years, and the ages of the joint ventures are in between. No change in the pattern of ownership can be discerned during recent years. Less than a third of the firms were established during the five years that immediately preceded the survey, and more of these were wholly Haitian-owned than were wholly foreign-owned; almost a third of the total were owned jointly.

Most of the companies interviewed do assembly work for U.S. principals. Those that are not wholly owned by U.S. manufacturers usually work with machinery and equipment supplied by U.S. principals. About three quarters of the forty-seven respondents to the survey said that the principals usually furnish the machinery and two-thirds said that the principals always do so. This arrangement, which is a special form of an arm's-length relationship, reduces the risks faced by both the subcontractor and the principal. The former does not have to venture his own capital, and having his customer's machinery is some assurance that the principal will not run out on the contract. The principal, on the other hand, can have his production abroad without having to establish a subsidiary company and without having to make a large capital investment. The machinery that he lends to his subcontractor is usually secondhand, often fully depreciated, and in any event he can take it back when a contract has been completed. That there is Haitian entrepreneurship is demonstrated by the fact that about three quarters of the local firms reported that they initiated the relationship with the principal. In many instances the Haitian enterprise has a relationship with more than one principal. The overall average is three U.S. principals per Haitian firm, and there may be as many as an average of nine for local firms in the electronics sector.

It should be noted that, aside from Haitian entrepreneurship, a strong reason for local ownership of assembly plants may be that government bureaucracy and red tape keep foreign firms from establishing their own subsidiaries in the country. This means that, as foreign enterprises become used to doing business with Haiti and learn the ropes—not to speak of any improvement coming about in the bureaucracy—more of them will invest directly in Haitian assembly production rather than subcontract with independent Haitian firms.

Size of Firms

As might be expected, foreign firms are larger than locally owned assembly companies. While the labor force is about equally distributed among Haitian, U.S., and jointly owned enterprises, the average number of workers per foreign enterprise is about 480, the average per Haitian firm about a third less. Since some firms have more than one plant, employment per plant is significantly lower than the average of about 400 workers per firm in the sample. The best available estimate is about 300 workers per plant, which is many more than in other manufacturing enterprises in Haiti. About two-thirds of the employment in assembly plants is concentrated in the one-third of the enterprises that employ more than 400 workers each. Assembly firms that produce for export are therefore the largest employers in the country.

By far the largest plants, both in employment and in square feet, can be found in baseball assembly, followed by electronics, in which they are only slightly larger than the average, and by textiles, in which they are somewhat smaller than the average. The smallest plants are to be found in the miscellaneous category.

Capital investment in assembly is extremely low. An average figure is difficult to arrive at because, as indicated earlier, many of the local firms receive from their principals machinery and equipment, the value of which could not be included in the survey. Taking only those firms which responded that the principals never provided the machinery, the average capital invested was about $740,000 in 1979. The average investment was nearly three times as large in electronics, but only about a third as large in textiles. The capital requirements of most local assembly companies are, of course, much less because the principals furnish most of the machinery and equipment. Capital is therefore not a great restraint upon a Haitian who wants to enter the assembly business. The difficulty of finding a principal in the United States is probably a greater barrier.

Similarly the ratio of capital to labor is very low. The approximate average of $1,500 per worker for assembly operations is far less than the $25,000 found in Haitian import-substituting enterprises. On the other hand, average utilization of capacity of about 75 percent in assembly plants is twice that in industries that produce for the domestic market. Part of the reason is that assembly plants—whether Haitian-owned or subsidiaries of foreign companies—have a regular relationship with their principals abroad

Wages

Haiti, like most countries, has minimum-wage laws. No matter how low the minimum wage may be in a developing country, the temptation is for economists to claim that it is too high, even in very poor countries, as long as there is a vast pool of underemployed persons, as there is in Haiti. That the minimum wage is higher than the opportunity cost of labor in Haiti can be demonstrated by the fact that the daily earnings of persons who work at home under contract to more formal enterprises—the "put out" system—are usually between half and a tenth of the official minimum wage.

It might therefore be claimed that more capital-intensive methods are employed in Haiti than would be warranted by the true opportunity cost of labor—the shadow price—particularly in industries that work for the domestic market, where, as indicated earlier, the average capital per worker is more than $25,000. On the other side of the coin, however, capital is not heavily subsidized in Haiti, as it is in many developing countries. Long-term financing is generally not available, and there are neither special interest rates nor special exchange rates or licenses for the import of capital. It was indicated that in the assembly-for-export sector, moreover, the capital-labor ratio is less than a tenth of the average in the industries that produce for the home market.

Average wages reported by the enterprises in the survey were more than a quarter higher than the official minimum wage of $51 a month, if a five-day week, the usual work week in assembly operations, is taken as the norm (see table 3). Although the implications of the law are not clear, strict interpretation of it might mean a legal monthly minimum of $69 a month, including nonworking Sunday pay, rather than the average wage in the enterprise sample of $65 a month. According to the World Bank, "the law stipulates a normal working week of 48 hours, 8 hours per day." Foreign firms pay a little more, local firms somewhat less than the average of $65 a month found in the survey, and, as expected, wages are higher in electronics than in textiles (for details and explanations, see table 3).[1]

In the survey of workers the respondents reported receiving about $10 a month less in the various categories than the figures given by the managers in the enterprise survey. It should be noted that the wages reported by the worker are the take-home pay and are therefore less than the wages reported by the firms. Nevertheless, the average monthly wage of $55 reported by the workers was almost 8 percent higher than the official minimum, assuming a five-day week. The average monthly wage paid in Port-au-Prince in 1975–76, on the other hand, was $45 in clothing manufacture and $49 in industries similar to other assembly operations. The workers in the sample who had had previous experience reported an average monthly

Table 3
Comparison between Wages Reported in Surveys of Firms and Workers in Haiti and Official Minimum Wages, December 1979[a]

U.S. dollars

Sector	Monthly wages reported by firm	Monthly wages reported by workers	Minimum monthly wages				Official minimum daily wage
			Twenty-two days per month[b]	Twenty-four days per month[c]	Twenty-six days per month[d]	Thirty days per month[e]	
Textiles	64	54	48	53	57	66	2.20
Electronics	72	59	57	62	68	78	2.60
Baseballs	60	51	53	58	62	72	2.40
Miscellaneous	63	59	51	55	60	69	2.30
Average	65	55	51	55	60	69	2.30

Sources: Leslie Delatour and Karl Voltaire, "International Subcontracting Activities in Haiti," paper prepared especially for the present study, tables IV-11 and V-4; World Bank, Economic Memorandum on Haiti, Report 3931-HA (Washington, D.C.: IBRD, May 25, 1982), table 2.5, corrected by the authors.

a. Detailed information on the Haitian minimum-wage law could not be obtained. One provision of the law is for nonworking Sunday pay under certain circumstances, principally after a worker has worked forty-eight hours in a six-day week. Workers in most assembly plants, however, work only a five-day week. Monthly minimum wages, therefore, were calculated on the basis of four different assumptions, as noted in the four columns under that heading.

b. Five-day week.

c. Five-and-a-half-day week.

d. Six-day week.

e. Seven-day week.

wage of $38 in their previous jobs, about 70 percent of the wages they reported in their present jobs.

There is no question that their wages put assembly workers in the upper income groups in Haiti. According to a 1976 survey, 70 percent of the households living in Port-au-Prince had incomes of less than $40 a month. Even allowing for inflation this would still be less than the average take-home pay reported by the assembly workers in the survey.

Given that fact and the existence of the put-out system in the subcontracting industry in Haiti, it might be expected that a majority of the firms would consider the minimum wage too high. The survey did not show this. In reference to the productivity of the workers, only 10 percent of the managers considered the minimum wage too high, while 73 percent considered it adequate and 17 percent considered it too low. In reply to another question in the enterprise survey, more than 80 percent of the respondents said that they would not hire additional workers if the minimum wage were abolished. This included even a majority of those who considered the minimum wage too high.

This does not mean that the minimum wage is irrelevant. It only suggests that the existence of the minimum wage in 1979–80 did not noticeably affect the level of employment. It should be noted that almost 60 percent of the assembly subcontractors in the sample charge their principals piece rates that are the piece rates paid to the workers plus overhead and a profit markup. The number of jobs created, therefore, is determined to a great extent by the demand for Haitian labor by the principals in the United States. That demand, in turn, largely depends on the difference between U.S. and Haitian wages, differences in productivity, tariffs, taxes, transportation costs, and so on. According to the responses in the survey, about half the firms reported that their productivity was about equal to that of U.S. firms with comparable product lines and activities, while about a fifth said that it was superior. The available evidence seems to support these findings.

As long as the difference between labor costs in the United States and those in Haiti greatly exceeds the sum of transportation costs and the U.S. tariffs on Haitian value added, there will be an economic incentive for U.S. companies to subcontract the labor-intensive parts of their production processes in Haiti. This does not imply that minimum wages, if enforced, can be pushed higher and higher as long as the difference is large enough to cover the other costs mentioned. First of all, to push minimum wages continually upward might cause distortions in the Haitian economy, such as large-scale substitution of capital for labor, which would have serious implications for a labor-abundant, poverty-stricken economy. Second, if the difference were not large enough, it might still cover transportation, tariff, and other direct costs yet not be sufficient to offset the bureaucratic and political risks in Haiti, which are perceived as formidable. Assembly business might thus be lost if wages rose to much higher levels, even if they remained lower than those of international competitors.

The Labor Force

Haiti is not different from other countries in having a high proportion of women in assembly plants. The 75 percent found in the survey is slightly smaller than the proportion in Mexico. In Haiti, however, women seem to constitute a much larger share of the total labor force than in Mexico and in many other countries. Abject poverty forces most adult women to go to work. It is also reported that in Port-au-Prince women outnumber men by 30 percent.

Before working in the assembly plants in which they were interviewed, more than half the women held other jobs or were looking for work. Well over 90 percent said that they would remain in the labor force if they were to lose their present jobs. Their responses were not substantially different from those given by men. The principal difference was that almost a quarter of the women workers—but half the men—were attending school before taking assembly jobs.

In Haiti as in other countries, two main reasons are offered for the predominance of women in assembly plants: women, unlike men, possess the manual dexterity needed in assembly operations, and they are more docile and less militant than men. Yet in the survey men were found to be more productive than women. Since workers are paid piece rates, their wages should reflect their output. Average wages of men were nearly 10 percent higher than those of women. Regarding docility, labor militancy is still unknown in Haiti. Indeed, more than three-fourths of the workers did not know what a labor union is, and in none of the fifty-one firms in the sample was labor organized. The high proportion of women in the assembly work force in Port-au-Prince is probably due in part to the more abundant supply of female labor there and in part to cultural factors already referred to in the study of Mexico.

Almost three quarters of the assembly workers were born outside Port-au-Prince, yet the assembly industries have not been a particularly strong stimulus to migration. At the time of the survey, the migrant worker had been in Port-au-Prince an average of fourteen years, long before the assembly industries emerged as a significant force in the economic development of Haiti.

The existence of a large labor pool for assembly work is demonstrated by the selectivity exercised in hiring practices. Since the minimum wage is higher than what needs to be paid for unskilled labor in the informal sector, assembly managers give preference to persons with some skills, such as literacy and previous experience. In the face of a literacy rate of only a quarter of the adult population in Haiti, more than half the enterprises—in electronics more than three quarters—always require literacy and generally pay higher wages than those that never require it as a condition of employment. Vocational training, surprisingly, was not required by many firms; the highest level required is in textiles, where only a quarter of the companies insist on it. Although about half the firms—more in textiles, fewer in electronics—require prior experience, they do not pay higher wages than firms that do not require it.

Whereas few assembly enterprises say that they require vocational training, about half the workers in the sample reported that they had had such training. About a third of the workers interviewed had had previous assembly experience and another 5 percent had worked in nonassembly jobs. While neither vocational training nor previous experience seemed to make much difference in workers' earnings, education is positively correlated with assembly wages. More than 90 percent of the assembly workers reported that they had had at least one year of schooling. A third had had some high school, and 44 percent had completed elementary school. These levels are far higher than those for the country as a whole and also significantly higher than those for Port-au-Prince, where the educational level is much higher than in the rest of the country. Reflecting the difference in wages, the highest educational level was found in electronics and the lowest in baseball assembly.

Compared with Mexico and other countries in which assembly activities take place, the average age of assembly workers in Haiti is high—about 29 years. It is somewhat lower in electronics and a little higher in the other assembly categories. Age, unexpectedly, is not correlated with wages. This finding is consistent with another that indicates that seniority of workers in the factories is unrelated to their earnings. Surprisingly, female workers in electronics are younger than their male colleagues yet earn more, while in textiles they are older and earn less. These findings appear to say that learning on the job is not rewarded, that there is no learning, or that productivity declines with time spent on the job. The last may be a consequence of boredom with routine tasks or a decline in efficiency with age. In either case, since piece rates prevail the result would be lower wages. It may be too early, however, for any pattern to emerge, since the average seniority is just under four years, and about half of the workers have been in their present jobs less than three years.

Linkages

Much is made of the apparently weak linkages in developing countries between assembly activities and the rest of the economy. In Haiti, however, the rudimentary status of overall industrial development combined with the predominance of imported inputs would lead to the assumption that intersectoral linkages are weak in general.

Although assembly plants in Haiti are not isolated from the rest of the economy as they are in other places through in-bond and free-zone restrictions, the output of sophisticated assembled intermediate goods, such as computer harnesses and integrated circuits, could hardly serve as inputs for Haitian industries in a forward linkage, nor, for that matter, could the output of assembled finished goods, such as baseballs, wigs, stuffed toys, and relatively expensive clothing, find a ready market in the country.

In respect to backward linkages, however, significant progress has been made. In the assembly production of cassettes, for example, the plastic shells are purchased from a local producer; in baseballs and softballs, the core is now fabricated in Haiti and the glue is supplied locally; some threads are also purchased on the domestic market even though the relation of quality to price may not be up to international standards. All these items were formerly imported.

Despite the increased use of local materials in baseballs, there was no increase in the share of dutiable value—that is, value added in Haiti—as a percent of total value of U.S. baseball imports from Haiti under U.S. tariff item 807.00. Value added in Haiti averaged about a third of total value of assembled products exported to the United States, according to data obtained from the U.S. International Trade Commission. In some product groups there was a noticeable expansion of the Haitian share of assembly export value after the mid 1970s, perhaps indicating both a greater use of Haitian materials and an increase in local wages and profits (see table 4).

If the overall proportion of Haitian value added in total U.S. imports from Haiti is examined through a number of years, a decline can be observed until 1974, then a gradual increase until 1981 (see table 1). Part of the swing is a result of changes in the composition of output. The early decline appeared to be attributable to the use of more expensive U.S. components, as U.S. producers became more confident of Haitian assembly capabilities. The subsequent rise in Haitian value added might be attributed to higher relative wages and to the greater attention being paid to quality control in Haitian assembly plants, which increased the payroll component of value added through the employment of more highly paid testing personnel, inspectors, and supervisors. This seems to be true of the assembly of office machines, the dutiable value of which rose from 6 percent of total U.S. 807 imports from Haiti in 1973, the first year of production, to about 30 percent in 1982 and 1983, and of assembly of miscellaneous electrical products and parts, of which the dutiable value rose from less than 10 percent in the mid 1970s to 31 percent in 1981. In almost all categories more than a quarter of total export value is added in Haiti; in gloves it has been about half since 1977, probably because of the increased use of local leather (see table 4).

The most important indirect linkage of assembly operations to the Haitian economy is through the consumption expenditures of assembly workers. If the U.S. embassy figure of 60,000 workers and the $55 average monthly wage reported by the workers in the survey are used, the result is a total annual wage bill of almost $40 million. If most of this amount were spent on food, shelter, simple clothing, and transportation, items that would have a very low import content, the effect on Haitian economic development would be substantial. Profits and the incomes of managers might have a lesser effect, since substantial portions would be spent either abroad or on imported luxury items.

In respect to transfer of technology, there is a flow of technical assis-

Table 4
Total U.S. Imports of Selected Products from Haiti under Tariff Item 807 and Dutiable Value as Percent of Total Value, 1969–81

Total values in millions of dollars; dutiable values in percent

Year	Textiles Total 807 value	Textiles Dutiable value	Baseballs Total 807 value	Baseballs Dutiable value	Toys and dolls Total 807 value	Toys and dolls Dutiable value	Office machines Total 807 value	Office machines Dutiable value	Equipment for circuits Total 807 value	Equipment for circuits Dutiable value
1969	1.0	46.4	2.0	38.6	0.1	44.2	a	a	0.1	25.2
1970	1.5	38.3	3.2	34.7	0.2	37.7	a	a	0.1	21.4
1971	1.8	36.2	5.0	36.7	0.3	38.4	a	a	0.4	18.2
1972	4.1	35.0	8.0	31.8	0.4	40.0	a	a	0.8	17.1
1973	10.1	29.9	9.2	29.7	1.2	33.0	1.7	5.9	1.0	27.0
1974	19.7	25.3	12.4	29.8	2.2	28.9	4.6	4.4	1.9	21.2
1975	19.4	25.2	13.6	31.8	3.0	28.3	3.1	6.9	1.1	25.5
1976	27.1	23.8	16.8	37.6	6.1	31.4	6.3	10.4	2.2	23.6
1977	30.7	24.7	17.4	29.2	7.2	30.3	5.5	13.7	2.3	30.6
1978	39.7	27.6	24.4	25.6	7.6	29.6	5.5	17.6	3.1	25.8
1979	46.2	29.9	28.4	24.9	10.0	28.5	6.3	26.8	8.1	25.8
1980	53.0	30.8	29.5	26.8	8.5	28.8	3.9	56.1	11.4	21.1
1981	61.0	31.9	31.5	29.9	7.4	33.3	1.5	37.9	17.6	23.1

Source: Special magnetic tapes from the ITC.
n.a. Not available.
a. Zero or less than $50,000.

tance between the foreign principals and the Haitian assembly plants. Well over half the firms in the survey reported that foreign principals always provide technical assistance and an additional few indicated that they sometimes do. Much of the assistance comes in the form of foreign technicians sent by the principal on a temporary basis. About 60 percent of the assembly firms that responded said that foreign technicians visit regularly, and well over a third reported that they regularly send their own technicians abroad for training. Since the majority of assembly factories are in Haitian hands, new working methods and production techniques can easily be transferred to other sectors of the Haitian economy, particularly because some Haitians simultaneously operate other businesses as well as assembly plants.

Stability

Subcontracting is generally perceived to be a volatile business, so assembly operations are expected to be unstable. Two possible reasons for instability in these activities are sensitivity to the external business cycle, particularly in the United States, and the involvement of footloose U.S. industries and fly-by-night enterprises out to make a fast profit and run.

Table 4 (continued)

Total values in millions of dollars; dutiable values in percent

	Capacitors		Electric motors		Gloves		Miscellaneous electrical products		Luggage and handbags		Miscellaneous other machinery	
	Total 807 value	Dutiable value	Total 807 value	Dutiable value	Total 807 value	Dutiable value	Total 807 value	Dutiable value	Total 807 value	Dutiable value	Total 807 value	Dutiable value
	a	a	a	a	a	a	a	a	n.a.	n.a.	n.a.	n.a.
	a	a	0.1	10.1	a	a	a	a	n.a.	n.a.	n.a.	n.a.
	a	a	0.1	13.8	0.1	38.6	a	a	n.a.	n.a.	n.a.	n.a.
	a	a	0.1	13.3	0.1	40.8	0.1	28.6	n.a.	n.a.	n.a.	n.a.
	a	a	0.2	16.4	0.3	38.3	0.3	14.9	n.a.	n.a.	n.a.	n.a.
	0.3	24.0	0.2	24.8	1.1	40.9	1.1	10.4	n.a.	n.a.	n.a.	n.a.
	2.8	21.1	0.5	39.0	1.2	35.7	1.3	7.5	n.a.	n.a.	n.a.	n.a.
	6.0	23.9	1.3	31.4	1.4	44.3	2.0	9.8	n.a.	n.a.	n.a.	n.a.
	4.5	27.2	1.8	33.1	1.5	50.0	2.6	11.8	n.a.	n.a.	n.a.	n.a.
	3.5	20.7	2.0	35.6	1.7	53.2	4.3	13.4	0.9	29.1	1.7	26.8
	1.9	35.3	2.0	45.5	2.1	50.7	6.1	15.8	3.3	30.6	2.1	19.1
	5.7	48.1	2.7	35.7	2.1	47.1	7.2	22.7	4.7	31.8	2.5	29.4
	6.7	28.3	5.3	28.1	2.9	48.6	6.6	31.0	8.2	37.8	3.0	34.6

Regarding the first, a look at table 1 shows that Haitian assembly exports to the United States increased by leaps between 1969 and 1983. The only interruption of the monotonic expansion occurred in 1975, when there was a 3 percent decline in total export value after it had doubled in 1974 its value in the preceding year. This slight break can probably be attributed to the U.S. recession of 1974–75 and was caused entirely by a drop of 7 percent in the value of U.S. components used for assembly. Haiti, however, was not adversely affected; to the contrary, assembly activities generated a rise of more than 7 percent in Haitian value added in 1975. In textiles in 1975 there was a barely noticeable hesitation in the steady increase, and the only product group in which a significant decline can be found is equipment for electric circuits, the total decline in value added of which hardly amounted to $250,000. The U.S. recession of 1980–82 did not have a negative effect on Haitian assembly operations for the U.S. market. Growth in value added in Haiti slowed from 1981 to 1982, though the decline was less than 3 percent in real terms when deflated by the U.S. consumer price index excluding food. But in 1983 it was almost 50 percent greater than in prerecession 1979.

The assembly firms in Haiti seem to have a sanguine outlook regarding the fragility of their operations. Nearly half of them believe that a U.S. recession will not affect them, according to the survey taken at the beginning of 1980, when a business downturn abroad threatened. Textile companies are the least optimistic of the assembly firms; the most optimistic are those that assemble stuffed toys. Depending on the industry, either a supply-side or a demand-side explanation is given. The supply-side assumption is that in a business downturn the U.S. principals will have an additional incentive to reduce production costs and therefore will move more operations abroad; the demand-side assumption is that consumers will shift from expensive products to cheaper ones—that is, that instead of buying fancy electronic gadgets for children, adults will give them stuffed toys.

In this respect, it is interesting to note that in Haiti light manufactured exports that use local materials appear to have been deeply affected by the U.S. recession of 1974–75. There was a decline of about a quarter in these exports from 1974 to 1975 and a further decline in the following year, so the value of 1976 exports was almost a third lower than that of 1974. Only in 1978 were 1974 export levels reached again in nominal terms. Exports of light manufacture also suffered on account of the U.S. recession of 1980–82.

In respect to the argument that footloose industries impart instability, the record shows that only a few products assembled abroad have emerged and disappeared, while most product lines have increased steadily. Because most of Haiti's assembly plants require small capital investments, it is surprising that so few industries have disappeared. This does not mean that there have been no fly-by-night operators and that no subcontractors in Haiti were hurt, but this kind of instability appears to have had only small effects on the Haitian economy.

Furthermore, Haitian assembly operators have learned to protect themselves against the ups and downs of subcontracting. Almost all the firms in the sample work with more than one U.S. enterprise and the majority subcontract with more than two principals each. This often tends to smooth fluctuations in the total orders, thereby reducing the risk. It will be especially true if the local enterprise assembles several different product lines, each of which may be subject to diverse seasonal and cyclical variations.

The low capital requirements in this business permit the local entrepreneur to diversify his assembly operations so that he can work for U.S. firms in different industrial sectors, either simultaneously or by shifting his work force from one assembly production to another according to the external demand. It was observed in the survey that the low capital-output ratio enables some firms, with only minor modifications in their factories and equipment, to move from baseballs to cassettes or from electronics to textiles, as they lose contracts for one product but find them for another. Thus, it is not the foreign principal that is footloose but rather the Haitian firm, which displays high mobility between foreign principals or production sectors. It is this flexibility, a manifestation of perceptive Haitian entrepreneurship, that has given stability to the growth of assembly activities in Haiti.

Conclusions

Assembly production has become an integral part of economic activities in Haiti. It is not a marginal appendage, as in some countries that are engaged in these operations, but a principal contributor to the economic development of Haiti and the leading edge of its industrialization. It has been the most buoyant sector of the Haitian economy for the past decade. The best estimate is that value added in assembly activities now constitutes more than a quarter of all income generated in the manufacturing sector and provides almost a third of the country's export earnings. While the employment it creates, estimated by the authors at 60,000 persons in 1980, may seem small in relation to the total labor force of more than 2.5 million, it is probably as large as employment in all the rest of the modern manufacturing sector in the country.

The strong linkages of assembly plants to foreign companies do not seem to have introduced extraordinary instability into the Haitian economy. While there is a dependence on foreign orders or contracts, Haitian entrepreneurs have learned to protect themselves against the extremes of this dependency, often by spreading the risk among several foreign firms and assembly products.

The foreign principal, in turn, apparently wants to keep the dependency relationship to a minimum: while prepared to provide machinery and equipment to the Haitian assembly plant, the principal does not provide much financing. Only 7 percent of the firms interviewed in the survey reported that they were in debt to the principal. Nearly all the rest borrowed from private banks in Haiti.

It would be difficult to support the assertion that, as in Mexico, assembly activities from an economic enclave in Haiti. First of all, Haiti is at the bottom of the development ladder, so linkages among its economic sectors are weak and it is still far from having an integrated economy. Second, assembly production is concentrated in the economic and population center of the country, metropolitan Port-au-Prince, and not in a remote region far removed from the capital, as in Mexico. There are no laws that explicitly restrict assembly plants to a free-trade zone or to in-bond operations. There is, however, a formal legal barrier to selling assembled items with imported components on the local market. At present, assembly firms have no strong economic incentives to sell at home, where the demand for their products is weak in relation to export demand.

The lack of tight integration of assembly operations into the rest of the economy is not too different from the relatively weak linkages among other industries that use imported materials. This is particularly true of many of the new import-substituting industries that depend on imported inputs.

The basic problem of national economic integration is the low level of economic development. As income levels rise and better public policy helps ease supply constraints, the various economic sectors, including assembly industries, are tied together more closely, thereby increasing their contribution to the economic development of the country. In the meantime, the linkages in assembly production have not been trivial; they compare favorably with linkages in other economic activities in the country. Not only do the incomes generated by assembly production create a significant demand for local goods and services, but assembly production has also provided a substantial stimulus to the banking, transport, and communications sectors, as well as to food services in or near the factories for the assembly workers. While some local components, such as materials for baseball cores, leather for shoes and handbags, plastic cases for cassettes, and corrugated cartons and other packing materials are used in assembly production, any substantial increase in the use of local inputs must await the elimination of supply problems in other sectors, particularly in agriculture.

In the context of an underdeveloped economy such as that of Haiti, assembly operations help significantly in the transfer of technology. The mere existence of these activities, which constitute perhaps half the country's modern sector, introduces workers to factory discipline and to new equipment and working methods. Even if, as the World Bank reports, the average training period for assembly work is only two months, it provides skill levels far higher than the mean of the Haitian labor force. As shown earlier, there is a considerable interchange of technicians between the Haitian plants and the U.S. principals. And the mere fact that assembly operations and other production activities are often carried on within the same firm makes an unusual economic and technological isolation of the former highly unlikely.

Barring a deterioration of the political situation in Haiti, indications are that the dynamism of assembly production will remain alive. The vast majority of the firms interviewed—84 percent of them—expected a further growth of subcontracting activities and said that they planned to expand their own operations. At least until 1983, the optimism of the managers has been borne out by the facts: value added generated in assembly production in Haiti in 1980 was nearly a quarter greater than in the preceding year, and growth continued in 1981, 1982, and 1983, despite a dramatic decline in GNP and in other exports.

Notes

1. The labor cost to the employer is 32 percent higher than the wage in order to cover legal fringe benefits and social security. For details see ADIH, "The Industrial Sector in Haiti," appendix II, p. 94. The figures given were adjusted by the authors. If workers work more than forty-eight hours or at night, they are legally entitled to time-and-a-half pay. They are also entitled to a paid rest day (ADIH, "The Industrial Sector in Haiti," p. 33). Often, however, the wage laws are not enforced in Haiti, and workers receive the regular rate of pay for overtime and do not receive Sunday pay. It should be recalled that most production workers in Haiti (all those in the underlying sample) are paid on a piece-rate basis. The piece rate is

determined by dividing the minimum wage by a productivity norm set on the basis of time-and-motion studies.

Ideas for Discussion

1. After studying this article, does it seem strategically sound to encourage U.S. manufacturers to assemble products in Haiti? Why? Why not?
2. Looking back to "The Enduring Logic of Industrial Success" could you argue U.S. investment in Haiti might make both countries globally competitive?
3. List the characteristics of the Haitian workforce.

Topics for Writing

1. Write a proposal for or against your company's plan to move assembly to Haiti.
2. After looking at an article such as "Detroit South," *Business Week*, March 16, 1992: 98–103, write an expanded comparison of Haiti and Mexico as centers for manufacturing assemblies.
3. Write a report for a new manager on how to treat the Haitian workforce.
4. Write a memo announcing your company's intention to move manufacturing to Haiti. Write that same memo as a letter to your company's board of directors. In a memo to your instructor, explain the changes you made from one format to another.
5. As the Ministry for Commerce in Haiti, prepare a promotional packet to attract business to your country.
6. Write a sales proposal or a feasibility study for doing business in Haiti.

Gary Hamel and C. K. Prahalad
Strategic Intent

Gary Hamel teaches at the London Business School. His coauthor, C. K. Prahalad, is a professor at the University of Michigan, offering courses in international business and corporate strategy. This essay, first published in Harvard Business Review (May–June 1989), reflects their joint interest in theories of industrial competitiveness.

TODAY MANAGERS in many industries are working hard to match the competitive advantages of their new global rivals. They are moving manufacturing offshore in search of lower labor costs, rationalizing product lines to capture global scale economies, instituting quality circles and just-in-time production, and adopting Japanese human resource practices. When competitiveness still seems out of reach, they form strategic alliances—often with the very companies that upset the competitive balance in the first place.

Important as these initiatives are, few of them go beyond mere imitation. Too many companies are expending enormous energy simply to reproduce the cost and quality advantages their global competitors already enjoy. Imitation may be the sincerest form of flattery, but it will not lead to competitive revitalization. Strategies based on imitation are transparent to competitors who have already mastered them. Moreover, successful competitors rarely stand still. So it is not surprising that many executives feel trapped in a seemingly endless game of catch-up—regularly surprised by the new accomplishments of their rivals.

For these executives and their companies, regaining competitiveness will mean rethinking many of the basic concepts of strategy.[1] As "strategy" has blossomed, the competitiveness of Western companies has withered. This may be coincidence, but we think not. We believe that the application of concepts such as "strategic fit" (between resources and opportunities), "generic strategies" (low cost vs. differentiation vs. focus), and the "strategy hierarchy" (goals, strategies, and tactics) have often abetted the process of competitive decline. The new global competitors approach strategy from a perspective that is fundamentally different from that which underpins Western management thought. Against such competitors, marginal adjustments to current orthodoxies are no more likely to produce competitive revitalization than are marginal improvements in operating efficiency.

Few Western companies have an enviable track record anticipating the moves of new global competitors. Why? The explanation begins with the way most companies have approached competitor analysis. Typically, competitor analysis focuses on the existing resources (human, technical, and financial) of present competitors. The only companies seen as a threat are those with the resources to erode margins and market share in the next planning period. Resourcefulness, the pace at which new competitive advantages are being built, rarely enters in.

In this respect, traditional competitor analysis is like a snapshot of a moving car. By itself, the photograph yields little information about the cars speed or direction—whether the driver is out for a quiet Sunday drive or warming up for the Grand Prix. Yet many managers have learned through painful experience that a business's initial resource endowment (whether bountiful or meager) is an unreliable predictor of future global success.

Think back. In 1970, few Japanese companies possessed the resource base, manufacturing volume, or technical prowess of U.S. and European industry leaders. Komatsu was less than 35% as large as Caterpillar (measured by sales), was scarcely represented outside Japan, and relied on just one product line—small bulldozers—for most of its revenue. Honda was smaller than American Motors and had not yet begun to export cars to the United States. Canon's first halting steps in the reprographics business looked pitifully small compared with the $4 billion Xerox powerhouse.

If Western managers had extended their competitor analysis to include these companies, it would merely have underlined how dramatic the resource discrepancies between them were. Yet by 1985, Komatsu was a $2.8 billion company with a product scope encompassing a broad range of earth-moving equipment, industrial robots, and semiconductors. Honda manufactured almost as many cars worldwide in 1987 as Chrysler. Canon had matched Xerox's global unit market share.

The lesson is clear: assessing the current tactical advantages of known competitors will not help you understand the resolution, stamina, and inventiveness of potential competitors. Sun-tzu, a Chinese military strategist, made the point 3,000 years ago: "All men can see the tactics whereby I conquer," he wrote, "but what none can see is the strategy out of which great victory is evolved."

Companies that have risen to global leadership over the past 20 years invariably began with ambitions that were out of all proportion to their resources and capabilities. But they created an obsession with winning at all levels of the organization and then sustained that obsession over the 10- to

20-year quest for global leadership. We term this obsession "strategic intent."

On the one hand, strategic intent envisions a desired leadership position and establishes the criterion the organization will use to chart its progress. Komatsu set out to "Encircle Caterpillar." Canon sought to "Beat Xerox." Honda strove to become a second Ford—an automotive pioneer. All are expressions of strategic intent.

At the same time, strategic intent is more than simply unfettered ambition. (Many companies possess an ambitious strategic intent yet fall short of their goals.) The concept also encompasses an active management process that includes: focusing the organization's attention on the essence of winning; motivating people by communicating the value of the target; leaving room for individual and team contributions; sustaining enthusiasm by providing new operational definitions as circumstances change; and using intent consistently to guide resource allocations.

Strategic intent captures the essence of winning. The Apollo program—landing a man on the moon ahead of the Soviets—was as competitively focused as Komatsu's drive against Caterpillar. The space program became the scorecard for America's technology race with the USSR. In the turbulent information technology industry, it was hard to pick a single competitor as a target, so NEC's strategic intent, set in the early 1970s, was to acquire the technologies that would put it in the best position to exploit the convergence of computing and telecommunications. Other industry observers foresaw this convergence, but only NEC made convergence the guiding theme for subsequent strategic decisions by adopting "computing and communications" as its intent. For Coca-Cola, strategic intent has been to put a Coke within "arm's reach" of every consumer in the world.

Strategic intent is stable over time. In battles for global leadership, one of the most critical tasks is to lengthen the organization's attention span. Strategic intent provides consistency to short-term action, while leaving room for reinterpretation as new opportunities emerge. At Komatsu, encircling Caterpillar encompassed a succession of medium-term programs aimed at exploiting specific weaknesses in Caterpillar or building particular competitive advantages. When Caterpillar threatened Komatsu in Japan, for example, Komatsu responded by first improving quality, then driving down costs, then cultivating export markets, and then underwriting new product development.

Strategic intent sets a target that deserves personal effort and commitment. Ask the chairmen of many American corporations how they measure their contributions to their companies' success and you're likely to get an answer expressed in terms of shareholder wealth. In a company that possesses a strategic intent, top management is more likely to talk in terms of global market leadership. Market share leadership typically yields shareholder wealth, to be sure. But the two goals do not have the same motivational impact. It is hard to imagine middle managers, let alone blue-collar employees, waking up each day with the sole thought of creating more shareholder wealth. But mightn't they feel different given the challenge to "Beat Benz"—the rallying cry at one Japanese auto producer? Strategic intent gives employees the only goal that is worthy of commitment: to unseat the best or remain the best, worldwide.

Many companies are more familiar with strategic planning than they are with strategic intent. The planning process typically acts as a "feasibility sieve." Strategies are accepted or rejected on the basis of whether managers can be precise about the "how" as well as the "what" of their plans. Are the milestones clear? Do we have the necessary skills and resources? How will competitors react? Has the market been thoroughly researched? In one form or another, the admonition "Be realistic!" is given to line managers at almost every turn.

But can you *plan* for global leadership? Did Komatsu, Canon, and Honda have detailed, 20-year "strategies" for attacking Western markets? Are Japanese and Korean managers better planners than their Western counterparts? No. As valuable as strategic planning is, global leadership is an objective that lies outside the range of planning. We know of few companies with highly developed planning systems that have managed to set a strategic intent. As tests of strategic fit become more stringent, goals that cannot be planned for fall by the wayside. Yet companies that are afraid to commit to goals that lie outside the range of planning are unlikely to become global leaders.

Although strategic planning is billed as a way of becoming more future oriented, most managers, when pressed, will admit that their strategic plans reveal more about today's problems than tomorrow's opportunities. With a fresh set of problems confronting managers at the beginning of every planning cycle, focus often shifts dramatically from year to year. And with the pace of change accelerating in most industries, the predictive horizon is becoming shorter and shorter. So plans do little more than project the present forward incrementally. The goal of strategic intent is to fold the future back into the present. The important question is not "How will next year be different from this year?" but "What must we do differently next year to get closer to our strategic intent?" Only with a carefully articulated and adhered to strategic intent will a succession of year-on-year plans sum up to global leadership.

Just as you cannot plan a 10- to 20-year quest for global leadership, the chance of falling into a leadership position by accident is also remote. We don't believe that global leadership comes from an undirected process of intrapreneurship. Nor is it the product of a skunkworks or other techniques for internal venturing. Behind such programs lies a nihilistic assumption: the organization is so hidebound, so orthodox ridden that the only way to innovate is to put a few bright people in a dark room, pour in some money, and hope that something wonderful will happen. In this "Silicon Valley" approach to innovation, the only role for top managers is to retrofit their corporate strategy to the entrepreneurial successes that emerge from below. Here the value added of top management is low indeed.

Sadly, this view of innovation may be consistent with the reality in many large companies.[2] On the one hand, top management lacks any particular point of view about desirable ends beyond satisfying shareholders and keeping raiders at bay. On the other, the planning format, reward criteria, definition of served market, and belief in accepted industry practice all work together to tightly constrain the range of available means. As a result, innovation is necessarily an isolated activity. Growth depends more on the inventive capacity of individuals and small teams than on the ability of top management to aggregate the efforts of multiple teams towards an ambitious strategic intent.

In companies that overcame resource constraints to build leadership positions, we see a different relationship between means and ends. While strategic intent is clear about ends, it is flexible as to means—it leaves room for improvisation. Achieving strategic intent requires enormous creativity with respect to means: witness Fujitsu's use of strategic alliances in Europe to attack IBM. But this creativity comes in the service of a clearly prescribed end. Creativity is unbridled, but not uncorralled, because top management establishes the criterion against which employees can pretest the logic of their initiatives. Middle managers must do more than deliver on promised financial targets; they must also deliver on the broad direction implicit in their organization's strategic intent.

Strategic intent implies a sizable stretch for an organization. Current capabilities and resources will not suffice. This forces the organization to be more inventive, to make the most of limited resources. Whereas the traditional view of strategy focuses on the degree of fit between existing resources and current opportunities, strategic intent creates an extreme misfit between resources and ambitions. Top management then challenges the organization to close the gap by systematically building new advantages. For Canon this meant first understanding Xerox's patents, then licensing technology to create a product that would yield early market experience, then gearing up internal R&D efforts, then licensing its own technology to other manufacturers to fund further R&D, then entering market segments in Japan and Europe where Xerox was weak, and so on.

In this respect, strategic intent is like a marathon run in 400-meter sprints. No one knows what the terrain will look like at mile 26, so the role of top management is to focus the organization's attention on the ground to be covered in the next 400 meters. In several companies, management did this by presenting the organization with a series of corporate challenges, each specifying the next hill in the race to achieve strategic intent. One year the challenge might be quality, the next total customer care, the next entry into new markets, the next a rejuvenated product line. As this example indicates, corporate challenges are a way to stage the acquisition of new competitive advantages, a way to identify the focal point for employees' efforts in the near to medium term. As with strategic intent, top management is specific about the ends (reducing product development times by 75%, for example) but less prescriptive about the means.

Like strategic intent, challenges stretch the organization. To preempt Xerox in the personal copier business, Canon set its engineers a target price of $1,000 for a home copier. At the time, Canon's least expensive copier sold for several thousand dollars. Trying to reduce the cost of existing models would not have given Canon the radical price-performance improvement it needed to delay or deter Xerox's entry into personal copiers. Instead, Canon engineers were challenged to reinvent the copier—a challenge they met by substituting a disposable cartridge for the complex image-transfer mechanism used in other copiers.

Corporate challenges come from analyzing competitors as well as from the foreseeable pattern of industry evolution. Together these reveal potential competitive openings and identify the new skills the organization will need to take the initiative away from better positioned players. The exhibit "Building Competitive Advantage at Komatsu," illustrates the way challenges helped that company achieve its intent.

For a challenge to be effective, individuals and teams throughout the organization must understand it and see its implications for their own jobs. Companies that set corporate challenges to create new competitive advantages (as Ford and IBM did with quality improvement) quickly discover that engaging the entire organization requires top management to:

Create a sense of urgency, or quasi crisis, by amplifying weak signals in the environment that point up the need to improve, instead of allowing inaction to precipitate a real crisis. (Komatsu, for example, budgeted on the basis of worst case exchange rates that overvalued the yen.)

Develop a competitor focus at every level through widespread use of competitive intelligence. Every employee should be able to benchmark his or her efforts against best-in-class competitors so that the challenge becomes personal. (For example, Ford showed production-line workers videotapes of operations at Mazda's most efficient plant.)

Provide employees with the skills they need to work effectively—training in statistical tools, problem solving, value engineering, and team building, for example.

Give the organization time to digest one challenge before launching another. When competing initiatives overload the organization, middle managers often try to protect their people from the whipsaw of shifting priorities. But this "wait and see if they're serious this time" attitude ultimately destroys the credibility of corporate challenges.

Establish clear milestones and review mechanisms to track progress and ensure that internal recognition and rewards reinforce desired behavior. The goal is to make the challenge inescapable for everyone in the company.

It is important to distinguish between the process of managing corporate challenges and the advantages that the process creates. Whatever the actual

Building Competitive Advantage at Komatsu

Corporate Challenge	Protect Komatsu's home market against Caterpillar	Reduce cash while maintaining quality	Make Komatsu an international enterprise and build expert materials	Respond to external shocks that threaten markets	Create new products and materials
Programs	early 1960s: Licensing deals with Cummins Engine, International Harvester and Bucyrus-Erie to acquire technology and establish benchmarks	1965: CD (Cost Down) program	early 1960s: Develop Eastern bloc countries	1975: V-10 program to reduce costs by 10% while maintaining quality; reduce parts by 20% to nationalize manufacturing system	late 1970s: Accelerate product development to expand product line
	1961: Project A (for Ace) to advance the product quality of Komatsu's small- and medium-sized bulldozers above Caterpillar's	1966: Total CD program	1967: Komatsu Europe marketing subsidiary established	1977: V 180 program to budget companywide for 180 yen to the dollar when exchange rate was 240	1979: Future and Frontiers program to identify new businesses based on society's needs and company's know-how
	1962: Quality Circles companywide to provide training for all employees		1970: Komatsu America established	1979: Project E to establish teams to redouble cost and quality efforts in response to crisis	1981: EPOCHS program to reconcile greater product variety with improved production efficiencies
			1972: Project B to improve the durability and reliability and to reduce costs of large bulldozers		
			1972: Project C to improve payloaders		
			1972: Project D to improve hydraulic excavators		
			1974: Establish presales and service department to assist newly industrializing countries in construction projects		

challenge may be—quality, cost, value engineering, or something else—there is the same need to engage employees intellectually and emotionally in the development of new skills. In each case, the challenge will take root only if senior executives and lower level employees feel a reciprocal responsibility for competitiveness.

We believe workers in many companies have been asked to take a disproportionate share of the blame for competitive failure. In one U.S. company, for example, management had sought a 40% wage-package concession from hourly employees to bring labor costs into line with Far Eastern competitors. The result was a long strike and, ultimately a 10% wage concession from employees on the line. However, direct labor costs in manufacturing accounted for less than 15% of total value added. The company thus succeeded in demoralizing its entire blue-collar work force for the sake of a 1.5% reduction in total costs. Ironically, further analysis showed that their competitors' most significant cost savings came not from lower hourly wages but from better work methods invented by employees. You can imagine how eager the U.S. workers were to make similar contributions after the strike and concessions. Contrast this situation with what happened at Nissan when the yen strengthened: top management took a big pay cut and then asked middle managers and line employees to sacrifice relatively less.

Reciprocal responsibility means shared gain and shared pain. In too many companies, the pain of revitalization falls almost exclusively on the employees least responsible for the enterprise's decline. Too often, workers are asked to commit to corporate goals without any matching commitment from top management—be it employment security, gain sharing, or an ability to influence the direction of the business. This one-sided approach to regaining competitiveness keeps many companies from harnessing the intellectual horsepower of their employees.

Creating a sense of reciprocal responsibility is crucial because competitiveness ultimately depends on the pace at which a company embeds new advantages deep within its organization, not on its stock of advantages at any given time. Thus we need to expand the concept of competitive advantage beyond the scorecard many managers now use: Are my costs lower? Will my product command a price premium?

Few competitive advantages are long lasting. Uncovering a new competitive advantage is a bit like getting a hot tip on a stock: the first person to act on the insight makes more money than the last. When the experience curve was young, a company that built capacity ahead of competitors, dropped prices to fill plants, and reduced costs as volume rose went to the bank. The first mover traded on the fact that competitors undervalued market share—they didn't price to capture additional share because they didn't understand how market share leadership could be translated into lower costs and better margins. But there is no more undervalued market share when each of 20 semiconductor companies builds enough capacity to serve 10% of the world market.

Keeping score of existing advantages is not the same as building new advantages. The essence of strategy lies in creating tomorrow's competitive advantages faster than competitors mimic the ones you possess today. In the 1960s, Japanese producers relied on labor and capital cost advantages. As Western manufacturers began to move production offshore, Japanese companies accelerated their investment in process technology and created scale and quality advantages. Then as their U.S. and European competitors rationalized manufacturing, they added another string to their bow by accelerating the rate of product development. Then they built global brands. Then they deskilled competitors through alliances and outsourcing deals. The moral? An organization's capacity to improve existing skills and learn new ones is the most defensible competitive advantage of all.

To achieve a strategic intent, a company must usually take on larger, better financed competitors. That means carefully managing competitive engagements so that scarce resources are conserved. Managers cannot do that simply by playing the same game better—making marginal improvements to competitors' technology and business practices. Instead, they must fundamentally change the game in ways that disadvantage incumbents—devising novel approaches to market entry, advantage building, and competitive warfare. For smart competitors, the goal is not competitive imitation but competitive innovation, the art of containing competitive risks within manageable proportions.

Four approaches to competitive innovation are evident in the global expansion of Japanese companies. These are: building layers of advantage, searching for loose bricks, changing the terms of engagement, and competing through collaboration.

The wider a company's portfolio of advantages, the less risk it faces in competitive battles. New global competitors have built such portfolios by steadily expanding their arsenals of competitive weapons. They have moved inexorably from less defensible advantages such as low wage costs to more defensible advantages like global brands. The Japanese color television industry illustrates this layering process.

By 1967, Japan had become the largest producer of black-and-white television sets. By 1970, it was closing the gap in color televisions. Japanese manufacturers used their competitive advantage—at that time, primarily, low labor costs—to build a base in the private-label business, then moved quickly to establish world-scale plants. This investment gave them additional layers of advantage—quality and reliability—as well as further cost reductions from process improvements. At the same time, they recognized that these cost-based advantages were vulnerable to changes in labor costs, process and product technology, exchange rates, and trade policy. So throughout the 1970s, they also invested heavily in building channels and brands, thus creating another layer of advantage, a global franchise. In the late 1970s, they enlarged the scope of their products and businesses to amortize these grand investments, and by 1980 all the major players—Matsushita, Sharp, Toshiba, Hitachi, Sanyo—had established related sets of businesses that could support global marketing investments. More recently, they have been investing in regional manufacturing and design centers to tailor their products more closely to national markets.

These manufacturers thought of the various sources of competitive advantage as mutually desirable layers, not mutually exclusive choices. What some call competitive suicide—pursuing both cost and differentiation—is exactly what many competitors strive for.[3] Using flexible manufacturing technologies and better marketing intelligence, they are moving away from standardized "world products" to products like Mazda's mini-van, developed in California expressly for the U.S. market.

Another approach to competitive innovation—searching for loose bricks—exploits the benefits of surprise, which is just as useful in business battles as it is in war. Particularly in the early stages of a war for global markets, successful new competitors work to stay below the response threshold of their larger, more powerful rivals. Staking out underdefended territory is one way to do this.

To find loose bricks, managers must have few orthodoxies about how to break into a market or challenge a competitor. For example, in one large U.S. multinational, we asked several country managers to describe what a Japanese competitor was doing in the local market. The first executive said, "They're coming at us in the low end. Japanese companies always come in at the bottom." The second speaker found the comment interesting but disagreed: "They don't offer any low-end products in my market, but they have some exciting stuff at the top end. We really should reverse engineer that thing." Another colleague told still another story. "They haven't taken any business away from me," he said, "but they've just made me a great offer to supply components." In each country, their Japanese competitor had found a different loose brick.

The search for loose bricks begins with a careful analysis of the competitor's conventional wisdom: How does the company define its "served market"? What activities are most profitable? Which geographic markets are too troublesome to enter? The objective is not to find a corner of the industry (or niche) where larger competitors seldom tread but to build a base of attack just outside the market territory that industry leaders currently occupy. The goal is an uncontested profit sanctuary, which could be a particular product segment (the "low end" in motorcycles), a slice of the value chain (components in the computer industry), or a particular geographic market (Eastern Europe).

When Honda took on leaders in the motorcycle industry, for example, it began with products that were just outside the conventional definition of the leaders' product-market domains. As a result, it could build a base of operations in underdefended territory and then use that base to launch an expanded attack. What many competitors failed to see was Honda's strategic intent and its growing competence in engines and power trains. Yet even as Honda was selling 50cc motorcycles in the United States, it was already racing larger bikes in Europe—assembling the design skills and technology it would need for a systematic expansion across the entire spectrum of motor-related businesses.

Honda's progress in creating a core competence in engines should have warned competitors that it might enter a series of seemingly unrelated industries—automobiles, lawn mowers, marine engines, generators. But with each company fixated on its own market, the threat of Honda's horizontal diversification went unnoticed. Today companies like Matsushita and Toshiba are similarly poised to move in unexpected ways across industry boundaries. In protecting loose bricks, companies must extend their peripheral vision by tracking and anticipating the migration of global competitors across product segments, businesses, national markets, value-added stages, and distribution channels.

Changing the terms of engagement—refusing to accept the front runner's definition of industry and segment boundaries—represents still another form of competitive innovation. Canon's entry into the copier business illustrates this approach.

During the 1970s, both Kodak and IBM tried to match Xerox's business system in terms of segmentation, products, distribution, service, and pricing. As a result, Xerox had no trouble decoding the new entrants' intentions and developing countermoves. IBM eventually withdrew from the copier business, while Kodak remains a distant second in the large copier market that Xerox still dominates.

Canon, on the other hand, changed the terms of competitive engagement. While Xerox built a wide range of copiers, Canon standardized machines and components to reduce costs. Canon chose to distribute through office-product dealers rather than try to match Xerox's huge direct sales force. It also avoided the need to create a national service network by designing reliability and serviceability into its product and then delegating service responsibility to the dealers. Canon copiers were sold rather than leased, freeing Canon from the burden of financing the lease base. Finally, instead of selling to the heads of corporate duplicating departments, Canon appealed to secretaries and department managers who wanted distributed copying. At each stage, Canon neatly sidestepped a potential barrier to entry.

Canon's experience suggests that there is an important distinction between barriers to entry and barriers to imitation. Competitors that tried to match Xerox's business system had to pay the same entry costs—the barriers to imitation were high. But Canon dramatically reduced the barriers to entry by changing the rules of the game.

Changing the rules also short-circuited Xerox's ability to retaliate quickly against its new rival. Confronted with the need to rethink its business strategy and organization, Xerox was paralyzed for a time. Xerox managers realized that the faster they downsized the product line, developed new channels, and improved reliability, the faster they would erode the company's traditional profit base. What might have been seen as critical success factors—Xerox's national sales force and service network, its large installed base of leased machines, and its reliance on service revenues—instead became barriers to retaliation. In this sense, competitive innovation is like judo: the goal is to use a larger competitor's weight against it. And that happens not by matching the leader's capabilities but by developing contrasting capabilities of one's own.

Competitive innovation works on the premise that a successful competitor is likely to be wedded to a "recipe" for success. That's why the most effective weapon new competitors possess is probably a clean sheet of paper. And why an incumbent's greatest vulnerability is its belief in accepted practice.

Through licensing, outsourcing agreements, and joint ventures, it is sometimes possible to win without fighting. For example, Fujitsu's alliances in Europe with Siemens and STC (Britain's largest computer maker) and in the United States with Amdahl yield manufacturing volume and access to Western markets. In the early 1980s, Matsushita established a joint venture with Thorn (in the United Kingdom), Telefunken (in Germany), and Thomson (in France), which allowed it to quickly multiply the forces arrayed against Philips in the battle for leadership in the European VCR business. In fighting larger global rivals by proxy, Japanese companies have adopted a maxim as old as human conflict itself: my enemy's enemy is my friend.

Hijacking the development efforts of potential rivals is another goal of competitive collaboration. In the consumer electronics war, Japanese competitors attacked traditional businesses like TVs and hi-fis while volunteering to manufacture "next generation" products like VCRs, camcorders, and compact disc players for Western rivals. They hoped their rivals would ratchet down development spending, and in most cases that is precisely what happened. But companies that abandoned their own development efforts seldom reemerged as serious competitors in subsequent new product battles.

Collaboration can also be used to calibrate competitors' strengths and weaknesses. Toyota's joint venture with GM, and Mazda's with Ford, give these automakers an invaluable vantage point for assessing the progress their U.S. rivals have made in cost reduction, quality, and technology. They can also learn how GM and Ford compete—when they will fight and when they won't. Of course, the reverse is also true: Ford and GM have an equal opportunity to learn from their partner-competitors.

The route to competitive revitalization we have been mapping implies a new view of strategy. Strategic intent assures consistency in resource allocation over the long term. Clearly articulated corporate challenges focus the efforts of individuals in the medium term. Finally, competitive innovation helps reduce competitive risk in the short term. This consistency in the long term, focus in the medium term, and inventiveness and involvement in the short term provide the key to leveraging limited resources in pursuit of ambitious goals. But just as there is a process of winning, so there is a process of surrender. Revitalization requires understanding that process too.

Given their technological leadership and access to large regional markets, how did U.S. and European companies lose their apparent birthright to dominate global industries? There is no simple answer. Few companies recognize the value of documenting failure. Fewer still search their own managerial orthodoxies for the seeds for competitive surrender. But we believe there is a pathology of surrender that gives some important clues.

It is not very comforting to think that the essence of Western strategic thought can be reduced to eight rules for excellence, seven 5's, five competitive forces, four product life-cycle stages, three generic strategies, and innumerable two-by-two matrices.[4] Yet for the past 20 years, "advances" in strategy have taken the form of ever more typologies, heuristics, and laundry lists, often with dubious empirical bases. Moreover, even reasonable concepts like the product life cycle, experience curve, product portfolios, and generic strategies often have toxic side effects: They reduce the number of strategic options management is willing to consider. They create a preference for selling businesses rather than defending them. They yield predictable strategies that rivals easily decode.

Strategy "recipes" limit opportunities for competitive innovation. A company may have 40 businesses and only four strategies—invest, hold, harvest, or divest. Too often strategy is seen as a positioning exercise in which options are tested by how they fit the existing industry structure. But current industry structure reflects the strengths of the industry leader; and playing by the leader's rules is usually competitive suicide.

Armed with concepts like segmentation, the value chain, competitor benchmarking, strategic groups, and mobility barriers, many managers have become better and better at drawing industry maps. But while they have been busy map making, their competitors have been moving entire continents. The strategist's goal is not to find a niche within the existing industry space but to create new space that is uniquely suited to the company's own strengths, space that is off the map.

This is particularly true now that industry boundaries are becoming more and more unstable. In industries such as financial services and communications, rapidly changing technology, deregulation, and globalization have undermined the value of traditional industry analysis. Map-making skills are worth little in the epicenter of an earthquake. But an industry in upheaval presents opportunities for ambitious companies to redraw the map in their favor, so long as they can think outside traditional industry boundaries.

Concepts like "mature" and "declining" are largely definitional. What most executives mean when they label a business mature is that sales growth has stagnated in their current geographic markets for existing products sold through existing channels. In such cases, it's not the industry that is mature, but the executives' conception of the industry. Asked if the piano business was mature, a senior executive in Yamaha replied, "Only if we can't take any market share from anybody anywhere in the world and still make money. And anyway, we're not in the 'piano' business, we're in the 'keyboard' business." Year after year, Sony has revitalized its radio and tape recorder businesses, despite the fact that other manufacturers long ago abandoned these businesses as mature.

A narrow concept of maturity can foreclose a company from a broad stream of future opportunities. In the 1970s, several U.S. companies thought that consumer electronics had become a mature industry. What could possibly top the color TV? they asked themselves. RCA and GE, distracted by opportunities in more "attractive" industries like mainframe computers, left Japanese producers with a virtual monopoly in VCRs, camcorders, and compact disc players. Ironically, the TV business once thought mature, is on the verge of a dramatic renaissance. A $20 billion-a-year business will be created when high-definition television is launched in the United States. But the pioneers of television may capture only a small part of this bonanza.

Most of the tools of strategic analysis are focused domestically. Few force managers to consider global opportunities and threats. For example, portfolio planning portrays top management's investment options as an array of businesses rather than as an array of geographic markets. The result is predictable, as businesses come under attack from foreign competitors, the company attempts to abandon them and enter others in which the forces of global competition are not yet so strong. In the short term, this may be an appropriate response to waning competitiveness, but there are fewer and fewer businesses in which a domestic-oriented company can find refuge. We seldom hear such companies asking: Can we move into emerging markets overseas ahead of our global rivals and prolong the profitability of this business? Can we counterattack in our global competitors' home markets and slow the pace of their expansion? A senior executive in one successful global company made a telling comment: "We're glad to find a competitor managing by the portfolio concept—we can almost predict how much share we'll have to take away to put the business on the CEO's 'sell list.' "

Companies can also be overcommitted to organizational recipes, such as strategic business units and the decentralization an SBU structure implies. Decentralization is seductive because it places the responsibility for success or failure squarely on the shoulders of line managers. Each business is assumed to have all the resources it needs to execute its strategies successfully, and in this no-excuses environment, it is hard for top management to fail. But desirable as clear lines of responsibility and accountability are, competitive revitalization requires positive value added from top management.

Few companies with a strong SBU orientation have built successful global distribution and brand positions. Investments in a global brand franchise typically transcend the resources and risk propensity of a single business. While some Western companies have had global brand positions for 30 or 40 years or more (Heinz, Siemens, IBM, Ford, and Kodak, for example), it is hard to identify any American or European company that has created a new global brand franchise in the last 10 to 15 years. Yet Japanese companies have created a score or more—NEC, Fujitsu, Panasonic (Matsushita), Toshiba, Sony, Seiko, Epson, Canon, Minolta, and Honda, among them.

General Electric's situation is typical. In many of its businesses, this American giant has been almost unknown in Europe and Asia. GE made no coordinated effort to build a global corporate franchise. Any GE business with international ambitions had to bear the burden of establishing its credibility and credentials in the new market alone. Not surprisingly, some once-strong GE businesses opted out of the difficult task of building a global brand position. In contrast, smaller Korean companies like Samsung, Daewoo, and Lucky Gold Star are busy building global-brand umbrellas that will ease market entry for a whole range of businesses. The underlying principle is simple: economies of scope may be as important as economies of scale in entering global markets. But capturing economies of scope demands interbusiness coordination that only top management can provide.

We believe that inflexible SBU-type organizations have also contributed to the deskilling of some companies. For a single SBU, incapable of sustaining investment in a core competence such as semiconductors, optical media, or combustion engines, the only way to remain competitive is to purchase key components from potential (often Japanese or Korean) competitors. For an SBU defined in product-market terms, competitiveness means offering an end product that is competitive in price and performance. But that gives an SBU manager little incentive to distinguish between external sourcing that achieves "product embodied" competitiveness and internal development that yields deeply embedded organizational competences that can be exploited across multiple businesses. Where upstream component manufacturing activities are seen as cost centers with cost-plus transfer pricing, additional investment in the core activity may seem a less profitable use of capital than investment in downstream activities. To make matters worse, internal accounting data may not reflect the competitive value of retaining control over core competence.

Together a shared global corporate brand franchise and shared core competence act as mortar in many Japanese companies. Lacking this mortar, a company's businesses are truly loose bricks—easily knocked out by global competitors that steadily invest in core competences. Such competitors can co-opt domestically oriented companies into long-term sourcing dependence and capture the economies of scope of global brand investment through interbusiness coordination.

Last in decentralization's list of dangers is the standard of managerial performance typically used in SBU organizations. In many companies, business unit managers are rewarded solely on the basis of their performance against return on investment targets. Unfortunately, that often leads to denominator management because executives soon discover that reductions in investment and head count—the denominator—"improve" the financial ratios by which they are measured more easily than growth in the numerator—revenues. It also fosters a hair-trigger sensitivity to industry downturns that can be very costly. Managers who are quick to reduce investment and dismiss workers find it takes much longer to regain lost skills and catch up on investment when the industry turns upward again. As a result, they lose market share in every business cycle. Particularly in industries where there is fierce competition for the best people and where competitors invest relentlessly, denominator management creates a retrenchment ratchet.

The concept of the general manager as a movable peg reinforces the problem of denominator management. Business schools are guilty here because they have perpetuated the notion that a manager with net present value calculations in one hand and portfolio planning in the other can manage any business anywhere.

In many diversified companies, top management evaluates line managers on numbers alone because no other basis for dialogue exists. Managers move so many times as part of their "career development" that they often do not understand the nuances of the businesses they are managing. At GE, for example, one fast-track manager heading an important new venture had moved across five businesses in five years. His series of quick successes finally came to an end when he confronted a Japanese competitor whose managers had been plodding along in the same business for more than a decade.

Regardless of ability and effort, fast-track managers are unlikely to develop the deep business knowledge they need to discuss technology options, competitors' strategies, and global opportunities substantively. Invariably, therefore, discussions gravitate to "the numbers," while the value added of managers is limited to the financial and planning savvy they carry from job to job. Knowledge of the company's internal planning and accounting systems substitutes for substantive knowledge of the business, making competitive innovation unlikely.

When managers know that their assignments have a two- to three-year time frame, they feel great pressure to create a good track record fast. This pressure often takes one of two forms. Either the manager does not commit to goals whose time line extends beyond his or her expected tenure. Or ambitious goals are adopted and squeezed into an unrealistically short time frame. Aiming to be number one in a business is the essence of strategic intent; but imposing a three- to four-year horizon on the effort simply invites disaster. Acquisitions are made with little attention to the problems of integration. The organization becomes overloaded with initiatives. Collaborative ventures are formed without adequate attention to competitive consequences.

Almost every strategic management theory and nearly every corporate planning system is premised on a strategy hierarchy in which corporate goals guide business unit strategies and business unit strategies guide functional tactics.[5] In this hierarchy, senior management makes strategy and lower levels execute it. The dichotomy between formulation and implementation is familiar and widely accepted. But the strategy hierarchy undermines competitiveness by fostering an elitist view of management that tends to disenfranchise most of the organization. Employees fail to identify with corporate goals or involve themselves deeply in the work of becoming more competitive.

The strategy hierarchy isn't the only explanation for an elitist view of management, of course. The myths that grow up around successful top managers—"Lee Iacocca saved Chrysler," "De Benedetti rescued Olivetti," "John Sculley turned Apple around"—perpetuate it. So does the turbulent business environment. Middle managers buffeted by circumstances that seem to be beyond their control desperately want to believe that top management has all the answers. And top management, in turn, hesitates to admit it does not for fear of demoralizing lower level employees.

The result of all this is often a code of silence in which the full extent of a company's competitiveness problem is not widely shared. We interviewed business unit managers in one company, for example, who were extremely anxious because top management wasn't talking openly about the competitive challenges the company faced. They assumed the lack of communication indicated a lack of awareness on their senior managers' part. But when asked whether they were open with their own employees, these same managers replied that while they could face up to the problems, the people below them could not. Indeed, the only time the work force heard about the company's competitiveness problems was during wage negotiations when problems were used to extract concessions.

Unfortunately, a threat that everyone perceives but no one talks about creates more anxiety than a threat that has been clearly identified and made the focal point for the problem-solving efforts of the entire company. That is one reason honesty and humility on the part of top management may be the first prerequisite of revitalization. Another reason is the need to make participation more than a buzzword.

Programs such as quality circles and total customer service often fall short of expectations because management does not recognize that successful implementation requires more than administrative structures. Difficulties in embedding new capabilities are typically put down to "communication" problems, with the unstated assumption that if only downward communication were more effective—"if only middle management would get the message straight"—the new program would quickly take root. The need for upward communication is often ignored, or assumed to mean nothing more than feedback. In contrast, Japanese companies win, not because they have smarter managers, but because they have developed ways to harness the "wisdom of the anthill." They realize that top managers are a bit like the astronauts who circle the earth in the space shuttle. It may be the astronauts who get all the glory, but everyone knows that the real intelligence behind the mission is located firmly on the ground.

Where strategy formulation is an elitist activity it is also difficult to produce truly creative strategies. For one thing, there are not enough heads and points of view in divisional or corporate planning departments to challenge conventional wisdom. For another, creative strategies seldom emerge from the annual planning ritual. The starting point for next year's strategy is almost always this year's strategy. Improvements are incremental. The company sticks to the segments and territories it knows, even though the real opportunities may be elsewhere. The impetus for Canon's pioneering entry into the personal copier business came from an overseas sales subsidiary—not from planners in Japan.

The goal of the strategy hierarchy remains valid—to ensure consistency up and down the organization. But this consistency is better derived from a clearly articulated strategic intent than from inflexibly applied top-down plans. In the 1990s, the challenge will be to enfranchise employees to invent the means to accomplish ambitious ends.

We seldom found cautious administrators among the top managements of companies that came from behind to challenge incumbents for global leadership. But in studying organizations that had surrendered, we invariably found senior managers who, for whatever reason, lacked the courage to commit their companies to heroic goals—goals that lay beyond the reach of planning and existing resources. The conservative goals they set failed to generate pressure and enthusiasm for competitive innovation or give the organization much useful guidance. Financial targets and vague mission statements just cannot provide the consistent direction that is a prerequisite for winning a global competitive war.

This kind of conservatism is usually blamed on the financial markets. But we believe that in most cases investors' so-called short-term orientation simply reflects their lack of confidence in the ability of senior managers to conceive and deliver stretch goals. The chairman of one company complained bitterly that even after improving return on capital employed to over 40% (by ruthlessly divesting lackluster businesses and downsizing others), the stock market, held the company to an 8:1 price/earnings ratio. Of course the market's message was clear: "We don't trust you. You've shown no ability to achieve profitable growth. Just cut out the slack, manage the denominators, and perhaps you'll be taken over by a company that can use your resources more creatively." Very little in the track record of most large Western companies warrants the confidence of the stock market. Investors aren't hopelessly short-term, they're justifiably skeptical.

We believe that top management's caution reflects a lack of confidence in its own ability to involve the entire organization in revitalization—as opposed to simply raising financial targets. Developing faith in the organization's ability to deliver on tough goals, motivating it to do so, focusing its attention long enough to internalize new capabilities—that is the real challenge for top management. Only by rising to this challenge will senior managers gain the courage they need to commit themselves and their companies to global leadership.

Notes

1. Among the first to apply the concept of strategy to management were H. Igor Ansoff in *Corporate Strategy: An Analytic Approach to Business Policy for Growth and Expansion* (New York: McGraw-Hill, 1965) and Kenneth R. Andrews in *The Concept of Corporate Strategy* (Homewood, Ill.: Dow Jones-Irwin, 1971).
2. Robert A. Burgelman, "A Process Model of Internal Corporate Venturing in the Diversified Major Firm," *Administrative Science Quarterly*, June 1983.
3. For example, see Michael E. Porter, *Competitive Strategy* (New York: Free Press, 1980).
4. Strategic frameworks for resource allocation in diversified companies are summarized in Charles W. Hofer and Dan E. Schendel, *Strategy Formulation Analytical Concepts* (St. Paul, Minn.: West Publishing, 1978).
5. For example, see Peter Lorange and Richard F. Vaneil, *Strategic Planning Systems* (Englewood Cliffs, N.J.: Prentice-Hall, 1977).

Ideas for Discussion

1. What is "strategic intent"?
2. What role(s) do(es) the numerous examples serve in the article?
3. What assumptions do the authors make about their audience? How do you know?

Topics for Writing

1. In an essay discuss why there is no "recipe" for success.
2. Create an abstract for this article.
3. Write a letter requesting Hamel and Prahalad to speak at your university.

4. Write a proposal detailing how to make your company globally competitive.
5. In a letter to upper management, define strategic intent and explain persuasively how attention to it will enhance your company's operations.

Arthur Pound
Pouring Ideas into Tin Cans

Arthur Pound contributed this essay on the Continental Can Company to a 1936 collection of articles on American manufacturing titled Industrial America. *In this piece, Pound profiles the invention of the tin can; how it was mass produced; and what the tin can would mean to the future of the supermarket consumer.*

IF ONE WERE searching for examples of the old and new to illustrate the swift changes of our common life through the past century, he need not look beyond the oldest business in the world—retail merchandising, and particularly the grocery store, around which revolves community life at its simplest. Fifty years ago the food and general supply store kept its staple goods in wooden barrels, with an overflow stock in a dingy back room. Barrels of molasses, vinegar, sugar, flour, pickles, kerosene, crackers, and vegetables stood about on the floor. Over these would bend a clerk probably wearing a soiled apron, with a ladle or scoop in his hand, presently to return with his order in a paper bag or wooden dish over which he placed a piece of wrapping paper. Almost the only packaged goods on the shelves were flavoring extracts, cans of salmon, meat, and condensed milk, together with a few articles in wooden or paper boxes.

Enter a similar store to-day. No barrels in sight. Even that familiar human type, the cracker-barrel philosopher, has gone. More than half the store's merchandise is on shelf display, packaged to be handed out in the twinkling of an eye. The contents have been protected by a container against most of the insanitary accidents of transportation and handling and against adulteration or substitution by unscrupulous dealers. The container is the key to modern merchandising and the basis of its guarantees. Each year sees more packaged goods consumed and more sorts of goods vended in containers and steady progress made toward improvement of both the package and its contents, a trend certain to continue because the public finds in it increasing satisfaction, confidence, and convenience. Try to think of any article which, once packaged, has been returned to bulk handling. In thousands of instances the public has voted with its nickels and dimes for container merchandising.

If an observer could step from the 1880's to the present, he would be struck by the presence in amazing variety and quantity of an object now so common that it is taken for granted—the tin can. The few packaged products found in the old-time stores were for the most part in bottles, cartons, and boxes. Now the tin can dominates the food trade, appears decisively in many other lines, including paints, varnishes, pharmaceuticals, cosmetics, polishes, and so forth, and is being adapted to use on more and more products of common sale. Our old-timer, transplanted from one era to another, would fall to wondering how such quantities of tin cans were made, and especially how they could be manufactured with such unvarying success that perfect protection of their contents could be guaranteed.

The spick-and-span store of the present is possible only because in spick-and-span factories, spaced widely over the country, tin cans are being turned out with almost lightning rapidity from tin plate, which consists of a thin sheet of steel coated with pure tin. The cleanest metal-working operation known to an industrial age must surely be that of a can factory. Let us visit one of Continental's modern can plants. A door is opened and the visitor steps through into a sunny room, the air of which seems full of blinding beauty and deafening clatter. Thousands of brightly shining, eloquently tinkling tin cans chase each other through automatic machines and along overhead conveyor systems. Each contributes its small voice to the rattling clamor; each adds its glinting surface to a kaleidoscopic motion picture. The frame of the pattern holds steady, the lines march unbroken alow and aloft, but each scintillating little soldier dances and sings as it moves, step by tiny step, through this amazing factory.

It is a scene to delight a modernist painter, alert for new combinations of light and shade in flashing motion. Not soon to be forgotten is the orderly beauty of simple, everyday objects, multitudinously on the march and singing as they travel automatically toward their utilitarian destiny. No hand touches them after they pass from slitting machine to storehouse; in fact, no hand has touched the inside of one of them since the tin plate sheets, two tons to a bundle, left the mills.

Comparatively few persons supervise the automatized operation of this plant. This must be "the can maker's paradise." On each of fifteen lines 325 cans may be made every minute of the working day, with 7,000,000 a day a mathematical possibility under continuous operation. Some of Continental's thirty-two can-making plants are larger than this one in volume, some smaller, but all reveal a high efficiency combined with speed. Celerity is the word which from first to last characterizes can making.

In a specialized plant like this one a high degree of automatization is attained. The cans are cut, shaped, soldered, sealed on one end, and tested under pressure by machinery more sensitive to flaws than human eye or hand. The last of the hand processes to yield to machinery was soldering. An expert task in craftsmanship was this one of soldering the outside of the seam of an open cylinder as it swung in front of the operator; occasionally one sees this skilled hand operation in a "general-line" plant, when a wide variety of orders, some of which are too small in quantity to justify the high initial cost of extreme automatization, has to be filled. Whereas a specialized plant may make only one or two shapes, the general-line plant often has a hundred shapes and sizes in process at once, ranging from 110-pound coffee drums and 50-pound lard cans to tiny sample containers.

Greatest volume is reached in manufacturing cans for packers of foodstuffs, greatest variety in general-line cans for other goods. Continental's thirty-two can plants may be classified as follows: twelve packers' can plants; fifteen general-line plants; and five combination plants.

The can opener in the kitchen is still the foundation of the business in spite of growth in other lines. Packers' cans are made in or near railroad centres from which transportation lines radiate to important food-growing areas. Both for canners and for can makers, even slight freight-rate advantages are important; as a result, the industrial map of canning operations and can making is dictated largely by the geography and timing of plant growth. Food packing and can making tend toward locations near the growing or producing areas suitable for quantity packs. The largest of these packs consists of evaporated and condensed milk; pineapples, pears, peaches, apricots, cherries, and grapefruit among the fruits; corn, peas, tomatoes, and beans in the vegetable world; soups in all varieties; and salmon, the most popular product of our fisheries. The average reader may be surprised to learn that although Maine, the original corn-canning state, still holds a proud place, the large-volume packs come from Iowa, Indiana, Ohio, Maryland, Minnesota, and Illinois; that California's peaches, better eating after canning than before, are used almost exclusively for commercial canning—very limited quantities of the celebrated Michigan and Georgia peaches being available in the package-goods market; that the largest quantities of tomatoes come from widely separated areas—California, Missouri, Arkansas, Indiana, Ohio, Kentucky, Tennessee, and the Atlantic seaboard area of southern New Jersey and the Delaware peninsula, composed of Delaware and the eastern portions of Maryland and Virginia. Wisconsin and New York are the largest producers of peas for canning.

This picture changes with weather and climate variations, discovery of new insecticides, development of fertilizers, opening of new lands to packable crops, improvement of seeds, and acceptance by growers of changes from traditional to scientific methods of agriculture. In the canning industry the plants are usually located in the small villages or towns close to a favored crop area. The can maker tends to follow the packer, spacing factories widely instead of concentrating operations. Both are answering the call of the land and the dictation of seasonal swings beyond the control of man.

Neither the canner nor the can maker, however, accepts Nature fully or blindly. The former stirs his contract farmers to improve yields in quantity and quality. His agricultural experts advise farmers and market gardeners on crop rotation, tillage, drainage, spraying, and seed selection. Many canners grow and distribute seed to growers, conduct long-range experiments in plant variations, inspect fields of ripening crops, and fix the day and hour when the crop should be gathered to ensure first quality. The old belief that only cull crops go into cans has become a myth; the fact is that the yields of the best acres go with the least possible delay to the plant and into tin cans. It could hardly be otherwise, with processors guaranteeing their products and profits depending on high-speed, continuous runs through short periods, conditions that make the best product infallibly the cheapest.

Another element reënforcing geographical spread is the stark need for prompt, almost instant, service. The food packer cannot wait; it is now or never with him. If his delivered crop exceeds his estimates, he must get containers by rush order. His commercial life hangs on the outcome of his shout for cans. In the case of general-line plants the need for supplies on the instant is not so acute. Nevertheless, canners want containers when they want them—*pronto* and no fooling! As a result, Continental Can Company, Inc., has packers' can plants in ten states, general-line plants in ten states, and combination plants in four states.

Organized with the modest capital of $500,000 in the fall of 1904, Continental acquired the patents and good will of United Machinery Company of Rochester, New York, which had developed a line of improved can-making machinery. With these machines it equipped plants at Syracuse and Chicago, natural centres for can manufacturing because at these points radiating railways tap rich agricultural sections. Shortly thereafter the Company established its third plant at Baltimore. In 1909, seeking assured supply of raw materials, the Company bought the Standard Tinplate Company at Canonsburg, Pennsylvania. In these five years occurred a technical change of great importance, the shift from the soldered end to the sanitary or double-seamed end now in general use. Under the leadership of its founder and first President, T. G. Cranwell, Continental quickly reëquipped its plants to meet the new situation. Later growth to present proportions has been partly through internal expansion and partly through merger.

Thus far Continental operations had been confined to packers' cans, a highly seasonal business with acute demand from April to October, followed by a comparatively slack season of almost six months. General-line business, on the other hand, is steadier; and corporate stability could hardly be gained without it. So Continental entered the general-line field in 1912 at Chicago.

In 1917, with canned goods in demand by the army, there arose the plant at Clearing, Illinois, near Chicago, now the Company's largest operation. Another development of that year was doubling the factory facilities at Syracuse, New York, for the manufacture of can-making and can-closing machinery, the latter designed for installation in packing plants for closing the cans after filling. This meant that Continental could offer its customers machines progressively adapted to every improvement in can design and built to meet particular needs. It became possible to increase the guarantee ratio, now surprisingly high. Continental guarantees 998 out of every 1000 packers' cans it delivers to fruit and vegetable canners.

The war induced a marked change in the food habits of the people, the net effect of which benefited the canning industry. After hostilities ceased, the letdown in this industry was less marked than in some others. The public had become educated to the use and convenience of canned foods, with the result that, with the return of peace, these foods continued in popular favor.

In contrast to the highly standardized production in packers' cans, general-line can making develops many variations. The desirable container in this field is one which will help sell the merchandise it holds, and its success depends not only on economy and safety, but also on ease of use, better display value, and all-round attractiveness. A chief consideration is to develop cans which catch and please the eye, while meeting adequately the practical needs established by the nature of the product to be sold.

The Development Department of Continental studies customer container problems from two angles. Does the present container provide ideal protection? Does it help to sell the merchandise? Failure in either respect means that Continental suffers with its customer, while success benefits all parties. Merely changing the shape of a container from flat to upright may double sales, because the new article is no longer hidden on the merchant's shelves. Adding an easy pouring spout to a tooth-powder container may help a manufacturer over his marketing barriers. In some instances the Development Department conducts surveys of retail conditions as they affect a certain product, takes the results to a manufacturer, and with his staff works out a new approach to the market.

Recently Continental's experts, in coöperation with the Dairy Division of the Department of Agriculture, undertook experiments looking toward a new method of marketing Cheddar cheese, as one way of relieving overproduction of milk. Cheddar has suffered in trade because the usual large, round units develop waste and deteriorate through rapid drying and rind formation. The quality which gives Cheddar its snappy taste, a certain gassiness, rendered ordinary containers useless. By the invention of a valve-vented can, an entirely new departure in can making, the problem has been solved. In the top of the can is a vent, so arranged that it releases gas at a low pressure without admitting air. This unique package of cheese will soon be on the market, the six months' aging process being now under way. A new delicacy becomes available in a package which ensures dependable freshness without waste and makes possible a complete guarantee of quality and trade identification. Other gas-forming products may eventually be marketed in this unique valved container.

Following the trend toward greater demand for tooth-powder containers, Continental designed a new can with an improved dispenser, making it an outstanding package in this field. The trial order for the above package was for only 50,000 containers, since when the sale has run into the millions. Many similar requests flow into the Development Department from buyers, with the result that this division must be alert to the trends in modern packaging. In the ever-changing picture of general-line can manufacture, a novel departure which catches on means substantial and steady gains.

One of Continental's new ideas, so new that it is not yet in assured production, is the window-top can, long a will-o'-the-wisp arousing keen pursuit but only lately brought into the realm of practical attainment. In the glass-top can Continental unites the protective qualities of the tin can with the visibility afforded by the glass top, producing a container which, while judged too expensive for universal use, is effective as an inspection sample. Through this method of display, sales of canned foods in certain test campaigns have been stimulated 300 to 400 per cent.

An innovation which rose to impressive volume within two years is canned motor oil. Progressive refiners realized that there was a vast market of buyers unable to distinguish between various grades or brands of oil. Taught by experience and advertising to ask for their favorite oils, they could not be sure of getting what they wanted. The fruits of extra care in refining, painstaking oil research, and strenuous advertising remained in jeopardy unless lubricants went to the customer in sealed and tamper-proof containers, the contents being drained into the crankcase under the eye of the purchaser.

In this conclusion not all refiners agreed, but it did not take long to convert most of the skeptics. The first canned oil came to the market early in 1933, yet within two years over 200 oil companies were packaging a sizable proportion of their motor lubricants. More than 65 per cent of oil dealer stations now handle canned oil. In 1934, the second year, authoritative estimates indicate that approximately 350,000,000 quarts of motor oil were sold in cans, about one fifth of total production.

In cans the motorist gets the oil he wants and pays for—in grade, measure, and brand. Dealers find these advantages: fewer complaints, cleaner and easier handling, and a reduced investment in stock as between a 55-gallon drum and a case of 24 quart tins, or a wider stock of popular brands on the same capital. Lithographed cans, bright with color, offer the first chance to display oil attractively. Canned oil has sharpened competition by enabling the refiner to place his goods in thousands of new outlets, but at the same time has lifted the motor-oil trade to a higher plane.

A primary reason why canned oil caught on so quickly is this: Continental took its advantages directly to the public through national advertising. Converted to the new idea, motorists came forward to demand canned oil, and the innovation became an assured success almost overnight.

National advertising by the can maker to the ultimate consumer was itself a novelty. Hitherto, can makers had relied almost entirely on trade papers, concentrating on potential can buyers and overlooking ultimate consumers, leaving the latter to be cultivated by the National Canners Association (composed of food canners) and individual packers. Continental joined forces fully with the canners in these joint efforts to educate housewives to the advantages and merits of the more than three hundred kinds of foods and delicacies available at all seasons in tin cans and to break down the old and unfounded prejudices against their use. While continuing its coöperation with the Association, Continental undertook to carry this work of education further on its own account. Advertising texts and illustrations emphasized the point that health and freshness are sealed in cans which preserve indefinitely the food values of fruits and vegetables canned within a few hours after leaving the field. During summer months Continental called public attention to the fact that, while Nature was at its best, products of farms, gardens, and orchards otherwise unavailable were being canned for winter use.

Acquainting the public with Continental as a service organization, helpfully joining its own interests with those of the many great industries which it serves, is another objective of Continental advertising. As business develops in size and complexity, producers of hitherto unregarded elements in an assembled product are more and more seeking a direct approach to a public mind appreciative of value, cleanliness, and design, and increasingly interested in the practical aspects of industrial coöperation. All these objectives have been realized; canned oil has already become a standard product, housewives reacted favorably to the messages on the uses and merits of canned foods, and the public understands better than it did the part which the tin container plays in the nation's mercantile and industrial processes.

In its consumer advertising, as in other departures from standard practice, the controlling thought is benefit to Continental's customers, as Continental's progress goes hand in hand with that of its patrons. Alive to the wisdom of change, and alert in research and developing new products, in machine improvements, in anticipating customer requirements, and in advertising, Continental drives ahead with vigor in its efforts to stimulate the broad array of American industries which use its products in ever-growing quantities.

Unless all signs fail, the ultramodern store of to-day will seem hopelessly antiquated fifty years hence, as out-of-date as an old-fashioned country general store is to-day. But one thing is certain: the store of the future will hold containers in ever greater variety than at present, and the tin can will without doubt still be an important factor in the commercial and domestic life of America.

Ideas for Discussion

1. Chart the development of the tin can's manufacture.
2. Comment on the essay's tone and style. Who is this essay intended for? Do you detect any bias on the author's part?

Topics for Writing

1. Research the role of government support for new industry in the early twentieth century.

2. Rewrite this essay with an emphasis on strategic intent not the narration of an invention.
3. In an essay compare the history of the tin can's development with that of Ford's car and McCormick's reaper.
4. Prepare a press release for this new product, "the tin can."
5. Write a marketing proposal for the tin can.

Stuart Blumin

Black Coats to White Collars: Economic Change, Nonmanual Work, and the Social Structure of Industrializing America

Stuart Blumin earned his Ph.D. from The University of Pennsylvania in 1968. He has been a professor of history at Skidmore College, Massachusetts Institute of Technology, and Cornell University, since 1977. His research interests include American social history and the U.S. city. Blumin has written several articles on nineteenth-century America, and his contributions include "Mobility in the Nineteenth-Century City" (1973); and "Rip Van Winkle's Grandchildren: Family and Household in the Hudson Valley 1800–1860" (1975). This selection first appeared in Stuart Bruchey's Small Business in American Life *(Columbia University Press, 1980).*

IT IS INCREASINGLY evident that the white collar–blue collar dichotomy that has served so long as the primary symbolic distillation of this culture's concept of its class structure (and particularly that point in the structure where the "middle class" is distinguished from the "working class") is today breaking down under the force of economic change. Several decades of successful wage negotiations by large and fully institutional labor unions, combined with the emergence of a huge tertiary sector of clerical and service workers, have produced a broad belt of middle income earners who confound the old categories by weakening the connection between type of work and income. Today it will no longer do to assume that a person who wears a "white collar" at his nonmanual desk job enjoys a higher income, better life-style, more respect within his community, more rewarding work experience, or greater opportunity for advancement than does his "blue collar" counterpart on the assembly line or behind the wheel of a television repair truck. To be sure, nearly all of those who are above or below this broad belt of moderate incomes conform quite well to the old categories. But within it there are too many, ever-more-numerous exceptions. "Postindustrial society," whatever that may come to be, will require its own shorthand vocabulary to describe class.

But if the "industrial" vocabulary and its attendant perceptions and values are destined to have an end, we may presume that they also had a beginning (a prehistory) and even an evolving life (a history). Clearly, the industrial age dawned in a world that could have made little use of white collar–blue collar dichotomy, even in the little urban economies that were mere islands in a sea of rural life and work, and even if we translate our anachronistic sartorial imagery into that of the black coat of the clerical worker and the leather apron of the urban artisan. For in the immediate preindustrial era (as, increasingly, in our own day), the difference between manual and nonmanual work did not correspond even remotely to that critical disjuncture between the "middling sorts" and "meaner sorts" within the urban class structure. The figure most responsible for this disjuncture of disjunctures was, of course, the master craftsman, the artisan-shopkeeper, who was both the producer and the retailer (and sometimes the wholesaler) of his goods, and a solid member of the city's middle orders. As long as all, or even most, of the consumer and capital goods made in the city were produced in the small shops of the independent artisan, by the master himself working alongside his journeymen and apprentices, and as long as the master's citizenship was that of the city or nation as a whole and not just its "producing classes," there could be no social stigma attached to the leather apron—certainly not, at least, by those who wore the black coat.

Industrialization, as we know very well, changed all this, first by eroding the independent craftsman's economic position, and then by eliminating him entirely from most lines of production. And as production was transferred from the small shop to the large shop, the "manufactory," and ultimately the mechanized factory, the distinctions between nonmanual and manual work, and between middle-class and working-class status, came increasingly into alignment. Those who "worked with their hands" were socially demoted into, or confirmed within, the "working class." Those who "worked with their heads" were promoted into, or confirmed within, the "middle class." Eventually, the evolution of a salaried, office-bound component of the latter group would bring to life the symbolic shorthand of the industrial era—"blue collar" and "white collar."

If this sketch seems too simple and too abstract, it is probably because it joins two lines of historical inquiry that have too seldom converged—the evolution of status distinctions in industrializing societies, and the *specific* economic events that underlay this evolving concept of social position and worth. What it suggests, therefore, is the need for a social and cultural history of industrializing societies (focused, in the current instance, on changing concepts of social class) that pays far closer attention to the relevant details of economic history. What specific structural, or merely quantitative, changes in economic life (in economic *relationships*) provoked men and women into altering their view of the social order? When did these changes occur? How should they affect our understanding of both the character and the timing of the effects of industrialization on society?

I do not mean to suggest that nothing of this sort has even been attempted. On the contrary, historians of the working class in America have been proceeding in something of this fashion ever since John R. Commons (beleaguered prophet!) provided the model in his long article on American shoemakers in 1909—though one might have hoped for more progress in industries other than boots and shoes since that date. But historians of the working class, even at their best, can be expected to do only part of the job. What about the experiences of those who were not in the working class? Retailers, wholesalers, brokers, bankers, managers, clerks, salesmen, and many others were affected by the economic changes of the industrial era; and these changes in turn affected both their position within, and their concept of, the class structure. Yet we have no history of the nonmanual work force (no *Making of the American Middle Class*) that we may place alongside those histories that follow the laborers, journeymen, and degraded master craftsmen into the factories and the new working class. Until we do, even a fully developed history of the industrial working class will tell us only half of the story of the changing economic foundations of social class in the industrial era.

Actually, the contribution of the labor historians is more than "half," for the history of the nonmanual work force in the nineteenth century is in many respects the reciprocal of the history of the emerging working class, and the historian of the middle class must call upon many of the same events already detailed by the historians of labor—if only to look at "the other side." Thus, the increasing association of nonmanual work with middle-class status may be said to be a product of the social degradation of manual work that accompanied the transfer of production from the artisan's shop to the factory. The nonmanual middle-class was formed, according to this line of reasoning, by the evaporation of the manual component—the gradual disappearance of the artisan-entrepreneur from the economy and class structure. And since the labor historian has already described this phenomenon from the artisan's point of view, the middle-class historian is left with what appears to be the simple and humble task of deducing its consequences to those who were *not* demoted to the working class.

On the other hand, it is not clear that the middle class was entirely "made" by "the making of the working class," or that the middle-class historian's role is necessarily a residual one. Events occurring within the world of nonmanual work—events parallel to, but different from, those already catalogued by the labor historians—may have helped shape and give consciousness to the emerging "white collar" middle class. What might these events have been? How shall we write their history? These are difficult questions, made still more difficult by the historical neglect of the small-scale, nonmanual entrepreneur and worker. As Arno Mayer has recently pointed out, the lower middle class (to use his phrase), unlike either the working class or the large-scale merchants and industrialists, has never been an appealing subject to historians. And in the American setting, at least, it has been particularly unappealing from just that point in history when it lost its artisan component! Thus, there are studies of colonial craftsmen as middle-class entrepreneurs and even as middle-class revolutionaries, but there is no analogous study of middle-class entrepreneurialism or political activity for the retail shopkeepers, dealers, agents, and salesmen of the industrial era. Studies of commercial life in both eras do exist, but they are confined to the large-scale merchants, whose wealth and standing placed them among the urban elite—or at least at the upper levels of the middle class—and are quite unrevealing about the nature of work and life in the smaller shops. Clearly the kind of analysis that I am proposing—one that isolates those economic changes in nonmanual work that contributed to the development of the white collar–blue collar dichotomy—must be built largely on its own foundation of ideas, evidence, and actual research. In this article, I shall attempt to lay a part of that foundation by proposing several ideas, discussing a body of evidence that seems especially pertinent to the subject, and analyzing selections from that evidence in a way that I hope will suggest what might be accomplished by further research on these particularly anonymous Americans.

I shall begin in the most basic way possible—with numbers. In some dichotomous distillations of the social structure—those which focus on the inequalities of wealth and power between such groups as lords and peasants, or capitalists and industrial workers—the size of the two classes is significant, because it emphasizes the inequity of the domination of a very large class by a very small one. It is the imbalance that is important, and it may be said that the dominant group is visible as a class, not because it is large, but precisely because it is small. The white collar–blue collar dichotomy that has generally prevailed in modern American perceptions, however, is quite different in this respect. The dichotomy itself, though hierarchical, does not purport to explain the mechanisms of domination and exploitation in society: indeed, it implicitly denies their importance by leaving room for the existence of elites within, or even beyond, the white-collar class, by simultaneously denying the fundamental importance of these elites, and by substituting a large, amorphous (if uniformly attired) middle class for Marx's capitalists in the historical *pas de deux* with the working class. Where Marxism in effect posits a small upper class and a large proletariat, the white collar–blue collar dichotomy allows for a small but essentially insignificant upper class, and posits large middle and lower classes, between which there lies a rather more narrow chasm of power and wealth. The point that needs emphasis here is the *size* of the class on the upper side of the basic dichotomy. Not accorded any inherent structural significance, the white-collar middle class acquires visibility and importance by growing in size, to a point where sheer numbers allow it to be placed in apposition to the blue-collar working class.

How large, then, was the nonmanual work force at various points in the industrial era? Did the growth of this work force, or particular components of it, help establish in people's minds the significance of the apposition of nonmanual and manual work? Certainly, the growth of the salaried, office-working component in the latter-nineteenth and early-twentieth centuries was crucial, as is revealed by the very imagery of the dichotomy as it finally evolved. But what of earlier changes that may have laid the foundation by suggesting the significance of the difference between working with one's "head" or one's "hands," well before the collective eye was redirected to one's shirt collar?

In the aggregate, the nonmanual work force seems to have changed very little in the decades before the proliferation of salaried office workers. My own study of the Philadelphia work force, for example, indicates that the proportion of merchants, storekeepers, professionals, clerks, and others who "worked with their heads" rose slightly between 1820 and 1860, from about 27 percent to just under 30 percent, while Thernstrom's study of Boston places the nonmanual work force at 32 percent in 1880 and 38 percent in 1900 (all of these figures refer only to males). What these calculations suggest is a steady but very gradual growth of the nonmanual sector of the urban work force through the nineteenth century—a trend so gradual that it was probably not even noticed by those who lived through it, much less perceived as a change of consequence to the class structure. But these calculations of aggregate change may be misleading. The more rapid growth of certain segments of the nonmanual sector during the early phases of industrialization may have provided the foundation for just the sort of alteration of perspective as occurred later, during the "white collar" era. In particular, the expansion of *retailing* as a distinct activity of distinct persons and firms, and the proliferation of *services* not before offered within the urban economy, would seem to have had a large potential impact on the perceptions of the city dweller. It is clear that both the retailing of consumer goods and the provision of such services as banking, insurance, local and nonlocal transportation, advertising, organized entertainment, and formal instruction of various kinds, changed rapidly in character during the early industrial era. How were these changes related to changing perceptions of the class structure?

Let us look first at retailing in the preindustrial American city. According to most accounts, there was but a small "retail sector" in the American towns of the eighteenth century. Most goods were sold to the public by the artisans who made them, by the farmers who grew them and carted them to town, and by the importers who imported them, sold them wholesale to country merchants, and only incidentally retailed them to individuals. Other goods were sold not in stores but by peddlers who carted or carried them through the streets. Such retail stores as did exist, moreover, were often very small, and yielded small incomes to their proprietors, many of whom were spinsters, widows, and seafarers' wives—the Aunt Hepzibahs of the colonial era. By the middle of the nineteenth century, this picture had altered greatly. American-made goods had largely supplanted the imports that had given the large-scale merchants and their cluttered wharves such an important role in the urban economies of the colonial period, and the men and women who made these goods engaged far less often in their sale to the public. Retailers—of hats and shoes, dry goods and hardware, pianos and paintings—filled the vacuum. Farmers' markets continued to exist, as they do in our own day, but were gradually losing custom to the full-time grocers. Perhaps, too, storekeepers of various kinds were forcing the peddlers into such lines as oysters and pepper-pot soup, and the collecting of rags and bones. Not all of these changes would have produced an increase in the overall proportions of nonmanual workers in the urban work force. The separation of sales from production may have increased the numbers of retail stores, but the numbers of storekeepers and sales clerks who worked in these stores were easily offset by the enlarged numbers of manual wage earners in the growing industrial sector. Nor did the transfer of retail activity from the importers to nonimporting storekeepers require a massive increase in nonmanual personnel. Yet, both changes were highly significant, for they enlarged that portion of the nonmanual work force— the retail storekeepers and their store clerks—who begged comparison, as the importers never did, with the increasingly dependent (i.e., wage-earning) working class. The retailers were becoming what the skilled workers once were—small-scale, independent businessmen. And the increasing *numbers* of these small businessmen, who sold goods they neither manufactured nor imported, made the comparison with the dependent manual workers that much more compelling.

The growth of the service sector contributed to the manual-nonmanual dichotomy too, though its major impact was reserved for a much later period. The number of nonmanual jobs in the "tertiary sector" of the nineteenth-century urban economy was still small, but it was larger than it had been in the eighteenth century. Institutional innovations introduced bankers, insurance salesmen, and credit investigators into the urban economy, while the general maturation of the city provided new opportunities for lawyers, real-estate agents, employment bureaus, Italian dancing teachers, and others outside the flow of tangible commerce. To the more radical workers and intellectuals, these men, even more than the industrialists and merchants, were the "accumulators" who produced nothing, but gained comfortable livings from other people's labor. To those of a more conservative hue, they were simply additions to the nonmanual middle class. From either perspective, the increase in the numbers of such nonmanual positions in the service sector strengthened, however modestly, the emerging manual-nonmanual dichotomy.

But how shall we examine and specify these quantitative shifts in the nonmanual work force through the course of the industrial era? The question is not as easily answered as might first appear, largely because the city directories and census manuscripts that constitute the basic sources for historical studies of the occupational structure are usually insufficiently precise, particularly in the early stages of industrialization. Seldom do they distinguish between owners of large and small enterprises, or between wholesalers and retailers (or artisans and manufacturers), and they often fail even to distinguish between employers and employees. The opportunities for combining these sources with others that help resolve these ambiguities, as Clyde and Sally Griffen have recently done with such effectiveness in Poughkeepsie, seem confined to the period after 1850. How shall we contend with the crucial period that precedes that date? One possibility is to treat the business advertisements appearing in a particular city's newspaper as a kind of business directory for that city, a method that has the advantage of being applicable to a period stretching as far back as the mid-eighteenth century for several American cities. The disadvantages and pitfalls inherent in this method, however, are all too obvious. First, the advertisements are not a complete directory, or a representative one, and the degree and direction of bias in their inclusiveness of a city's economic units may well have changed over time. Second, the advertisements, though usually far more informative than the simple labels one finds in an occupational list, often to not reveal enough about the character of the businesses that placed them—whether they are large or small, wholesalers or retailers, or even whether they are regular businesses or just private parties placing what we would now call a classified ad. For some purposes, as I shall argue shortly, the newspaper ads constitute a valuable source, but for calculating the changing proportions of different kinds of businesses in the urban economy (much less different kinds of workers) they are suspect at best, and are probably worthless. There may, in fact, be no way of reliably reconstructing the urban economies of the early industrial era in a way that will reveal the timing of the emergence and complete development of the retail and service sectors. (The New York state census tabulates 9,617 retail stores in New York County in 1855. How many stores were there in New York in 1800?) But there should be little doubt that such a phenomenon accompanied, and gave further significance to, the more familiar events that shaped the careers of the industrial workers.

Perhaps, though, this point is better expressed not in terms of numbers, but in terms of the changing character of the nonmanual sector. As I have already noted, most retail goods were sold in the colonial city by men whose main activity lay in their production or importation, by other men—the peddlers—whose modest scale of operations placed them below the artisans in the social and economic scale, and by women, who found in shopkeeping a substitute for, or supplement to, a husband's wage. Keeping a retail store was not one of the ordinary ambitions of young men of middling rank, and the term "merchant" was reserved for the large-scale import-exporters. By the mid-nineteenth century, retailing had become respectable, and many men were able to fulfill in the retail sector ambitions that formerly would have been fulfilled only in the crafts or in some form of extra-local commerce. Undoubtedly, this enhancement of the respectability of retail trade was underpinned by a significant increase in the incomes that were earned in the retail sector. There were a few spectacular examples of this, such as A. T. Stewart of New York, who opened a dry goods store in 1823 and went on to become one of the wealthiest men in America, and in every city there were less spectacular but still very visible examples of retail merchants (as they could now be called) whose stores brought them not merely comfort but prosperity. The changing fortunes of more ordinary retailers, however, are difficult, if not impossible, to determine. As is the case with the changing size of the retail sector, the evidence is as elusive as the argument is compelling.

Fortunately, one aspect of retailing in the industrializing American city—one that bears on the changing character of retail institutions, and the work performed in them, if not on the incomes earned in selling—may be examined with more confidence, and it is here that the greatest opportunities for research in this area probably lie. I refer to a specialization and regularization of retail activity that may be regarded, in my view, as a form of institutional development analogous to the new forms of production that were transforming the industrial sector. During the colonial era, product specialization—and the close knowledge of particular products that specialization implies—belonged mainly to the artisans. Commerce, whether wholesale or retail, was a highly diversified and episodic affair, in which individual firms offered for sale whenever goods happened to be arriving in the town from foreign parts, the coastal trade, or the back country. For the larger firms in the larger towns, and for most of the firms in smaller towns, it was also a two-way affair, involving the purchase of hinterland produce at the same time as, and often in exchange for, imported goods. The evolution of a far different system, in which the sellers of goods offered specialized product lines from regularly replenished inventories, while leaving the purchase of hinterland produce to other specialized dealers, was a phenomenon of the early industrial era. And it was a phenomenon that may be traced with some reliability through the very newspaper advertisements that proved unreliable in estimating the changing sizes of the various sectors of the local economy. The newspapers still provide a biased "sample" of local firms, of course, but for the purpose of inferring structural changes in commercial activity (as opposed to inferring the changing numbers of different types of businesses) the bias is probably quite constant. In all eras, it is the more prosperous, more fully rationalized, and more cosmopolitan firms of each type (wholesaler, retailer, and so on) that we expect to find making regular use of existing channels of advertising. For this segment of the urban economy at least, therefore, we can reliably compare the merchandising practices of commercial firms of the preindustrial and industrial eras, to the extent that these practices are revealed in the newspaper ads.

To accomplish such a comparison, I selected four cities of differing regions, sizes, and character, whose surviving newspapers stretch back to at least the late eighteenth century. I then selected eight relatively normal economic years spanning the preindustrial and early industrial era, and specified which newspaper from each city and each year was to be used for the analysis. A coding scheme for products and certain other aspects of each ad was developed, and, according to this scheme, information was recorded from every advertisement appearing in four issues (one issue each from January, April, July, and October) of each specified newspaper in each year. The cities are Philadelphia, New York, Charleston, and Hartford, and the years are 1772, 1792, 1805, 1815, 1825, 1835, 1845, and 1855. The product codes and the list of newspapers are detailed in the appendixes to this essay. The newspapers utilized were those available at Olin Library, Cornell University, and because of the incompleteness of this collection, there are a number of gaps, but for each city it was possible to construct a time series covering much of the period.

The most interesting and telling results from this analysis are those pertaining to the specificity of the products offered for sale in the four cities' commercial firms. In each city there is a remarkably clear pattern, visible at several levels, of increasing product specialization. Table 1 divides the retailers and wholesalers of each city into four categories representing various levels of specialization: (1) general merchants, who offered goods from three or more of the product categories defined in the coding scheme; (2) slightly less general merchants, who offered goods from two different categories (but not including those who clearly specialized in one type of product while tacking on a few articles of another type); (3) merchants offering a variety of goods within a single product category; and (4) highly specialized merchants, offering for sale a restricted and specific product line. In all four cities, the proportion of advertisers in the first category declined dramatically over the period, with the exception that in Charleston it leveled off at 10 to 15 percent after declining from 38 percent in 1772. In Philadelphia, where the pattern is most clear, the general merchants declined continuously, from 65 percent in 1772, to 12 percent in 1805, and to a mere 1 percent in 1855. Clearly, the general merchant, that central figure of the colonial urban economy, was on the way out. Nor was he replaced by those at the next level of specialization. Advertisers offering goods from two product categories also declined between 1772 and 1855, although the decline was more gradual and seems to have begun somewhat later in the period. In all probability, this second category of table 1 represents a transitional stage of specialization and includes a number of once-general merchants who were in the process of shedding certain product lines while gradually coming to focus on others. Ultimately, these firms too were shaken out of the urban economy—in the three seaport cities, they constituted only 2 percent of the commercial advertisers in 1855.

The third level of specialization in table 1—firms offering two or more types of products within a single product category—reveals no consistent pattern of change. In Philadelphia, this type of firm increased relative to other advertisers between 1772 and 1792, and then maintained a fairly stable proportion through the rest of the period. In New York (where gaps in the data make the timing of change most difficult to observe) there was a significant decline between 1772 and 1855. In Charleston the proportion remained fairly stable through 1835, then rose in 1845, then fell drastically

Table 1
Product Specialization, All Commercial Advertisers[a]

Philadelphia	1772	1792	1805	1815	1845	1855
General merchants	64.9%	28.6%	11.7%	9.0%	2.1%	1.3%
Merchants offering goods in two product categories	10.8	9.5	9.1	10.3	3.6	1.7
Merchants offering a variety of goods in one product category	10.8	24.9	19.5	20.5	21.6	18.2
Specialized merchants	13.5	37.0	59.8	60.3	72.7	78.8
Total[b]	100.0%	100.0%	100.0%	100.0%	100.0%	100.0%
n	37	189	77	78	473	527
New York						
General merchants	23.5%				0.0%	0.4%
Merchants offering goods in two product categories	12.6				7.8	2.1
Merchants offering a variety of goods in one product category	39.5				12.6	10.4
Specialized merchants	24.3				79.5	87.1
Total[b]	99.9%				99.9%	100.0%
n	119				301	241
Charleston	1772		1825	1835	1845	1855
General merchants	37.6%		15.2%	10.6%	13.3%	15.2%
Merchants offering goods in two product categories	6.5		6.1	8.3	3.5	2.3
Merchants offering a variety of goods in one product category	35.5		30.3	28.0	39.8	16.7
Specialized merchants	20.5		48.5	53.1	43.4	65.9
Total[b]	100.1%		100.1%	100.0%	100.0%	100.1%
n	93		66	132	113	132
Hartford	1792	1805	1815	1825	1835	1845
General merchants	46.9%	49.0%	25.5%	13.8%	8.1%	5.5%
Merchants offering goods in two product categories	18.8	14.0	7.3	11.3	11.5	13.1
Merchants offering a variety of goods in one product category	18.8	29.0	30.9	34.4	27.7	27.4
Specialized merchants	15.5	8.0	36.4	40.4	52.8	54.0
Total[b]	100.0%	100.0%	100.1%	99.9%	100.1%	100.0%
n	32	100	110	247	235	274

[a] Excludes those deemed to be artisans, factories, providers of services, and private parties.
[b] Rounded off.

in 1855. In Hartford, it rose between 1792 and 1805, and then remained stable thereafter. Only at the fourth and final level of specialization, therefore, do we find a pattern reciprocal to that of the general merchants. In Philadelphia, these most specialized of merchants—hat stores, shoe stores, piano stores, and even pet shops—increased from 14 percent of all commercial advertisers in 1772 to 79 percent in 1855. In New York, they increased from 24 percent to 87 percent, in Charleston from 21 percent to 66 percent, and in Hartford from 16 percent to 54 percent. In all four cities, in short, there occurred a dramatic reversal in the relative proportions of the general and specialized merchants among commercial advertisers, with the latter becoming the dominant figure during the early industrial era. And as table 2 reveals, the emergence of the specialist was not restricted to one or two lines of goods. In each city, specialized sellers of dry goods, clothing, home furnishings, and many other consumer goods took up increasing amounts of newspaper advertising space, and so did firms selling particular types of equipment, construction materials, and other capital goods. The increasing specialization of wholesale and retail sales was a general and pervasive fact in the economies of the industrializing city.

Specialization took other forms, moreover, than the increasing focus on particular goods. For example, firms that formerly engaged in a two-way exchange of goods, and that thereby served two often radically different markets, gave way to firms that facilitated the flow of goods in only one direction, and therefore may be said to have specialized in a particular market. More tangibly, storekeepers who sold dry goods or hardware did not also become grain or cattle dealers as a secondary pursuit, because they dealt increasingly with cash or credit customers and not with exchanges of consumer goods for country produce. Requiring no knowledge of how to move grain or livestock to mills, slaughterhouses, or foreign ports, they could focus their attention on the market for dry goods or hardware, and become specialists in that sense as well.

Table 3 attempts to assess the magnitude of this change by recording the proportion of each city's commercial advertisements that specifically mention the advertisers' willingness to purchase or exchange goods. This method, I contend, greatly understates the increasing specialization in a single market, for many firms in the earlier period (but few, I believe, in the later period) engaged in two-way commerce without specifically mentioning that fact in their advertisements. Still, a pattern is evident in all the cities, and is quite pronounced in the inland city of Hartford, where a large number of advertisers did specify two-way exchanges in the earlier years. In Philadelphia, New York, and Charleston, the decline is from a small proportion in 1772 to a negligible proportion or nil in 1855. In Hartford, the decline (and hence increasing specialization) is from 34 percent in 1792 to only 4 percent in 1845.

The newspaper advertisements include at least one other datum suggestive of a significant change in the character of urban commerce in the early industrial era. In the colonial period, many commercial ads prominently mention the arrival of a specific shipment. Many, indeed, are in effect *announcements* of the arrival of a particular ship, with particular goods offered for sale. During subsequent decades, however, advertisements of the form "Just Arrived on the Sloop Elizabeth" gave way to those which make no mention of particular shipments, and which imply the regular replenishment of inventories by means that are of no intrinsic interest to the consumer. What this in turn implies is a more general regularization of

Table 2
Product Categories of Specialized Merchants (percent of all commercial advertisers)

Philadelphia	1772	1792	1805	1815	1845	1855
Dry goods and clothing	8.1	12.7	1.3	11.5	14.6	16.3
Home furnishings	2.7	2.1	0.0	3.9	7.4	12.0
Food products	0.0	12.2	10.4	9.0	6.1	7.6
Buildings and construction products	0.0	2.1	1.3	1.3	2.3	4.2
Equipment, hardware, and tools	2.7	3.2	5.2	6.4	5.9	7.0
Printing, publishing, paper	0.0	4.2	39.0	16.7	8.5	2.9
Drugs	0.0	0.0	0.0	6.4	9.9	3.4
Other goods	0.0	0.5	2.6	5.1	18.0	25.4
Total	13.5	37.0	59.8	60.3	72.7	78.8
New York						
Dry goods and clothing	5.0				22.3	23.2
Home furnishings	0.8				10.3	24.5
Food products	8.4				6.0	2.9
Buildings and construction products	0.0				4.3	10.0
Equipment, hardware, and tools	0.8				11.6	5.8
Printing, publishing, paper	4.2				9.0	11.6
Drugs	3.4				3.0	2.5
Other goods	1.7				13.0	6.6
Total	24.3				79.5	87.1
Charleston	1772		1825	1835	1845	1855
Dry goods and clothing	2.2		6.1	9.9	8.9	12.1
Home furnishings	0.0		0.0	1.5	2.7	6.1
Food products	7.5		16.7	10.6	4.4	6.8
Buildings and construction products	0.0		1.5	0.8	0.0	2.3
Equipment, hardware, and tools	1.1		1.5	0.8	3.5	3.0
Printing, publishing, paper	1.1		10.6	6.8	5.3	9.1
Drugs	1.1		9.1	7.6	7.1	17.4
Other goods	7.5		3.0	15.1	11.5	9.1
Total	20.5		48.5	53.1	43.4	65.9
Hartford	1792	1805	1815	1825	1835	1845
Dry goods and clothing	3.1	0.0	6.4	2.0	2.6	6.9
Home furnishings	0.0	0.0	0.9	0.0	3.8	6.2
Food products	6.2	2.0	8.2	14.6	11.5	4.4
Buildings and construction products	3.1	1.0	2.7	4.9	3.8	3.3
Equipment, hardware, and tools	3.1	2.0	1.6	2.8	6.8	4.0
Printing, publishing, paper	0.0	0.0	1.8	2.8	4.3	2.9
Drugs	0.0	1.0	0.9	1.6	3.4	13.5
Other goods	0.0	2.0	10.9	11.7	16.6	12.8
Total						

commercial activity, a phenomenon that should be expected of an age that was turning to regularly scheduled railroads, steamboats, and packet lines as the carriers of its goods. Of course, the "just arrived" form of advertising is based on other things besides the actual regularity of supply, so we should not expect to find that it completely disappeared in the railroad age.

Table 3
Percentage of Commercial Advertisements Offering to Purchase or Exchange Goods

	1772	1792	1805	1815	1825	1835	1845	1855
Philadelphia	6.4	2.0	0.0	0.0			1.0	0.7
New York	1.3						0.0	0.3
Charleston	6.1				1.3	0.0	2.6	0.0
Hartford		34.1	12.4	4.4	2.6	5.4	4.2	

Indeed, it is still with us. But as table 4 indicates, it did decline markedly between 1772 and 1855—most markedly in Philadelphia and Charleston (actually disappearing from the latter city's ads), rather less so in the smaller city of Hartford. The newspaper advertisements of all four cities, in sum, imply the development of a commercial life that was a good deal less episodic than the one that prevailed in the colonial era.

But how did these various changes in the process of merchandising promote the association between nonmanual work and respectable social status? It is reasonable to suggest that the specialization and regularization of commerce created a new *physical and institutional setting* in which nonmanual work was both more distinct, and more professional, than it had been before. It was certainly more distinct from manual work, for even the physical locus of the job of selling had become increasingly separated from the locus of production. Retail and wholesale firms that had no workshops provided both an institutional and physical boundary between the environments of manual and nonmanual work, but the separation of these environments of manual and nonmanual work, but the separation of these environments was no less significant within the producing firms themselves. An interesting glimpse of this is provided by . . . an 1850s lithograph of the Chestnut Street hat store of Charles Oakford of Philadelphia. The view is that of an elegant retail sales area, a store with marble floors, ornate counters topped by still more ornate gaslight fixtures, and, along each wall, glass-doored display cases filled with long rows of silk and beaver hats. Oakford's firm seems to have made the hats it offered for sale, but the workshop, and its journeymen hatters, are nowhere in sight. The lithograph suggests nothing of the process of hat making, of the existence of the men who made the endless rows of hats, or of the possibility that these men might ever communicate with the men who sold the hats from the glass display cases. There are several of the latter depicted in the lithograph, and each wears the fine suit (and white collar) that clearly tells the store's customers that these are men who do not "work with their hands." Yet, the fact that the display cases hold hats and nothing but hats tells them also that these men are specialists, who can advise them in an expert way about the quality of this or that beaver skin, about the superior way in which Oakford's hats are constructed, about fit, and about the proper time and place to wear a silk hat. These men are not sea captains' widows, or peddlers, or black-coated clerks, but specialists of a new kind—nonartisanal experts in a partic-

Table 4
Percentage of Commercial Advertisements Announcing a Particular Shipment of Goods

	1772	1792	1805	1815	1825	1835	1845	1855
Philadelphia	72.2	57.7	23.1	25.6			1.5	0.8
New York	70.0						11.6	32.2
Charleston	53.8				23.4	11.4	9.7	0.0
Hartford			46.9	57.0	45.5	41.2	30.2	28.5

ular product line. As such, they are the inheritors of the respect that formerly accrued to the master craftsmen themselves, even while they escape the stigma that now increasingly attached to the workshop, the leather apron, and manual labor.

In end analysis, it is the latter phenomenon—the demise of the artisan-entrepreneur—that must remain central to our understanding of the way in which the manual-nonmanual dichotomy came to dominate our view of the class structure. But, as I hope these pages have suggested, events occurring in the world of nonmanual work made their own, independent contribution to the increasing alignment of work-type and social status. Decades before the emergence of a large, salaried corps of office workers gave rise to the "new middle class" of the late nineteenth and early twentieth centuries, other economic events affecting the nonmanual sector helped produce what C. Wright Mills later called the "old middle class" of retail and wholesale storekeepers, salesmen, managers, and other nonmanual proprietors and workers of various types. The term suggests a static residual category, no more than a baseline from which interesting history can begin. But Mills's "old middle class" was itself once "new," and the events that created it may tell us more than we previously imagined about the evolution of the social structure of industrializing America.

Ideas for Discussion

1. What is the predominant stylistic feature of this essay?
2. What effect did specialization in product retailing mean to the development of American preindustrial economies?
3. What roles do the visual aids play in the effectiveness of the article?

Topics for Writing

1. Blumin offers a straightforward, historical treatment of preindustrial America. Select one of the cities he describes, and write a report on its industrial state today.
2. Write a letter to a merchant in one of the cities discussed advising him of the importance of retail specialization.

Beatrice Webb
Women and the Factory Acts

Beatrice Webb was a political writer who dedicated her efforts to social reform. The eighth of nine daughters of Richard Potter, a railroad entrepreneur, Beatrice became her father's business partner in 1882. She published her first book in 1891, The Co-Operative Movement in Great Britain, *which was derived from her work with immigrant communities in London's East End. She married Sidney Webb when she was thirty-four, in 1892. In 1905, Beatrice became a member of the Royal Commission on the Poor Law, and in 1913, she and her husband founded the* New Statesman. *Beatrice Webb continued to write on social issues, publishing such works as* Industrial Democracy *(1897),* The Decay of Capitalist Civilization *(1923),* My Apprenticeship *(1926), and* Soviet Communism: A New Civilization *(1935). This selection is characteristic of Webb's treatment of issues involving women as she argues for equal treatment based on fairness not gender. This is also a Fabian tract from February 1896 and reprinted from* Women's Fabian Tracts, *edited by Sally Alexander (1988).*

THE DISCUSSIONS on the Factory Act of 1895 raised once more all the old arguments about Factory legislation, but with a significant new cleavage. This time legal regulation was demanded, not only by all the organizations of working women whose labor was affected, but also by, practically, all those actively engaged in Factory Act administration. The four women Factory Inspectors unanimously confirmed the opinion of their male colleagues. Of all the classes having any practical experience of Factory legislation, only one—that of the employers—was ranged against the Bill, and that not unanimously. But the employers had the powerful aid of most of the able and devoted ladies who have usually led the cause of women's enfranchisement, and whose strong theoretic objection to Factory legislation caused many of the most important clauses in the Bill to be rejected.

The ladies who resist further legal regulation of women's labor usually declare that their objection is to special legislation applying only to women. They regard it as unfair, they say, that women's power to compete in the labor market should be "hampered" by any regulation from which men are free. Any such restriction, they assert, results in the lowering of women's wages, and in diminishing the aggregate demand for women's work. I shall, later on, have something to say about this assumed competition between men and women. But it is curious that we seldom find these objectors to unequal laws coming forward to support even those regulations which apply equally to men and to women. Nearly all the clauses of the 1895 Bill, for instance, and nearly all the amendments proposed to it, applied to men and women alike.

* * *

We often forget that the contract between employer and workman is to the employer simply a question of the number of shillings to be paid at the end of the week. To the workman it is much more than that. The wage-earner does not, like the shopkeeper, merely sell a piece of goods which is carried away. It is his whole life which, for the stated term, he places at the disposal of his employer. What hours he shall work, when and where he shall get his meals, the sanitary conditions of his employment, the safety of the machinery, the atmosphere and temperature to which he is subjected, the fatigue or strains which he endures, the risks of accident or disease which he has to incur: all these are involved in the workman's contract and not in his employer's. Yet about the majority of these vital conditions he cannot bargain at all. Imagine a weaver, before accepting employment in a Lancashire cotton mill, examining the quantity of steam in the shed, the strength of the shuttle-guards, and the soundness of the belts of the shafting; an engineer prying into the security of the hoists and cranes, or the safety of the lathes and steam hammers among which he must move; a dressmaker's assistant computing the cubic space which will be her share of the workroom, criticising the ventilation, warmth and lighting, or examining the decency of the sanitary accommodation; think of the woman who wants a job at the white lead works, testing the poisonous influence in the particular process employed, and reckoning, in terms of shillings and pence, the exact degree of injury to her health which she is consenting to incur. No sensible person can really assert that the individual operative seeking a job has either the knowledge or the opportunity to ascertain what the conditions are, or to determine what they should be, even if he could bargain about them at all. On these matters, at any rate, there can be no question of free contract. We may, indeed, leave them to be determined by the employer himself: that is to say, by the competition between employers as to who can most reduce the expenses of production. What this means, we know from the ghastly experience of the early factory system; when whole generations of our factory hands were stunted and maimed, diseased and demoralized, hurried into early graves by the progressive degeneration of conditions imposed on even the best employers by the reckless competition of the worst. The only alternative to this disastrous reliance on a delusive freedom is the settlement, by expert advice, of standard conditions of health, safety, and convenience, to which all employers, good and bad alike, are compelled by law to conform.

We see, therefore, that many of the most vital conditions of employment cannot be made subjects of bargain at all, whilst, even about wages, unfettered freedom of individual bargaining places the operative at a serious disadvantage. But there is one important matter which stands midway between the two. In manual work it is seldom that an individual can bargain as to when he shall begin or leave off work. In the most typical processes of modern industry, individual choice as to the length of the working day is absolutely impossible. The most philanthropic or easy-going builder or manufacturer could not possibly make separate arrangements with each of his workpeople as to the times at which they should come and go, the particular intervals for meals, or what days they should take as holidays. Directly we get machinery and division of labor—directly we have more than one person working at the production of an article, all the persons concerned are compelled, by the very nature of their occupation, to work in concert. This means that there must be one uniform rule for the whole establishment. Every workman must come when the bell rings, and stay as long as the works are open; individual choice there can be none. The hours at which the bell shall ring must either be left to the autocratic decision of the employer, or else settled by collective regulation to which ever workman is compelled to conform.

We can now understand why it is that the representative wage earner declares, to the astonishment of the professional man or the journalist, that a rule fixing his hours of labor, or defining conditions of sanitation or safety, is not a restriction in his personal liberty. The workman knows by experience that there is no question of his ever settling these matters for himself. There are only two alternatives to their decision by the employer. One is their settlement by a conference between the representatives of the employers and the representatives of the organized workmen; both sides, of course, acting through their expert salaried officials. This is the method of collective bargaining—in short, Trade Unionism. The other method is the settlement by the whole community of questions which affect the health and industrial efficiency of the race. Then we get expert investigation as to the proper conditions, which are enforced by laws binding on all. This is the method of Factory legislation.

No greater mistake can be made in comparing these two methods than to assume that Trade Unionism sacrifices the imaginary personal liberty of the individual workman to make his own bargain any less than Factory legislation. Take, for instance, the Oldham weaver. Here we see both methods at work. The rate of wages is determined entirely by Trade Unionism; the hours of labor and sanitary conditions are fixed by law. But there is no more individual choice in the one than in the other. I do not hesitate to say, indeed, that an employer or a weaver would find it easier and less costly to defy the Factory Inspector and work overtime, than to defy the Trade Union official and evade the Piecework List of Prices. Or, take the Northumberland coalminer. He, for particular reasons, objects to have his hours fixed by law. But we need be under no delusion as to his views on "personal liberty." If any inhabitant of a Northumberland village offered to hew coal below the rate fixed by the Trade Union for the whole country, or if he proposed to work two shifts instead of one, the whole village would rise against him, and he would find it absolutely impossible to descend the mine, or to get work anywhere in the county. It is not my business to-day either to defend or to criticise Trade Union action. But we cannot understand this question without fully realizing that Trade Unionism, in substituting for the despotism of the employer or the individual choice of the workman a general rule binding on all concerned, is just as much founded on the subordination of the individual whim to the deliberate decision of the majority as any law can be. If I had the time I could show you, by elaborate technical arguments, how the one method of over-riding the individual will is best for certain matters, and the other method more expedient in regard to other matters. Rates of wages, for instance, are best settled by collective bargaining; and sanitation, safety, and the prevention of overwork by fixed hours of labor are best secured by legal enactment.

But this question of the relative advantages of legislative regulation and Trade Unionism has unhappily no bearing on the women employed in the sweated industries. Before we can have Trade Union regulation we must build up strong Trade Unions; and the unfortunate women workers whose overtime it was proposed to curtail, and whose health and vigor it was proposed to improve, by Mr. Asquith's Bill of 1895, are without any effective organization. The Lancashire women weavers and card-room hands were in the same predicament before the Factory Acts. It was only when they were saved from the unhealthy conditions and excessive hours of the cotton mills of that time that they began to combine in Trade Unions, to join Friendly Societies, and to form Co-operative Stores. This, too, is the constant experience of men's trades. Where effective Trade Unions have grown up, legal protection of one kind or another has led the way. And it is easy to see why this is so. Before wage-earners can exercise the intelligence, the deliberation, and the self-denial that are necessary for Trade Unionism, they must enjoy a certain standard of physical health, a certain surplus of energy, and a reasonable amount of leisure. It is cruel mockery to preach Trade Unionism, and Trade Unionism alone, to the sempstress sewing day and night in her garret for a bare subsistence; to the laundrywoman standing at the tub eighteen hours at a stretch; or to the woman whose health is undermined with "Wrist-drop," "Potter's-rot," or "Phossy-jaw." If we are really in earnest in wanting Trade Unions for women, the way is unmistakable. If we wish to see the capacity for organization, the self-reliance, and the personal independence of the Lancashire cotton weaver spread to other trades, we must give the women workers in these trades the same legal fixing of hours, the same effective prohibition of overtime, the same legal security against accident and disease, the same legal standard of sanitation and health as is now enjoyed by the women in the Lancashire cotton mills.

So much for the general theory of Factory legislation. We have still to deal with the special arguments directed against those clauses of the 1895 Bill which sought to restrict the overtime worked by women in the sweated trades. If, however, we have fully realized the advantages, both direct and indirect, which the workers obtain from the legal regulation of their labor, we shall regard with a good deal of suspicion any special arguments alleged in opposition to any particular Factory Acts. The student of past Factory agitations sees the same old bogeys come up again and again. Among these bogeys the commonest and most obstructive has always been that of foreign competition, that is to say, the risk that the regulated workers will be supplanted by "free labor"—whether of other countries or of other classes at home. At every step forward in legal regulation the miner and the textile worker have been solemnly warned that the result of any raising of their standard of sanitation, safety, education or leisure would be the transference of British capital to China or Peru. And to my mind it is only another form of the same fallacy when capitalists' wives and daughters seek to alarm working women by prophesying, as the result of further Factory legislation, the dismissal of women and girls from employment, and their replacement by men. The opposition to Factory legislation never comes from workers who have any practical experience of it. Every existing organization of working women in the kingdom has declared itself in favor of Factory legislation. Unfortunately, working women have less power to obtain legislation than middle-class women have to obstruct it. Unfortunately, too, not a few middle-class women have allowed their democratic sympathies and Collectivist principles to be overborne by this fear of handicapping women in their struggle for employment. Let us, therefore, consider, as seriously as we can, this terror lest the capitalist employing women and girls at from five to twelve shillings a week, should, on the passage of a new Factory Act, replace them by men at twenty or thirty shillings.

First let us realize the exact amount of the inequality between the sexes in our Factory Acts. All the regulations with respect to safety, sanitation, employers' liability, and age apply to men and women alike. The only restriction of any importance in our Labor Code which bears unequally on men and women is that relating to the hours of labor. Up to now there has been sufficient influence among the employers, and sufficient prejudice and misunderstanding among legislators, to prevent them expressly legislating, in so many words, about the hours of labor of adult men. That better counsels are now prevailing is shown by the fact that Parliament in 1892 gave power to the Board of Trade to prevent excessive hours of work among railway servants, and that the Home Secretary has now a similar power in respect of any kind of manual labor which is injurious to health or dangerous to life and limb. I need hardly say that I am heartily in favor of regulating, by law, the hours of adult men, wherever and whenever possible. But although the prejudice is breaking down, it is not likely that the men in the great staple industries will be able to secure for themselves the same legal limitation of hours and prohibition of overtime that the women in the textile manufactures have enjoyed for nearly forty years. And thus it comes about that some of the most practical proposals for raising the condition of the women in the sweated trades must take the form of regulations applying to women only.

It is frequently asserted as self-evident that any special limitation of women's labor must militate against their employment. If employers are not allowed to make their women work overtime, or during the night, they will, it is said, inevitably prefer to have men. Thus, it is urged, any extension of Factory legislation to trades at present unregulated must diminish the demand for women's labor. But this conclusion, which seems so obvious, really rests on a series of assumptions which are not borne out by facts.

The first assumption is, that in British industry to-day, men and women are actively competing for the same employment. I doubt whether any one here has any conception of the infinitesimal extent to which this is true. We are so accustomed, in the middle-class, to see men and women engaged in identical work, as teachers, journalists, authors, painters, sculptors, comedians, singers, musicians, medical practitioners, clerks, or what not, that we almost inevitably assume the same state of things to exist in manual labor and manufacturing industry. But this is very far from being the case. To begin with, in over nine-tenths of the industrial field there is no such thing as competition between men and women: the men do one thing, and the women do another. There is no more chance of our having our houses built by women than of our getting our floors scrubbed by men. And even in those industries which employ both men and women, we find them sharply divided in different departments, working at different processes, and performing different operations. In the tailoring trade, for instance, it is often assumed that men and women are competitors. But in a detailed investigation of that trade I discovered that men were working at entirely separate branches to those pursued by the women. And when my husband, as an economist, lately tried to demonstrate the oft-repeated statement that women are paid at a lower rate than men, he found it very difficult to discover any trade whatever in which men and women did the same work. As a matter of fact, the employment of men or women in any particular industry is almost always determined by the character of the process. In many cases the physical strength or endurance required, or the exposure involved, puts the work absolutely out of the power of the average woman. No law has hindered employers from engaging women as blacksmiths, steelsmelters, masons, or omnibus-drivers. The great mass of extractive, constructive, and transport industries must always fall to men. On the other hand, the women of the wage-earning class have hitherto been distinguished by certain qualities not possessed by the average working man. For good or for evil they eat little, despise tobacco, and seldom get drunk; they rarely strike or disobey orders; and they are in many other ways easier for an employer to deal with. Hence, where women can really perform a given task with anything like the efficiency of a man, they have, owing to their lower standard of expenditure, a far better chance than the man of getting work. The men, in short, enjoy what may be called a "rent" of superior strength and endurance; the women, on their side, in this preference for certain employments, what may be called a "rent" of abstemiousness.

I do not wish to imply that there are absolutely no cases in British industry in which men and women are really competing with each other. It is, I believe, easy to pick out an instance here and there in which it might be prophesied that the removal of an existing legal restriction might, in the first instance, lead to some women being taken on in place of men. In the book and printing trade of London, for instance, it has been said that if women were allowed by law to work all through the night, a certain number of exceptionally strong women might oust some men in book-folding and even in compositors' work. We must not overlook these cases; but we must learn to view them in their proper proportion to the whole field of industry. It would clearly be a calamity to the cause of women's advancement if we were to sacrifice the personal liberty and economic independence of three or four millions of wage-earning women in order to enable a few hundreds or a few thousands to supplant men in certain minor spheres of industry.

The second assumption is, that in the few cases in which men and women may be supposed really to compete with each other for employment, the effect of any regulation of women's hours is pure loss to them, and wholly in favor of their assumed competitors who are unrestricted. This, I believe, is simply a delusion. Any investigator of women's work knows full well that what most handicaps women is their general deficiency in industrial capacity and technical skill. Where the average woman fails is in being too much of an amateur at her work, and too little of a professional. Doubtless it may be said that the men are to blame here: it is they who induce women to marry, and thus divert their attention from professional life. But though we cannot cut at the root of this, by insisting, as I once heard it gravely suggested, on "three generations of unmarried women," we can do a great deal to encourage the growth of professional spirit and professional capacity among women workers, if we take care to develop our industrial organization along the proper lines. The first necessity is the exclusion of illegitimate competitors. The real enemies of the working woman are not the men, who always insist on higher wages, but the "amateurs" of her own sex. So long as there are women, married or unmarried, eager and able to take work home, and do it in the intervals of another profession, domestic service, we shall never disentangle ourselves from that vicious circle in which low wages lead to bad work, and bad work compels low wages. The one practical remedy for this disastrous competition is the extension of Factory legislation, with its strict limitation of women's hours, to all manufacturing work wherever carried on. It is no mere coincidence that the only great industry in which women get the same wages as men—Lancashire cotton weaving—is the one in which precise legal regulation of women's hours has involved the absolute exclusion of the casual amateur. No woman will be taken on at a cotton mill unless she is prepared to work the full factory hours, to come regularly every day, and put her whole energy into her task. In a Lancashire village a woman must decide whether she will earn her maintenance by working in the mill or by tending the home: there is no "betwixt and between." The result is a class of women wage-earners who are capable of working side by side with men at identical tasks; who can earn as high wages as their male competitors; who display the same economic independence and professional spirit as the men; and who are, in fact, in technical skill and industrial capacity, far in advance of any other class of women workers in the kingdom. If we want to bring the women wage-earners all over England up to the level of the Lancashire cotton weavers, we must subject them to the same conditions of exclusively professional work.

There is another way in which the extension of the Factory Acts to the unregulated trades is certain to advance women's industrial position. We have said that the choice of men or women as workers is really determined by the nature of the industrial process. Now these processes are constantly changing; new inventions bring in new methods of work, and often new kinds of machinery. This usually means an entire revolution in the character of the labor required. What to-day needs the physical strength or the life-long apprenticeship of the skilled handicraftsman may, to-morrow, by a new machine, or the use of motive power, be suddenly brought within the capacity of the nimble fingers of a girl from the Board School. It is in this substitution of one process for another that we discover the real competition between different classes or different sexes in industry. The tailoring trade, for instance, once carried on exclusively by skilled handicraftsmen, is now rapidly slipping out of their hands. But it is not the woman free to work all the night in her garret who is ousting the male operative. What is happening is that the individual tailor, man or woman, is being superseded by the great clothing factories established at Leeds, or elsewhere, where highly-paid skilled designers prepare work for the costly "cutting-out" guillotines, and hundreds of women guide the pieces through self-acting sewing and button-holing machines, to be finally pressed by steam power into the "smart new suit" of the City clerk.

Now this evolution of industry leads inevitably to an increased demand for women's labor. Immediately we substitute the factory, with its use of steam power, and production on a large scale, for the sweater's den or the domestic workshop, we get that division of labor and application of machinery which is directly favorable to the employment of women. It is to "the factory system, and the consequent growth of the ready-made trade," declares Miss Collet, that must "be traced the great increase in the number of girls employed in the tailoring trade." The same change is going on in other occupations. Miss Collet notices that the employment of female labor has specially increased in the great industry of boot and shoe making. But, as in the analogous case of the tailoring trade, the increase has not been in the number of unregulated women workers in the sweaters' dens. Formerly we had a man working in his own room, and employing his wife and daughter to help him at all hours. Some people might have argued that anything which struck at the root of this system would deprive women of employment. As a matter of fact, the result has been, by division of labor in the rapidly growing great boot factories, to substitute for these few hundreds of unpaid assistants, many thousands of independent and regularly employed women operatives. For we must remember that when these changes take place, they take place on a large scale. Whilst the Society for Promoting the Employment of Women is proud to secure new openings for a few scores or a few hundreds, the industrial evolution which I have described has been silently absorbing, in one trade or another, hundreds of thousands of women of all classes. It is therefore infinitely more important for the friends of women's employment to enquire how an extension of the Factory Acts would influence our progress towards the factory system, than how it would affect, say, the few hundred women who might be engaged in nightwork book-folding.

If there is one result more clearly proved by experience than another, it is that the legal fixing of definite hours of labor, the requirement of a high standard of sanitation, and the prohibition of overtime, all favor production on a large scale. It has been the employers' constant complaint against the Factory Acts that they inevitably tend to squeeze out the "little master." The evidence taken by the House of Lords' Committee on Sweating conclusively proved that any effective application of factory regulations to the workplaces of East London and the Black Country would quickly lead to the substitution of large factories. Factory legislation is, therefore, strenuously resisted by the "little masters," who carry on their workshops in the back slums; by the Jewish and other subcontractors who make a living by organizing helpless labor; and by all who cherish a sentimental yearning for domestic industries. But this sentiment must not blind us to the arithmetical fact that it is the factory system which provides the great market for women's labor. Those well-meaning ladies who, by resisting the extension of Factory legislation, are keeping alive the domestic workshop and the sweaters' den, are thus positively curtailing the sphere of women's employment. The "freedom" of the poor widow to work, in her own bedroom, "all the hours that God made"; and the wife's privilege to supplement a drunken husband's wages by doing work at her own fireside, are, in sober truth, being purchased at the price of the exclusion from regular factory employment of thousands of "independent women."

We can now sum up the whole argument. The case for Factory legislation does not rest on harrowing tales of exceptional tyranny, though plenty of these can be furnished in support of it. It is based on the broad facts of the capitalist system, and the inevitable results of the Industrial Revolution. A whole century of experience proves that where the conditions of the wage-earner's life are left to be settled by "free competition" and individual bargaining between master and man, the worker's "freedom" is delusive. Where he bargains, he bargains at a serious disadvantage, and on many of the points most vital to himself and to the community he cannot bargain at all. The common middle-class objection to Factory legislation—that it interferes with the individual liberty of the operative—springs from ignorance of the economic position of the wage-earner. Far from diminishing personal freedom, Factory legislation positively increases the individual liberty and economic independence of the workers subject to it. No one who knows what life is among the people in Lancashire textile villages on the one hand, and among the East End or Black Country unregulated trades on the other, can ever doubt this.

All these general considerations apply more forcibly to women wage-earners than to men. Women are far more helpless in the labor market, and much less able to enforce their own common rule by Trade Unionism. The only chance of getting Trade Unions among women workers lies through the Factory Acts. We have before us nearly forty years' actual experience of the precise limitation of hours and the absolute prohibition of overtime for women workers in the cotton manufacture; and they teach us nothing that justifies us in refusing to extend the like protection to the women slaving for irregular and excessive hours in laundries, dressmakers' workrooms, and all the thousand and one trades in which women's hours of work are practically unlimited.

Finally, we have seen that the fear of women's exclusion from industrial employment is wholly unfounded. The uniform effect of Factory legislation in the past has been, by encouraging machinery, division of labor, and production on a large scale, to increase the employment of women, and largely to raise their status in the labor market. At this very moment the neglect to apply the Factory Acts effectively to the domestic workshop is positively restricting the demand for women workers in the clothing trades. And what is even more important, we see that it is only by strict regulation of the conditions of women's employment that we can hope for any general rise in the level of their industrial efficiency. The real enemy of the woman worker is not the skilled male operative, but the unskilled and half-hearted female "amateur" who simultaneously blacklegs both the workshop and the home. The legal regulation of women's labor is required to protect the independent professional woman worker against these enemies of her own sex. Without this regulation it is futile to talk to her of the equality of men and women. With this regulation, experience teaches us that women can work their way in certain occupations to a man's skill, a man's wages, and a man's sense of personal dignity and independence.

Ideas for Discussion

1. What does Webb hope to accomplish in her tract?
2. Why does she use the pronoun "we" throughout her essay? How does that affect you as a reader?
3. Why has the need for women workers increased according to the essay?

Topics for Writing

1. Identify a need for improvement in people's working conditions and write a letter to a legislator urging reforms.
2. Find out about the Triangle Shirtwaist Factory fire or another manufacturing-related disaster. Compare the event to the relevant laws for factory safety and Webb's arguments.
3. Rewrite Webb's tract as a proposal for a manufacturer or business decision maker.
4. Write a memo to your employer about ways to improve employee morale.
5. Consider some visual aids to accompany this essay. In a memo to the graphics department, request some suitable illustrations for an oral presentation of the major points of Webb's article.

Alfred Pritchard Sloan
Co-Ordination by Committee

Alfred P. Sloan was born in New Haven, Connecticut, earned his degree from the Massachusetts Institute of Technology in 1895, and began his professional career as a draftsman at the Hyatt Roller Bearing Company. In 1901, at the age of twenty-six, he was company president, electing to sell to General Motors in 1916. He became president of General Motors in 1923, serving until 1937, when he became chairman. Under his leadership, which lasted until 1956, Sloan made the company a major force in the U.S. auto industry. He was also a founder of the Sloan-Kettering Institute for Cancer Research in New York City. A. P. Sloan died in 1966. This chapter from My Years with General Motors *(Doubleday, 1964) describes Sloan's approach to management by committee.*

THE ATMOSPHERE in the corporation in the fall of 1923 was one of excitement at the prospect that the industry's first four-million car-and-truck year had opened up, and there was a great desire for reconciliation of the organizational issues raised by the copper-cooled engine. The experience with that engine had a profound effect on General Motors. At the same time the power of the great demand for automobiles acted as a disciplinary force. It was clearly time to gather ourselves to meet the challenge of the boom twenties, and to gather meant to co-ordinate.

The problem of co-ordination was one of developing the practical means of relating the various functions of management. We had the principles of organization which were laid down in the "Organization Study" of 1919–20. We needed now concretely to co-ordinate such very different bodies in the corporation as the general office, the research staff, and the decentralized divisions. The divisions in General Motors are self-contained units combining the functions of engineering, production, and sales—the profit-creating activities, in other words. Corporation staff work in each of these functions cuts across these divisional units. For example, the function of staff engineering is potentially and sometimes immediately related to the engineering activities of any or all of the self-contained operating divisions. The junctures between staff and line are critical, as we had learned the hard way. The experience with the copper-cooled car showed what a paralyzing effect one of these junctures could have if it were turned into a battlefield.

The broad problem of co-ordination and decentralization began at the top of the corporation, and was now my responsibility. What I did about it I had already begun to do in the first period of the new administration. In my notes on the situation in the corporation, which I had written at the end of 1921, I introduced the question of decentralization in relation to the activity of the top executive group. First I set down a declaration of principle, as follows:

> . . . That I approached the matter [of organization] from the standpoint of a thorough belief in a decentralized organization. I still, after a year's experience, [am] just as firmly of the same belief that a decentralized organization is the only one that will develop the talent necessary to meet the Corporation's big problems, but certain things, notwithstanding a decentralized organization, must be recognized and I appreciate these much more than I did before . . .

The main questions, I said, looking forward to the liquidation of the emergency of 1921, were related to the Executive Committee itself, the highest body in the operating structure. These questions were: the role of the Executive Committee as a policy-making group, the representation from the operating side, and the need for authority in the person of the president. I wrote:

> a. That the Executive Committee [should] confine itself more particularly to principles which should be presented to it by the [operating organization], properly developed and thoughtfully carried out rather than to constitute itself as it is now, a group management.

This needs little explanation in view of what I have already related. But let me say that, though I have often been taxed, by people who do not know me, with being a committee man—and in a sense I most certainly am—I have never believed that a group as such could manage anything. A group can make policy, but only individuals can administer policy. At that time, and in particular in relation to the copper-cooled engine, the four of us on the Executive Committee were, in my opinion, trying to manage the divisions.

My next point was not specifically aimed at the lack of automotive experience, but at the need for integration of the top executive committee and the operating organization:

> b. The Operating side on the Executive Committee is not strongly enough represented. This should be corrected by the increase of the Executive Committee, and I suggest Messrs. Mott, McLaughlin and Bassett as additional members. The Executive Committee to meet not oftener than every two weeks and perhaps once a month.

I then proposed that the president should assume not less but more authority. This is not as surprising as it may seem at first sight, for it followed the principle that an individual and not a group should administer. In actual fact general operations had devolved on me at the time I was vice president of operations, and we had a situation of confused authority. I wrote as follows:

> c. Whoever is in charge of Operations should be designated with real authority to be used in case of an emergency. It will probably be best if the President of the Corporation could absolutely have charge of Operations. If this is not feasible, somebody should be so appointed and whoever has charge should develop a reasonable organization to contact with the Operations side as well as the Executive Committee.

I then gave examples of the distinction between policy and administration. Over-all pricing policy, I said, should be reserved to the Executive Committee. Obviously, since we were working with divisional price classes, we were not likely to want Cadillac to produce a car in the Chevrolet price class.

And on the question of Executive Committee action on the character and quality of product, I wrote as follows:

> It would seldom be suitable for the Executive Committee to approve the specifications of contemplated product, or even the principal characteristics unless they be of peculiar significance as involving the entry into new fields, or the possibility of undermining the position of existing profitable lines. The Executive Committee should treat with the question from the standpoint of policy and in the direction of regulating the general quality of product of divisions respectively so as to gain a wholesome distribution of product by class ranges, and the avoidance of undue interference between divisions. A carefully designed policy should be enunciated that will convey to each division a complete understanding of the general quality of product that should be attained or maintained and all major alterations of design should be submitted to the Executive Committee for approval from this standpoint. The Executive Committee should not attempt to pass upon the mechanical features, but must rely upon some competent individual or body in the operating organization.
>
> In general, the activity of the Executive Committee should be guided along the lines of establishing policies and laying the same down in such clear cut and comprehensive terms as to supply the basis of authorized executive action . . .

I cannot recall how Mr. du Pont expressed himself on these proposals. I think he must have agreed, for he co-operated in bringing them into effect. In 1922 he caused to be elected to the Executive Committee Mr. Mott and Fred J. Fisher, both experienced in operations. And later, in 1924, when I was its chairman, he concurred in adding Mr. Bassett, Mr. Brown, Mr. Pratt, Charles T. Fisher, and Lawrence P. Fisher, making ten in all, seven experienced in operations, two in finance, and Mr. du Pont himself. The Executive Committee thus achieved an identity with the operating organization, which, under one title or another, it has maintained ever since. Eventually the Executive Committee limited itself to policy matters, and left administration to the president.

Now the question of the relation of the staff, line, and general officers. I shall describe here the steps which, when completed, put form into the organization.

Two early steps, one in the area of purchasing, the other in advertising, assisted in pointing the way to a practical form of organization. The setting up of the General Purchasing Committee was a task I undertook in 1922. There are two things about this committee which it is important to consider. One was its value or lack of it in its own right; the other was its incidental value as a lesson in co-ordination, which is more germane to the story here.

Centralized purchasing was not an original idea with us. In those days it was considered to be an important industrial economy, and in some circumstances I believe it was. I had experience with volume economies at Hyatt as a supplier to Ford. But centralized purchasing, in which a single purchasing office executes contracts involving more than one division of an enterprise, was an oversimplified notion, as we discovered. The problem for General Motors, as I saw it in 1922, was to get the advantages of volume by buying on general contracts such items as tires, steel, stationery, rags, batteries, blocks, acetylene, abrasives, and the like, and at the same time to permit the divisions to have control over their own affairs. In a preliminary memorandum I argued that co-ordination of purchasing would save the corporation an estimated five to ten million dollars a year; that it would make it easier to control—especially to reduce—inventory; that in an emergency one division could obtain materials from another, and that the corporation's purchasing specialist could take advantage of price fluctuations. I conceded, however, the peculiar difficulties that arise "when one considers the extremely technical character of practically all the Corporation's product and recognizes that we are dealing with many personalities and viewpoints developed through years of contact with certain products as compared with others." In other words, it was a question of acknowledging the natural constraints of decentralization that were built into both the technology of the product and the minds of the managers. The latter were not long in making precisely this point when I first proposed to have a purchasing staff do the co-ordinating. They gave as argument their long experience, the variety of their requirements, and the loss of divisional responsibility in an area which could affect their ability to carry out their car programs.

To meet these objections I proposed the General Purchasing Committee, with a membership drawn mostly from the divisions. The divisions supported the proposition when they learned that they would be represented and would participate in deciding on policy and procedures for purchasing, determine specifications, and draw up contracts, and that their decisions in the committee would be final. Thus it was arranged that in the committee the representatives of the divisions had the opportunity to draw the balance between their special needs and the general interest. A corporation purchasing staff was to administer, but not dictate, the committee's decisions; that is, the relation of committee to staff was to be that of "principal and agent." The Purchasing Committee lasted about ten years and worked reasonably well during that time. But a number of limitations on its value arose:

The first was that quantities of any particular product needed for one division were generally large enough to justify the supplier giving the lowest possible price to that division.

The second was the question of administration. For instance, if the corporation made a contract available to all divisions it sometimes happened that a supplier who did not get the contract would go to one of the divisions and make a lower price, even though he had participated in the original offering. That would cause confusion and unhappiness.

Third, a large number of parts and supplies to be purchased had no common denominator. They were special items applicable to a particular engineering concept.

Therefore, I think the General Purchasing Committee itself cannot be cited as an unqualified success. It caused us, however, to make a strong effort to standardize articles where possible. This and the description of standardized production were very important matters. The General Purchasing Committee's real and lasting success was in the area of standardization of materials.

Also, this committee provided our first lesson in co-ordination. It was our first experience of interdivisional activity, combining line (at the level of a divisional function), staff (a general purchasing section), and general officers (I was the first chairman of the committee). Two years later, reviewing its work, I wrote:

> . . . The General Purchasing Committee has, I believe, shown the way and has demonstrated that those responsible for each functional activity can work together to their own profit and to the profit of the stockholders at the same time and such a plan of co-ordination is far better from every standpoint than trying to inject it into the operations from some central activity.

The next significant step toward co-ordination was in the area of advertising. I had had some consumer studies made in 1922, and we found that people throughout the United States, except at the corner of Wall and Broad streets, didn't know anything about General Motors. So I thought we should publicize the parent company. A plan submitted to me by Barton, Durstine, and Osborn, now BBDO, was approved by the Finance Committee and by our top executives. But since divisional matters would come into the picture, I asked both divisional personnel and other executives in Detroit for their viewpoints on the propriety of such a program. It was agreed that the plan was worthwhile, and Bruce Barton was given full responsibility for conducting the campaign. We then formed the Institutional Advertising

Committee, consisting of car-division managers and staff men, to assist Mr. Barton and "to effect the necessary co-ordination with other phases of the Corporation's publicity." I made a rule that if any advertising theme dealt with a particular division, it must have the approval of that division. It was another little lesson in divisional relationships.

The really big step forward in co-ordination, however, followed from the copper-cooled-engine experience. When that issue was concluded with the parties to it divided against one another—particularly the research engineers on the one hand and the divisional engineers on the other—something had to be done to heal the wounds, and to resolve this fundamental conflict between those seeking new concepts and those with the responsibility for producing automobiles. First of all, what was needed was a place to bring these men together under amicable circumstances for the exchange of information and the ironing out of differences. It seemed to me preferable that such a meeting of minds should take place in the presence of the general executives, who would in the end have to make or approve the big decisions on forward programs.

Rather than try to give the picture wholly through recollection, in which it might well be supposed that I would share in the human failing of making it seem more logical than it was, I shall quote here at length a proposal—the key statement, I believe, in the whole affair—which I wrote and circulated to a number of executives of the corporation and obtained approval of during September 1923:

> I have felt for a long time past that if a proper plan could be developed that would have the support of all those interested that a great deal could be gained for the Corporation by co-operation of an engineering nature between our various Operations, particularly our Car Divisions, dealing as they do in so many problems having the same general characteristics. Activities of this type have already been started in the way of purchasing and have been very helpful and I am confident that as time goes on will be justified in a great many different ways beside[s] resulting in very material profit to the Corporation. The activities of our Institutional Advertising Committee have been constructive and Mr. du Pont remarked to me the other day after one of those meetings that even if it was assumed that the value of the advertising was negligible the other benefits accruing to the Corporation by the development of a General Motors atmosphere and the working together spirit of all members of the Committee representing the various phases of the Corporation's activities . . . the cost was well justified. I am quite confident that we all agree as to these principles and assuming that is the case and there is no reason why the same principle does not apply to engineering, it appears to me to be well worth a serious attempt to put the principle into practical operation. I am thoroughly convinced that it can be made a wonderful success. I believe, therefore, that we should at this time establish what might be termed a General Technical Committee which Committee would have certain powers and functions which should be broadly defined at the start and amplified in various ways as the progress of the work seems to justify.

Before attempting to outline even the general principles upon which I believe a start could be made, I think it should be very clearly set forth and distinctly understood by all that the functions of this Committee would not in any event be to deal with the specific engineering activities of any particular Operation. According to General Motors plan of organization, to which I believe we all heartily subscribe, the activities of any specific Operation are under the absolute control of the General Manager of that Division, subject only to very broad contact with the general officers of the Corporation. I certainly do not want to suggest a departure even to the slightest degree from what I believe to be so thoroughly sound a type of organization. On the contrary I do believe and have believed for a long time that one of the great problems that faces the General Motors Corporation was to add to its present plan of organization some method by which the advantages of the Corporation as a whole could be capitalized to the further benefit of the stockholders. I feel that a proper balance can and must necessarily be established in the course of time between the activities of any particular Operation and that of all our Operations together and as I see the picture at the moment no better way or even as good a way has yet been advanced as to ask those members of each organization who have the same functional relationship to get together and decide for themselves what should be done where co-ordination is necessary, giving such a group the power to deal with the problem where it is felt that the power can be constructively applied. I believe that such a plan properly developed gives the necessary balance between each Operation and the Corporation itself and will result in all the advantages of co-ordinated action where such action is of benefit in a broader way without in any sense limiting the initiative of independence of action of any component part of the group.

Assuming that this is correct in principle, I might set forth specifically what the functions in the case of the General Technical Committee would be, although this discussion would, I think, apply equally well to other Committees dealing with all functions common to all manufacturing enterprises.

1. The Committee would deal in problems which would be of interest to all Divisions and would in dealing with such matters largely formulate the general engineering policies of the Corporation.

2. The Committee would assume the functions of the already constituted Patent Committee which would be discontinued and in assuming these functions would have the authority to deal with patent matters, already vested in the Patent Committee.

3. The Committee would not, as to principle, deal with the specific problems of any individual Operation. Each function of that Operation would be under the absolute control of the General Manager of that Division.

It is to be noted that the functioning of the Patent Section, Advisory Staff, differs materially from that of any other staff activity and is in a sense an exception to General Motors plan of organization in the fact that all patent problems come directly under the control of the Director of the Patent Section. In other words, all patent work is centralized. The Patent Procedure provides, however, for an Inventions Committee and for co-operation with the Director of the Patent Section and the dividing under certain conditions of responsibility in patent matters. In view of the fact that the personnel of the Inventions Committee must necessarily largely parallel that of the General Technical Committee it is thought advisable to consolidate the two for the sake of simplification.

There is also to be considered the functions of the General Motors Research Corporation at Dayton. I feel that up to the present time the [General Motors] Corporation has failed to capitalize what might be capitalized with a proper system of administration, the advantages that should flow from an organization such as we have at Dayton. In making this statement I feel that there are a number of contributing causes, the most important being a lack of proper administrative policy or, I might say, a lack of getting together which it is hoped that this program will provide not only, as just stated, for better co-ordination with the Research Corporation but better co-ordination also among the Operating Divisions themselves. I believe that we would all agree that many of our research and engineering problems in Dayton can only be capitalized through the acceptance and commercializing of same by the Operating Divisions. I fully believe that a more intimate contact with what the Research Corporation is trying to do will be all that is necessary to effect the desired result and strengthen the whole engineering side of the entire [General Motors] Corporation.

It is my idea that the General Technical Committee should be independent in character and in addition to developing through its Secretary, as hereafter described, a program for its meetings, which it is believed would be helpful and beneficial to all the members of the Committee, would conduct studies and investigations of such a character and scope as its judgment would dictate as desirable and for that purpose would use the facilities of the Research Corporation or of any Operating Division or of any outside source that in its judgment would lead to the most beneficial result. Projects of this character would be presented to the Committee by any member of the Committee itself, by the Research Corporation or by any member of the General Motors Corporation through the Committee's Secretary. Beginning January 1, 1924 the cost of operating the General Motors Corporation will be under the control of a budget system and funds will be provided in that budget to cover this purpose.

I have presented the above ideas at an Operations Committee meeting of which all the General Managers of the Car Divisions primarily interested in this matter and the Group Vice Presidents are members and they all seemed to think that the step was a constructive one and would have the support of all.

In order, therefore, that all the above may be crystallized in a few principal points which will be sufficient to form a starting point, I propose the following:—

1. That co-operation shall be established between the Car Divisions and the Engineering Departments within the Corporation, including the engineering and research activities of the General Motors Research Corporation and that co-operation shall take the form of a Committee to be established to be termed the General Technical Committee.

2. The Committee will consist as to principle, of the Chief Engineers of each Car Division and certain additional members . . .

Thus formed, the General Technical Committee became the highest advisory body on engineering in the corporation. It brought together the very persons who had parted over the copper-cooled engine: the divisional chief engineers, including, notably, Mr. Hunt; staff engineers, including, notably, Mr. Kettering; and a number of the general officers of the corporation, including myself as the committee's chairman. It was, as my proposal stated, an independent staff organization with its own secretary and budget. It held its first meeting on September 14, 1923. I was pleased to sit among those fine men—Mr. Kettering, who had the research responsibility; Mr. Hunt, who had a production-engineering responsibility at Chevrolet; Henry Crane, who was my assistant on engineering matters; and the others—all of whom met in a friendly atmosphere and entered afresh into the future development of the automobile.

The General Technical Committee raised the prestige of the engineering group in the corporation and supported its efforts to acquire more adequate facilities and personnel. Its activities emphasized the importance of product integrity as the basic requirement for the future success of the business. It had a remarkable effect in stimulating interest and action everywhere in the corporation in matters of product appeal and product improvement, and produced a free exchange of new and progressive ideas and experience among division engineers. In short, it co-ordinated information.

A number of specific functions were given to the General Technical Committee. For a while it dealt with patent matters, but these were soon turned over to a special New Devices Committee. More important was the committee's role as a kind of board of directors of the great new Proving Ground that we built at Milford, Michigan. Testing had clearly become a crucial question for the future of our products. The Proving Ground, with its controlled conditions, was the logical step away from testing on public roads, which the industry up to that time had practiced. The committee saw to it that the Proving Ground developed standardized test procedures and measuring equipment, and that it became the corporation's center for making independent comparisons of division products and the products of competition. Although engine testing was not assigned to the Proving Ground, the committee was charged with developing an engine test code that would produce uniformity in the engine-testing practices of the various divisions.[1]

And yet the General Technical Committee was the mildest kind of organization. Its most important role was that of a study group. It got to be known as a seminar. Its meetings usually were opened with the reading of one or two papers on a specific engineering problem or device, and these would then be the center of a general discussion. Sometimes the committee's discussion would conclude with the approval of a new device or method, or a recommendation on engineering policy and procedure, but more often the results were simply that information was transmitted from one to all. The members returned to their divisions with a broader understanding of new developments and current problems of automotive engineering and with knowledge of what their associates in other areas of the corporation were doing.

In its reports, papers, and discussions the General Technical Committee studied such short-range engineering problems as those concerning brakes, fuel consumption, lubrication, changes required in the steering mechanism as a result of the development of four-wheel brakes and "balloon" tires (this led to a subcommittee conferring with the rubber companies), and the condensation of products of combustion that resulted in internal rust and oil sludge (which was finally eliminated by proper crankcase ventilation). In 1924 and 1925 the committee gave attention to the education of the dealers and sales departments on the advertising and sales value of current engineering developments. I asked the committee to develop a series of criteria by which "car value" of the different makes and models might be objectively determined. In 1924, too, I gave the committee the task of setting up the broad specifications of the different cars to assist in our efforts to keep the several General Motors' cars distinct and separate products and in a proper price and cost relationship to one another.

Mr. Kettering's staff made most of the long-range investigations and submitted most of the reports during the early years of the committee. They discussed such matters as control of cylinder-wall temperature, cylinder heads, sleeve-valve engines, intake manifolds, tetraethyl lead for gasoline, and transmissions. Fundamentally the subject matters were fuels and metallurgy, the two areas which have furnished the most important improvements in the performance of the automobile since that time.

A meeting on September 17, 1924, in which the subject of transmissions was considered, is a good example of the committee at work. I rely on the minutes for this description. Mr. Kettering began by describing the relative merits and demerits of various types of transmissions. This was followed by a long discussion on the practicality of the inertia-type transmission from an engineering standpoint. Mr. Hunt discussed the different types from the "commercial angle." The growing traffic problem, he said, was calling for a car which "has real acceleration, and in addition it has to have real brakes." After some give and take around the table, I closed this part of the meeting by saying: "I take it that the sentiment of the Committee is something like this: First, that we should look to the ultimate, which is directly a Research problem, and that the inertia type [transmission] is the one which offers the greatest possibilities. This being strictly a Research problem, should not the Committee charge Mr. Kettering with doing everything possible toward its development? . . . Second, for the present we must have minimum inertia and minimum friction in our clutch and transmission elements at our various divisions, and this problem is their own."

In such manner we separated the function of the Research Corporation from that of the divisions. The divisions in those days, however, also had long-term projects; Chevrolet, for example, developed a six-cylinder low-priced car.

That summer I wrote Mr. Kettering about a session of the Technical Committee in Oshawa, Canada. This passage gives the general idea:

> . . . We had a splendid meeting not only so far as the meeting itself went but the boys stayed over Saturday and some of them Sunday and some went fishing and others played golf and that helps a lot in bringing men who are thinking in the same direction, more closely together. I can't help but feel, considering the magnitude of our picture and all that sort of thing, that this co-operation in engineering is working out just splendidly. We must be patient, but I am sure that as time goes on we are going to be fully repaid for the way we have handled it as compared with a more military style which I do not think would have ever put us anywhere.

The interdivisional committee, tried in a rudimentary way in purchasing and advertising, and applied more intensively in the General Technical Committee, was the first big idea for co-ordination in the corporation. We went on from the General Technical Committee to apply the concept to most of the principal functional activities of the divisions. The next interdivisional committee to be formed was in sales. The sales area was relatively unexplored, for the industry in the mid-twenties for the first time had entered its commercial phase. I therefore arranged to set up the General Sales Committee, made up of the sales managers of the car and truck divisions, sales-staff members, and general officers of the corporation. As its chairman, I opened its first meeting on March 6, 1924, with the following remarks:

> While General Motors is definitely committed to a decentralized plan of operation, it is nevertheless obvious that from time to time general plans and policies beneficial to the Corporation and its stockholders, as well as to the individual divisions, can best be accomplished through concerted effort.
>
> The necessity for concerted action on the broader phases of our activities is emphasized by the likelihood of some of our competitors merging their interests—perhaps in the near future. This, as you know, is the trend of the industry. Narrowing profits will add impetus to such a tendency and under the highly competitive conditions of the near future we may expect a decidedly different situation in the field.
>
> General Motors, as you know, has made quite a lot of progress in lining up its products into different price groups, which, relatively speaking, are non-competitive. From the standpoint of design and manufacture we have, through the cooperation of our Division Managers and Engineers, made wonderful progress in the direction of coordination.
>
> Much is to be gained through a similar coordination of sales activities. I think that we, in General Motors, have all got to recognize that the "neck of the bottle" is going to be the sales end. This is perfectly natural in any industry; it eventually gets down to the sales end, and certainly the automotive industry is beginning to reach that period—if it has not already arrived.

It is our idea that this Committee will take in hand all those major sales problems which [a]ffect the Corporation as a whole. It is your Committee. You can feel perfectly free to bring up any sales problems that seem to require general discussion and concerted effort. Whatever general policies or actions you may decide upon will be fully supported by the parent Corporation.

We should, I believe, confine our discussions in these meetings to problems of common interest [a]ffecting all divisions. Realizing that all of you men are extremely busy we will try to keep away from details—dealing only with the basic problems. We will do everything in our power to make the sessions business-like and to the point. No time will be taken to prepare papers, etc. unless in some instances you may wish it that way. Mr. [B. G.] Koether [director, Sales Section] will serve as Secretary of the Committee. He has a Staff which can be expanded if necessary, and whose services are entirely at your command.

We have not developed any definite programs for these meetings because we want to leave such matters to you, realizing that you are in a better position to know just what problems require the most urgent attention and while we may suggest a number of things from time to time it is entirely up to you to act upon such suggestions as you may see fit . . .

The chairmanship of the General Sales Committee was later given to Donaldson Brown, vice president of finance, because of the bearing of statistical and financial controls on production and sales problems. Co-ordination in sales thus extended to the Financial Staff.

After a study of the interdivisional type of committee by Mr. Pratt in late 1924 confirmed in the minds of everyone that this was the best form of co-ordination we had found up to that time, it was made more or less official and was extended to works managers and the power and maintenance staff. Something of the same sort of co-ordination was extended to the very top level of management—but with a difference.

The reader will recall that under Mr. Durant the Executive Committee was composed largely of division managers who campaigned there for the interests of their respective divisions. When we formed the new temporary Executive Committee of four, we placed the former members, mainly the division managers, in an advisory operations committee. For some time, while the emergency was being liquidated, this advisory committee was not regularly active. After I became president and the Executive Committee was enlarged again, it included at different times one or two division managers, depending on circumstance, or motivated by the thought that the largest car division should have representation there. But these were exceptions, not the rule, for I believed in principle that the top operating committee should be a policy group detached from the interests of specific divisions. In other words, it should contain only general executives. Holding this view, after I became president I felt that something should be done to bring the general managers into contact in a regular way with the members of the top operating policy group. I therefore reactivated the Operations Committee and had placed on it all the general operating officers on the Executive Committee and the general managers of the principal divisions, thus making it the major point of regular contact between the two types of executives. The Operations Committee was not a policy-making body but a forum for the discussion of policy or of need for policy. The Operations Committee would receive a full set of data on the performance of the corporation and would review that performance. The word "forum" may suggest something idle, but I assure you that in this case it does not mean that. In a large enterprise some means is necessary to bring about a common understanding. It is perhaps sufficient to note that, with all of the members of the top operating policy group present, an agreement on a policy, suggested say by a division manager, would be tantamount to acceptance on the operating side of the corporation.

In sum then, the whole picture of co-ordination in 1925 and for a number of years thereafter was as follows: The interdivisional-relations committees gave a measure of co-ordination to the functions of purchasing, engineering, sales, and the like. The Operations Committee, including the general managers, appraised the performance of the divisions. The Executive Committee, with contacts in all directions, made policy. It sat at the head of operations, responsible to the board of directors—indeed it was a committee of the board—but beholden to the Finance Committee for its larger appropriations. On the operating side the Executive Committee was supreme. Its chairman was the president and chief executive officer of the corporation; and he had all the authority he needed to carry out established policy. This was the new General Motors scheme of management from which developments down to this day, through much evolution, have been derived.

Notes

1. The inertia-type transmission did appear to have great possibilities technically, but in actual performance it did not prove to be sufficiently smooth or long-lived to warrant production.

Ideas for Discussion

1. What does Sloan mean by "Co-Ordination by Committee"? Does his approach seem effective in light of what you may know about organizational management?
2. How does the first person narrative style influence your response to his piece?
3. Can you think of any improvements to the format of this chapter which you would recommend if it seemed difficult for you to read?

Writing Topics

1. Create the appropriate visual aids to express the information in Sloan's text.
2. Re-write Sloan's chapter as a proposal to one of the committees Sloan describes having worked with.
3. Prepare a summary, abstract and outline for this selection.
4. Prepare a business meeting agenda on the topics presented to the General Purchasing Committee. If time allows, hold the meeting and discuss the ideas.

Paul R. Lawrence and Davis Dyer
Autos: On the Thin Edge

Paul R. Lawrence has been a professor of organizational behavior at the Harvard Business School since 1947. He is coauthor of Renewing American Industry *with Davis Dyer (1983) and the author of* Human Resource Management Trends and Challenges *(1985) and* Behind Factory Walls *(1990). This chapter from* Renewing American Industry *examines the shortcomings of the U.S. auto industry, paying special attention to Alfred Sloan's management of General Motors.*

THE SORRY CONDITION of the U.S. auto industry is news to no one. Chrysler's brush with bankruptcy, record losses and layoffs at Ford, the decision of the American Motors Corporation to sell controlling interest to a French company, staggering losses at once-mighty General Motors: all this has been front-page stuff for several years now. In a crisis of such magnitude, there is plenty of blame to go around, and there has been no shortage of critics (inside and outside the industry) offering opinions as to what went wrong. Some say government is at fault, exacting compliance with unreasonable and expensive safety, environmental, and fuel-efficiency standards too soon, and subsidizing the price of oil for too long. Others blame our trade partners, especially Japan, for a headlong drive to export autos in order to satisfy their domestic employment needs. Still others accuse leaders in industry management or labor of being short-sighted and greedy, sacrificing quality products, responsibility to the public, and productivity to fat margins or high wages and benefits.

There is truth in some of these analyses, but much of the current discussion is pointless because it is offered in an accusatory spirit and is seldom accompanied by useful recommendations for action. Nor has anyone (to date) attempted to view the current crisis against a broader understanding of the way our economy works and, in particular, of what conditions inside and outside the firm are most likely to promote continued economic growth and social responsibility. In short, what is missing is some attention to the theory of how organizations function.

We believe economic growth and social responsibility are most likely to be realized when organizations face intermediate amounts of uncertainty about their environments, when these organizations adopt a particular form with particular human resource practices, and when these organizations fix as their long-term aim the reconciliation of two often contradictory goals—efficiency and innovation. In this chapter we shall show how and why the U.S. auto industry once produced these outcomes, no longer does so, but might once again.

The U.S. Automobile Industry in the Early 1980s

In 1980, a memorable year for the industry, the major U.S. automakers—General Motors, Ford, Chrysler, and American Motors—collectively lost more than $4 billion, or about $12 million per day. Chrysler set a U.S. corporate record by losing $1.7 billion and was saved from bankruptcy only by the election-year beneficence of the government. Ford was scarcely in better shape, with losses totaling $1.5 billion; only its profitable operations overseas kept it from much worse results. By the end of 1980 one-fourth of the total industry labor force—more than 200,000 workers—had been laid off. Chrysler alone permanently closed ten manufacturing plants. 1981 brought little relief. Rather, the sales slump intensified and even the automakers' major efforts at retrenchment achieved only minor financial improvement, a collective loss of $1.3 billion.

The ripple effects of the auto crisis are enormous. It is estimated that about one-fifth of our gross national product springs from the automobile and related industries. Some 3,000 automobile dealers have gone out of business since 1979. Another 650,000 jobs were threatened in steel, rubber, and glass companies, machine toolmakers, and hundreds of small suppliers to the auto industry.

The American auto crisis has been easier to describe than to analyze. Several views have emerged in the popular press. First, industry managers have made an easy target. *Time* magazine, for instance, pronounced that "In the end . . . the auto industry's problems rest with Detroit's managers, who failed to plan for a new-car market after the 1974 oil embargo." Joseph Kraft, writing in the *New Yorker*, espoused a similar view when he described the industry's "downsizing decision" after 1974. Kraft, who documented General Motors' lead over the other American firms in building small cars implicitly suggested that if corporate leaders had made the decision to produce small cars in 1973 and had then acted swiftly on the decision, there would have been no crisis later on. On the other hand, William Tucker, an editor of *Harper's*, argued cogently that the Big Three automakers were encouraged to build large cars by consumers, who spurned compacts between 1974 and 1979, and by the government, which refused to lift price controls on oil in the mid-1970s.

A recent report from the government takes a more balanced view and even-handedly blames the industry, the labor force, and itself for the crisis. *The U.S. Automobile Industry, 1980,* an interagency study directed by Neil Goldschmidt, secretary of transportation in the Carter administration, pinpoints four major causes of the crisis: slowdown of demand, government mandated technological change, encroachment by imports, and high labor costs. First, the energy shocks of 1973–74 and 1978–79 have changed American taste in automobiles profoundly. Consumers of the 1980s are likely to prefer more fuel-efficient cars and to retain them longer than their earlier models. This behavior poses two problems for the industry: consumers are demanding a type of automobile that the American producers have never built in volume, while annual sales are growing much more slowly than in the past.

Federal regulation also has had a severe impact on the industry. Since the late 1960s, government has pressed strenuously for improvements in safety, pollution, and fuel efficiency, and the industry has been required to invest heavily in what some insiders consider "nonproductive innovation." In addition, the changing design of the automobile requires new process and product technologies. Robotics will be used increasingly on assembly lines to handle materials and to weld and paint the chassis. The new car itself will have freshly designed engines and transmissions and make greater use of lightweight materials. At present, the automakers have used up most of their working capital and debt capacity to finance this innovation, with dire consequences for profitability and their ability to attract new investment. It is estimated that the U.S. industry will require a capital investment of some $80 billion in the 1980s to produce the new automobiles competitively. In view of its present financial condition, it is difficult to see where this money will come from.

A third fundamental problem for the automakers is that increasing competition from abroad will prevent them from passing on their higher costs to consumers. The share of the American market captured by imports (especially Japanese) jumped from 15% in 1972 to 27% at the end of 1981 (Figure 1). Foreign automakers have been in a strong position to sell in the suddenly energy-conscious American market, because for years they have built small, light, fuel-efficient cars. Thus the foreign companies already possess production skills that American firms are now acquiring at great cost. Indeed, it is estimated that Japanese producers enjoy a cost advantage of some $1,500 to $2,000 per car. Some of this cost differential derives from wage costs, which run higher in the United States than in Japan, while American productivity has improved more slowly (Figure 1). U.S. autoworkers earn about twice the benefits of their Japanese counterparts, for example, while the Japanese use approximately half the man-hours to build an automobile. And, as Douglas Fraser, president of the United Auto Workers (UAW), points out, "the typical U.S. auto manager outearns his Japanese colleague by some 700%."

The federal government has taken various actions recently to counteract the automobile crisis. Secretary Goldschmidt, in his letter of transmittal accompanying the report, recommended that the government immediately negotiate trade restrictions with the Japanese and adopt fiscal policies to help the industry attract capital. For the longer term, Goldschmidt hoped

Figure 1 Market Share of Imported Automobiles 1960–1980
SOURCE: *Ward's Automotive Yearbook*, various years.

that management, government, and labor might establish a new "compact" to work together for the common benefit of each. The Reagan administration has pursued more traditional approaches. In 1981 the government negotiated a voluntary import restraint agreement with the Japanese; it also modified the tax laws to allow faster depreciation and removed or relaxed some safety and environmental regulations.

Such prescriptions are fine as far as they go. Amid all the verbiage on the auto crisis, however, there has been little analysis of the auto firms as *organizations* or of the automakers' ability to adapt to their changing environment. It is largely assumed that if the right public policy is found, or if the right business strategy is articulated and pursued, then the crisis will end. No one has asked whether the automakers have the kind of organization that can respond to the right public policies and internal strategies. Our analysis of organizational adaptation suggests that this is a crucial question indeed.

According to our model, the automakers currently face conditions of high resource scarcity (losses, lack of capital, regulatory compliance costs) and moderate to high information complexity (imports, demanding regulations, new technologies). The strategies and organizations that landed them in these circumstances are unlikely to get them out. While the automakers have clearly changed their strategies, it is less certain that their organizations will be able to respond effectively.

* * *

Strategy and Organization

In the evolution of the U.S. auto industry, organizational choice and internal structure were critical factors in determining not only the industry's shape but also the particular fate of each automaker. In the early years of competition, as we have seen, Ford's strategy of hard-driving production dominated all competitors. The company averaged better than 40% of the market between 1911 and 1925. By contrast, the combined market share of General Motors' auto assembly divisions oscillated between 12% and 20% in this period. However, the effects of GM's reorganization in the early 1920s, under Du Pont and Sloan, combined with the articulation of a definite strategy of segmenting the market, proved more enduring than Ford's

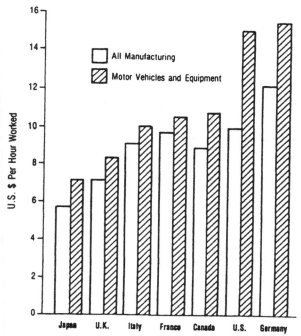

Figure 2 Average Production Worker Compensation, Mid-Year 1980: All Manufacturing and Motor Vehicles and Equipment
SOURCE: U.S. Department of Transportation, *The U.S. Automobile Industry, 1980* (Washington, D.C., 1981), p. 41.

initial success. It was in the 1920s that General Motors permanently passed Ford in sales and production, as the two giant companies acted out one of the great morality tales of American management.

Once automobiles became big business the advantages of good organization come to the fore. Henry Ford never understood this. In a famous essay entitled "Machines and Men," Ford observed:

> That which one has to fight hardest against in bringing together a large number of people to do work is excess organization and consequent red tape. To my mind there is no bent of mind more dangerous than that which is sometimes described as the "genius for organization."

"And so," he boasted proudly,

> the Ford factories and enterprises have no organization, no specific duties attaching to any position, no line of succession or of authority, very few titles, and no conferences. We have only the clerical help that is absolutely required; we have no elaborate records of any kind, and consequently no red tape.

The Ford Motor Company was run just so. A brilliant organizer of production, Ford made no effort to conceal his contempt for organized management. He himself asserted leadership sporadically but always with finality. He sometimes left management of the company to his son Edsel or to his assistants for months at a time before returning suddenly, imperiously, to make critical decisions. Aside from Edsel and a few vice-presidents, executives in the Ford Motor Company carried no titles and only vaguely defined responsibilities. Not surprisingly, turnover of top executives was high, especially among those in finance and sales.

Henry Ford's attitude toward professional management accorded with his general strategy of vertical integration. Rather than purchasing parts suppliers, as Durant had done, Ford went a step further and decided to build his own parts from raw materials. He bought enormous reserves of iron ore and coal and built a steelworks adjacent to his Model T assembly plant at the River Rouge in Dearborn. He also added a sawmill (after buying timber reserves) and a glassworks to the plant in the 1920s. He even purchased 2.5 million acres of land in Brazil, an estate christened Fordlandia, to plant rubber trees in hopes of eventually including a rubber factory at the River Rouge. Ford's notion, evidently, was to create a single monumental assembly line from raw minerals to the finished Model T. In the 1920s the company's slogan read: "From Mine to Finished Cars, One Organization." Management's task became essentially the first-line supervisor's task—to keep this assembly process moving at full speed.

The contrast between the Ford Motor Company and the General Motors of Du Pont and Sloan could hardly be greater. GM's head-to-head competition with Ford necessarily entailed some emulation of production techniques. All similarities ended there, however. GM shied away from full-scale vertical integration and concentrated instead upon weeding and sorting Durant's sprawling empire. It was Sloan's "genius for organization," so ridiculed by Henry Ford, that made GM a successful company in the 1920s. Sloan's formal association with Durant had begun in 1916 when Durant purchased Sloan's firm, Hyatt Roller Bearing. Disturbed by General Motors' chaotic organization, Sloan first proposed a plan to coordinate production and distribution in the company in 1920. Implementation of this plan, however, awaited Durant's fall and replacement by Du Pont, whose own explosives and chemical firm was simultaneously wrestling with similar problems of large-scale organization. Sloan's achievement was to combine the advantages of Durant's decentralized organization, which, after all, provided GM some strength in different markets, with strong financial coordination and scheduling of supplies common to the separate automaking units.

The organization in place at GM by 1924 embodied these principles exactly (Figure 3). Sloan gathered General Motors' subsidiaries together according to product or market criteria and created the concept of the group executive. The five automakers, the truck-making subsidiary, and various suppliers were established as autonomous divisions and treated as profit centers. This recognition of decentralization was balanced by the creation of a large general staff at headquarters, formed to monitor information and coordinate policy for the corporation as a whole. Other top executives, including John Lee Pratt and Donaldson Brown, supplemented Sloan's "organization plan" by inventing new management techniques at General Motors. Pratt worked out statistical procedures to provide accurate records and forecasts of inventories, while Brown developed "standards" and controls for production costs and anticipated sales. This organization and these refinements nicely supported GM's strategy, finally made explicit in 1921, of having "a car at every price position, just the same as a general conducting a campaign wants to have an army at every point he is likely to be attacked."

GM has long been recognized as a pioneer in the introduction of the "decentralized," or the "multidivisional," organization. Whereas most large businesses were organized with departments of engineering, manufacturing, sales, and so on, the multidivisional structure subordinated these functions to a division or product general manager. As Alfred Chandler describes the new structure:

> A general office plans, coordinates, and appraises the work of a number of operating divisions and allocates to them the necessary personnel, facilities, funds, and other resources. The executives in charge of these divisions, in turn, have under their command most of the functions necessary for handling one major line of products or set of services over a wide geographical area, and each of these executives is responsible for the financial results of his division and for its success in the market place.

Chandler's description and terminology have won wide acceptance in academic and management circles, yet it seems clear that GM's organization represents something more than simply a multidivisional form. Indeed, its central characteristic was not simply the balance between centralization and decentralization, as drawn on a chart, but also the practical manner in which this balance was maintained. Sloan called this "decentralization with co-ordinated control," a concept that "evolved gradually at General Motors as we responded to the tangible problems of management."

In addition to creating an active role for the central staff, Sloan grafted other coordinating mechanisms onto the multidivisional form, including committees at various levels. At the top two committees of the board of directors and senior managers focused on strategic policy and finance. Lower down—and here was a brilliant innovation—Sloan gradually introduced "interdivisional relations committees" to bridge the autonomous divisions on matters of common concern such as purchasing, advertising, technical engineering, operations, and so on. These committees served two principal purposes, education and persuasion. They were "not designed to function as administrative bodies," observed Donaldson Brown,

> but rather as a means of providing an opportunity for general discussion of problems of common interest. . . . Questions of policy are clarified; and through the operation of these several committees a means is provided of gaining a widely diffused knowledge and understanding of corporate policies, and a sympathetic compliance with them as they may bear upon the immediate problems of divisional management.

The General Motors of 1924 therefore represented a new species of corporate organization. It managed "co-ordinated decentralization" by balancing a high degree of differentiation (high D) in its product/market divisions and a high degree of integration (high I) through various committees. In our terms GM was not only one of the earliest multidivisional structures but the first high D and I form.*

The strategic and organizational differences between GM and Ford were reflected in their respective profits in the 1920s. As per capita income grew, the American public shed its allegiance to the Model T for GM's more stylish models. Ford's market share fell from 56% in 1921 to less than 40% in 1926 and the company lost money for the first time. Despite prices slashed as low as $260, Model Ts would not move in sufficient volume to keep the River Rouge plant operating in the black. Indeed, in 1927 Ford closed the plant for nine months while he retooled for a new car, the Model A. This opportunity permitted General Motors to surpass Ford in production, a lead it has maintained in all but one year since.

Despite GM's great success in the 1920 and 1930s, it was some time

---

*This integration distinguishes the organization of GM from that of the classic multidivisional structure established at Du Pont after World War I.

before "Sloanism" crept over to Ford and Chrysler. The latter company, as we have seen, became a full-line producer after 1928; it became more vertically integrated only after experiencing production and supply problems during World War II. Ford finally added a middle-priced line of cars in 1938 with the Mercury. Thereafter, each of the Big Three competed directly in most markets, from inexpensive coupes to luxury sedans. Each wooed customers with new colors and styles and occasionally an innovation such as automatic transmissions or power accessories. They no longer competed explicitly on production costs.

Ford and Chrysler adopted GM's strategy more readily than its structure. At Ford the continuing dominance of the family owners hindered the full development of a decentralized organization. The elder Henry Ford remained a crabby and unconventional despot presiding over the company until shortly before his death in 1947. Because of his devotion to the days when selling cars meant producing them at the lowest cost, his contempt for modern management, and his senile infatuation with strong-arm personnel policies, his company performed erratically in the 1930s and 1940s. The company did not really recover from the depression until after World War II. It was then that Henry Ford II asserted control and brought in a group of General Motors executives and the famous "Whiz Kids," recasting the corporation into a multidivisional structure with tight financial controls but without other integrating mechanisms.*

In the 1950s Ford made a more determined attempt to copy General Motors' organization more directly when it created a separate division to manage a new car in the middle price range, the Edsel. The Edsel failed for many reasons—poor styling and quality is irrelevant during a recession, and a confused marketing strategy. Its failure was also organizational. The Edsel division was not sufficiently integrated into the rest of the company and its managers failed to win the support and cooperation of executives in other auto divisions. The Edsel was discontinued after the 1960 model year.

Ford's failure to replicate GM's system of decentralization and committee management happened partly because the Ford family was reluctant to relinquish power to outside managers. Henry Ford II's much publicized troubles in maintaining good relations with his chief operating officers and establishing a clear line of succession make a case in point. Between 1960, when Henry II asserted full control, and 1980, when Philip Caldwell succeeded him as chief executive officer, the company passed through a half-dozen presidents; in the entire postwar era, only four men have held the number-two position for more than two years. These periodic shake-ups at the top prevented a stable differentiated and integrated organization from taking root at Ford.

---

*The Whiz Kids were a group of ten young systems analysts and management control specialists from the army air force who arrived at Ford in 1946. Six of the ten later became vice-presidents in the company, while two (Robert S. McNamara and Arjay Miller) served as president of Ford.

At Chrysler the centralized, functional organization endured until the 1970s. Walter Chrysler and his successor, K. T. Keller, kept close control over the company through the early 1950s. With centralization and traditional management styles deeply entrenched, the company resisted efforts to copy General Motors' structure when "Tex" Colbert (who followed Keller in the top job) tried to reorganize the company. The result was "sheer chaos," according to one company executive. "There were none of the mechanics for decision making and review, none of the accounting tools to support divisionalization." It was only in the early 1970s that the company finally adopted the multidivisional form. Even then bizarre organizational arrangements remained. When Lee Iacocca joined Chrysler in 1978 he was startled to find some staff officers claiming wide powers and running their departments as independent fiefdoms. Reporting relationships were defined with more regard to tradition than to organizational rationality. Corporate staff reported to the chairman or president with no clear understanding of which functions belonged to operations and which did not. Iacocca fully reorganized the company along more conventional lines in 1979 and 1980.

From the 1930s to at least the mid-1970s, then, there was remarkable continuity in strategy and performance among the Big Three automakers. Market shares and profitability held more or less constant, although there were significant organizational differences among the companies. GM took advantage of its larger size and more stable management to dominate the industry. All three companies, nonetheless, earned handsome rewards for their efforts. To all outward appearances, the Big Three possessed a sure formula for success.

The Readaptive Record

The U.S. automobile industry's long, stable tenure . . . produced some outcomes that generally pleased and rewarded both the industry and the public. For most of the period, the companies earned generous returns and offered excellent salaries and wages to attract and hold employees, while the consumer got a reasonably inexpensive and continually improving product. Over the years, however, the long equilibrium also reinforced some bad habits. Sloan's dominant strategy based on marketing and style subordinated technological innovation to production efficiency. There was little substantive innovation in the essential mechanics of autos after the 1920s. At the same time, as the business became better understood, managerial work became routine. The companies' concern for production efficiency also created tensions in the labor force, which erupted in a series of violent strikes in the 1930s and 1940s and culminated in the recognition of the United Automobile Workers (UAW) as a national bargaining agent. While unionization brought substantial material advancement to workers, the contracts struck between management and labor left control over production in the hands of the former. Over the years, management's implicit narrowing of focus on efficiency polarized its relationship to the work force.

Industry stability produced substantial material rewards for managers and workers. Sloan, the pioneer of so many innovations at General Motors, also developed a managerial compensation package that served as a model for the other automakers and for industry in general. General Motors started its annual bonus plan in 1918 in the belief

> that the interests of the corporation and its stockholders are best served by making key employees partners in the corporation's prosperity, and that each individual should be rewarded in proportion to his contribution to the profit of his own division and of the corporation as a whole.

Sloan's aim in adding this market incentive was to create a group of "owner-managers" inside General Motors whose personal ambitions would dovetail with the continued prosperity of the company. On the basis of annual performance, key managers received handsome bonuses in cash or stock options, which, over the years, made them among the highest paid executives in all of American business. In our terms, Sloan added market incentives to supplement bureaucratic ones. On the other hand, the plan reinforced short-term strategic thinking in the company because it was in the manager's best interests to maximize profits year by year. There was little incentive, for example, to forego profits and bonuses for several years to make a long-term investment that would pay off only in the future. This was not such a critical problem while the business expanded or while Sloan and his immediate successors ran GM, but it became especially troublesome in the 1960s and 1970s.

A second unforeseen consequence of Sloan's managerial system also surfaced in the 1960s and 1970s. As the automobile business became more certain, it also became less challenging. The routinization of managerial work was compounded by the tendency of some of the companies to inbreed their managers. As a result the companies gradually lost some of the advantages of internal differentiation. This problem was particularly acute at GM, where many senior managers spent their entire careers with the company, starting with their undergraduate education. General Motors has been the only corporation in America with its own accredited college as a subsidiary—the General Motors Institute (GMI). Although the company recently announced plans to divest GMI, most graduates traditionally received offers to join General Motors and most never worked anywhere else.

John Z. DeLorean's partisan memoir of life at the top in General Motors offers abundant testimony to the dangers of corporate conformism and committee management gone mad. Senior executives, claims DeLorean, "spent little time looking at the big picture, instead occupying themselves with minuscule matters of operation which should have been considered and disposed of in the divisions or much farther down. . . ." To DeLorean,

General Motors seemed to offer "a system which puts emphasis on form, style, and unwavering support for the decisions of the boss," and he blamed "system-men" for stifling initiative and creativity in the corporation. Once these system-men "get into power," said DeLorean, "they don't tamper with the system that promoted them. So a built-in method of perpetuating an imperfect management system is established." He also believed this explained why the company misread the market in the late 1960s and early 1970s.

If managerial work in the auto industry became "stultifying" to people like DeLorean, then what of the actual work of production? The story is similar: As the industry expanded, higher and higher wages were traded for more and more boring and enervating jobs. Henry Ford, of course, established the nature of work in the plants when he introduced the moving assembly line in 1913. Ford also, in the next year, saw to it that labor would be well compensated for its efforts when he introduced the eight-hour shift for the unheard-of pay of five dollars per day. For these rewards workers in the plants committed themselves to routines and schedules determined by the speed of conveyor belts. It was a costly bargain. One worker told Edmund Wilson in the 1930s:

> Ye get the wages, but ye sell your soul at Ford's—ye're worked like a slave all day, and when ye get out ye're too tired to do anything—ye go to sleep on the car comin' home. . . . A man checks his brains and his freedom at the door when he goes to work at Ford's.

The rise of the UAW in the 1930s and 1940s scarcely affected the nature of assembly line jobs. The establishment of the union solved the economic insecurity of the workers and provided a countervailing power to management. The UAW's great victories pertained chiefly to wages and benefits. In part, the union was deflected from addressing the enervating nature of the work itself by the very collective bargaining laws that gave it recognition and rights. The Wagner Act restricted negotiations between labor and management to "wages, hours, and other terms and conditions of employment." This clause was narrowly construed in the 1930s and 1940s and has opened up only gradually to include such items as supplementary unemployment benefits, work rules, and safe working conditions. The laws have seldom been read as permitting the labor force to achieve greater participation in the manufacturing process or more control over the substance of their own jobs.

The auto companies, moreover, severely resisted any proposals that threatened their "prerogatives" to manage and direct the work force. These issues came to a head in 1945–46 when the UAW struck GM for 113 days in a vain attempt to open the company's books in order to see how much of a wage hike the firm could afford without raising prices. General Motors resisted this demand successfully but mollified the union at the next contract by offering to improve wages by an annual cost-of-living adjustment plus an "annual improvement factor" reflecting productivity gains in the U.S. economy in general. The 1948 contract thus formally institutionalized rising real incomes and expectations for UAW members. The autoworkers have been among the highest paid hourly workers in America ever since.

Even so, there were signs that the tradeoff of wages for control of the workplace sanctioned by the labor laws, management, and the UAW was no happy bargain. A famous survey conducted in the 1950s showed that autoworkers resented the time-discipline of the assembly line and that they derived little emotional satisfaction from their daily toil. "The work isn't hard," said a worker quoted in Walker and Guest's *The Man on the Assembly Line* (1952), "it's the never-ending pace. . . . The guys yell 'hurrah' whenever the line breaks down . . . you can hear it all over the plant." At the time, however, the high wages and relative security of jobs in the plants were powerful incentives to preserve the status quo.

In the late 1960s and early 1970s, however, troubles in the workplace manifested themselves in disturbing ways. At General Motors, for instance, as a new generation of workers entered the plants, absenteeism rose 50% between 1965 and 1969, and ran as high as 20% on Mondays and Fridays. Over the same period employee turnover climbed 72%, the number of grievances lodged against supervisory personnel rose 38%, and disciplinary dismissals increased 44%. The most widely publicized of these troubles occurred in 1972 when workers struck GM's assembly plant in Lordstown, Ohio where the Chevrolet Vega was being assembled. Media reporting glossed over some of the principal causes of the strike including the reorganization of the plant, the merging of two local unions into one, and the elimination of hourly jobs. But working conditions were also at issue. Since the Vega was designed to be built with 43% fewer parts than most automobiles, Lordstown management planned to increase production from the normal rate of about sixty cars per hour to one hundred cars per hour. After several months of this frantic pace—some tasks had to be repeated every thirty-six seconds—some workers began sabotaging cars on the line while levels of absenteeism and grievances at the plant soared. Only after a three-week strike and the scrutiny of national attention did plant managers and workers redefine working conditions in the plant. By then, however, "the Lordstown syndrome" and "the blue-collar blues" had entered the language as symbols of the modern industrial worker's "alienation" from his daily labor.

The automakers' internal structures, power relationships, and human resource practices led to an unbalanced record of managing efficiency and innovation. The companies were more successful with efficiency. Economist Lawrence White calculates that labor productivity in the auto industry improved at a rate of 4% per annum during the 1950s and 1960s, well above the rate of manufacturing generally. This is an impressive record that is sometimes lost in comparisons with foreign automakers' productivity. Yet in the 1970s productivity declined substantially—by 25% or more as labor troubles surfaced.

At the same time, as discussed above, the automakers generated unusually high rates of return, which were paid out to stockholders rather than reinvested. Thus when lean times arrived the automakers were caught without the capital investment or the labor involvement to maintain historic improvements in productivity and efficiency.

The industry's record of efficiency came at the expense of innovation. While it is clear that automobiles improved steadily over the years, innovations (many of which originated at suppliers) tended to make driving simpler, faster, safer, and more comfortable. Automatic transmissions, for instance, came in the 1940s and 1950s, with power and disk brakes soon thereafter. However, American automakers did not spend much money on fundamental technology to change the power train until they were forced to do so by federal regulation and the energy crisis of the 1970s. Most of the pioneering efforts toward increased fuel efficiency, including advanced diesel engines, fuel injection, and stratified-charge engines, came from abroad.

Indeed, it has been argued that the auto industry achieved little substantive innovation after its earliest years. Writing from the vantage point of the early 1960s, Sloan admitted that "Great as have been the engineering advances since 1920, we have today basically the same kind of machine that was created in the first twenty years of the industry." Certainly the most prominent features of today's automobile date from long ago: the internal combustion engine (the nineteenth century); shock absorbers, electric headlamps, asbestos drum brakes, steering wheels, standardized parts and designs (by 1910); the electric self-starter (1911); and the closed chassis, low-pressure pneumatic tires, and synchromesh transmissions (1920s). In this sense, Sloan was right: The product had matured by the 1920s.

The readaptive record of the American automakers stands in marked contrast to that of their chief rivals in the 1980s, the Japanese. In the past two decades the Japanese have maintained astonishing levels of efficiency and innovation. It is generally agreed that Toyota is the low-cost auto producer in the world, as well as a leader in quality control. Fundamental product innovations originated at Honda and Toyo Kogyo (Mazda), although the Japanese have relatively small budgets for research and development. The Japanese are adept, moreover, at licensing and adopting technologies developed elsewhere when the market signals. Much of the Japanese automakers' success stems from their human resource practices, which involve workers in quality control and management of the workplace. These policies have been extremely successful in promoting a strong identity of purpose between workers and their employers.

The Coming of the Automobile Crisis

The U.S. automakers' failure to balance efficiency and innovation left them in an exposed position when a series of shocks disturbed the industry's competitive equilibrium. We shall do no more than note these problems here, as they are familiar to most readers. In the first place, the social value of the automobile, so long unquestioned in America, suddenly became controversial in the 1960s. Consumers had registered disapproval of Detroit's products from time to time in earlier days. In the late 1950s, for instance, imports of small cars earned about a 10% market share, as city drivers apparently grew disgusted with congested freeways and scarce parking spaces. In the next decade, however, consumer disenchantment grew by orders of magnitude in three distinct ways: agitation for safer automobiles, promoted by Ralph Nader's indictment of the Chevrolet Corvair and given momentum by General Motors' underhanded attempt to discredit Nader; the clamor to reduce pollution caused by the automobile; and the cry to improve public transportation at the expense of the auto companies and the highway lobby. Small cars and foreign cars became more attractive and the imports' market share grew substantially in the late 1960s, well before the energy crisis sent their fortunes climbing still higher. The root of the foreign success, especially the Japanese, was an ability to deliver high-quality vehicles to the American market at low cost.

Although the energy crisis of 1973-74 rocked the industry, federal energy policy muffled that signal of continuing change. The 1975 Energy Policy and Conservation Act retained price controls on oil and shielded consumers from the effects of the energy crunch. As a result, small cars did not sell as well as predicted in the mid-1970s, and even the Japanese automakers watched their inventories pile up on the West Coast docks. The market share of imports actually declined in 1975 and again in 1977-78 before the Iranian revolution reestablished the gravity of the energy crisis. (See Figure 1).

Twice then, in the late 1960s and in the late 1970s, Detroit was suddenly caught manufacturing the wrong product—overpowered and oversized vehicles with a growing reputation for low quality. Even when U.S. automakers attempted to compete directly with the imports the results were dismal. Indeed, the Ford Pinto and the Chevrolet Vega will be remembered in automotive history chiefly for the humiliations they caused their makers as objects of recalls and lawsuits, symbols of the blue-collar blues. The changing automotive reality of the late 1970s, born of trends in public policy, consumer attitudes, and demographics, now seems permanent. Pollster Daniel Yankelovich has captured the new mood:

> While Americans continue to cherish their cars, the psychocultural meaning of owning a car *is* changing. People feel that a car means freedom and independence and convenience—it takes you where you want to go when you want to. Our cars still make personal statements about us, but what they express is our taste, individuality, and autonomy, not our social status. Also, to increasing numbers of Americans a car is "just a car," a necessity of life that should be comfortable and styled to "suit me," but need not symbolize one's achievement in life.

Therefore bigness, newness, excess power, and the other elements of conspicuous consumption are not as important as they once were—not worth endless sacrifice. Ninety percent of the public say they are willing to do without annual model changes in automobiles.

Here in a nutshell is the backdrop to the automobile crisis of 1980. Caught between shifting public policies, especially with respect to energy, the industry compounded its troubles by misreading the American market. Indeed, the wonder is not that the auto crisis arrived, but that it was so long in coming.

Senility or Rejuvenation? The Case of General Motors

Juxtaposing this gloomy portrait and our earlier developmental model of the auto industry's history, we might be tempted to conclude that the industry has entered a fifth era, or stage, one of senility and onrushing death. Yet the experience of General Motors in the recent crisis—it suffered market share and dollar losses much less than its domestic rivals—suggests a weakness in any deterministic life-cycle model of industry evolution. GM has shown that the industry's decline may not be inevitable or irreversible. In part, GM owes its relatively steady performance to advantages of size. For example, it has absorbed regulatory compliance costs more easily than Ford or Chrysler. But GM was also quicker to recognize its changing environment and to adopt a long-term strategy which better balances innovation and efficiency. Moreover, GM revived the high D and I form through the adoption of project centers at the managerial level and Quality of Work Life (QWL) programs for hourly employees. As a result, GM is the only American auto company entering the 1980s with much reason for optimism.

GM's relative stability in the 1970s was a direct result of the company's ability to adapt to changing circumstances. After the humiliations of the Nader incident, Lordstown, and the Vega, GM began to reevaluate some of its policies and practices. The company seemed to understand that its mistakes could be traced to a lack of internal differentiation. In the late 1960s and early 1970s GM took several steps to change this. First, it began to proliferate internal committees to draw upon specialized points of view. In 1965–66 the company created an automotive safety engineering group to act as a bridge between the operating divisions and government policymakers. Similarly, in 1970 GM formed a research policy committee to study and consider the best technological response to the demands of the Clean Air Act. These steps were in keeping with a second change in GM's behavior, its decision to cooperate with government regulations on most issues rather than to fight them. One manager recalled that key decision makers put out the message in the early 1970s that "the [federal] standards are here to stay. Find out what the regulators are doing. Help them out. We have to learn to live with this."

Third, GM began to listen to what critics and outsiders were saying. After the failure of Ralph Nader's 1970 "Campaign GM" to place consumer representatives on the board of directors, the board itself added outsiders and formed a public policy committee to examine and oversee GM's public affairs positions. The outside directors were later responsible for influencing the company's response to two critical issues of the early 1970s, environmental policy and energy policy. In 1971 General Motors took the unusual step of hiring two senior managers from the outside: Ernest S. Starkman, an engineering professor at Berkeley, became vice-president for environmental activities, and Harvard Business School professor Stephen H. Fuller was appointed vice-president for personnel administration.

In the early 1970s senior management and the board together sensed that big changes were coming. In April 1973, the company switched from a five-year to a ten-year horizon in product planning. This long-term perspective enabled the company to see, however dimly, that the U.S. market would eventually become a battleground for global competition and that foreign automakers enjoyed certain advantages. Accordingly, GM began investing about $1 billion annually on research and development, designing new ways to reduce automobile weight and increase engine efficiency without sacrificing the company's image as a builder of roomy cars for families. This approach paid off. Between 1972 and 1978 GM's fleet's average mileage per gallon rose from America's worst (12 mpg) to America's best (17.8 mpg). After spending some $20 billion during the 1970s, GM has budgeted $80 billion for the 1980s, with plans to compete directly against imports and produce an electric-powered commuter vehicle by the end of the decade. Furthermore, GM became a more aggressive competitor in overseas markets. In 1972 it purchased a 34% interest in Isuzu, a Japanese firm.

In 1973 GM also committed itself to a new kind of organizational coordinating mechanism, the project center, to link corporate staff and line management in the car-making divisions. Traditionally, individual divisions, called "lead divisions," developed new products common to all of the corporation's models. For example, Pontiac was responsible for improvements in air conditioning, Chevrolet for frames, and Buick for brakes. Beginning in 1973 project centers took over some of the lead divisions' functions. The first center, formed to preside over the gradual shrinkage of the coming generation of autos, brought together representatives from each division as well as managers from the design, customer service, and marketing staffs. According to GM's president, this enabled the corporation "to get most of the advantages of divisionalization. . . . The development engineers were on a loan basis from all divisions. . . . We wanted a new design that would give each section of the car as much compatibility as possible across the full GM line." The utility of the project centers became clear later, when GM established separate centers to manage the development of the X-body and J-body cars.

Finally, in 1972 the board ordered that a study of federal energy policy be undertaken. Completed in the spring of 1973, it predicted that energy policy "would have a profound effect" on the corporation's business. When the first energy shock came that October, then, GM was strategically, organizationally, and psychologically prepared. GM rapidly committed itself to a new course by speeding up the development of the Chevette subcompact and the shrunken Cadillac Seville, as well as by "downsizing" its entire fleet in the later 1970s. As a result, GM brought its X-body, front-wheel-drive compacts to market more than a full year before Ford or Chrysler offered similar models.

In the later 1970s GM also strove to overcome the oppressive conformity that had alienated managers like DeLorean. It succeeded in restoring a comparatively high level of managerial commitment by its rejuvenation of the high D and I form and by its consensual decision making, although layoffs in 1974 and 1979–80 worked in the opposite direction. A writer for *Fortune* found it hard in 1978 "to find a top executive at GM who does not evidence enthusiasm for what he or the company is doing." GM insiders attributed this change of mood to the company's ability to generate ideas and decisions without creating conflict. As a marketing executive put it:

> In this company there is a real competition of ideas. We thrive on the adversary process. Anyone with a good idea can get heard. The top people have been around a long time. They know each other and their way down in the organization very well. They have had many opportunities to appraise people's actions, their capabilities, their judgments. In this kind of organization, there are rarely single instants of decision. I frequently don't know precisely when a decision is made in General Motors. I don't remember being in a committee meeting when things came to a vote. Usually someone will simply summarize a developing position. Everyone else either nods or states his particular terms of consensus.

The same *Fortune* writer, in an article written in 1981, described General Motors as "essentially a tribal organization," similar to its Japanese rivals. GM executives tended to agree. "Managers come to GM and somehow become part of it—and then it's like a religion," claimed erstwhile outsider Stephen Fuller. "We're more Japanese than we think."

GM became the first of the Big Three to apply the remedies of "tribal organization" to the blue-collar blues. GM reasoned that one solution to low levels of productivity and high levels of absenteeism, turnover, alcoholism, grievances, and industrial sabotage would be to transform the nature of jobs on the assembly line. Beginning in 1968, various experiments in job improvement and enrichment were tried in several GM plants with encouraging results. At an assembly plant in Lakewood, Georgia, for instance, first-line supervisors were trained to be more open and less formal with their subordinates. As the number of grievances dropped, hourly employees were encouraged to help rearrange work areas in order to make their job tasks more interesting and productive. To accomplish this, a plant manager recalled, "It was necessary to push the decision-making process to the lowest level of the organization."

In 1973 the UAW lent its formal support to what union vice-president Irving Bluestone dubbed GM's "Quality of Work Life" (QWL) efforts.* QWL proceeded in a decentralized fashion, with local plant managers and union representatives working together to define and implement the concept. The program has occasionally met opposition from local managers who see it as a surrender of their traditional prerogatives, and from suspicious workers who fear that it is a cover for speedups and heavier work loads. Nonetheless, participation in QWL programs has become mandatory for management and voluntary for labor, and the effort has become generally popular with both sides. Where QWL has won employee acceptance, the results have been impressive. An assembly plant in Tarrytown, New York, for instance, had a poor record of quality performance, as measured by inspection counts or dealer complaints. Quality performance improved so much under QWL that the plant went from being one of GM's worst to one of its best. At the same time, absenteeism and the filing of grievances dropped significantly.

A diesel engine plant in Bay City, Michigan, produced an immediate payoff under QWL. Before the plant opened, the general manager calculated that with an investment of about $350,000 in machinery and equipment, engine push rods could be built for 29 cents apiece. Before this investment was made, however, the manager met with the employees who would be producing the rods and, as he recalled,

> One operator suggested that he could handle six cut-off machines instead of four by modifying our material handling system which simplified his job. Another operator stated that he could run *two* welders instead of one with a minor rearrangement. The jobsetter said that with the rebuilt equipment he could service this job along with his current assignment. Another operator noted that he could use a current welder for both service and the diesel rod, thus saving the purchase of new equipment. In addition, there were several other proposals to improve the efficiency of the operation. The final result of this effort is the elimination of [some] investment and piece cost reduction of five cents. This cost reduction will result in a savings to the corporation of sixteen cents per engine.

QWL has gained the firm support of GM's top management and most of the directors of the UAW. At the end of 1980, 84 of 155 GM bargaining plants had some sort of QWL program in place. The program has not been an unqualified success, however. Several plants, including, ironically, Lakewood, Georgia, have moved away from QWL back to the old way of doing business. Furthermore, serious obstacles must be overcome. The shrinkage of the industry in the coming decade will force economic trade-offs between

---

*Ford and Chrysler lagged General Motors in QWL efforts. Ford formally established its Employe Involvement program as part of its contract with the UAW in 1979. Less is known about Chrysler's efforts, although the corporation has had a memorandum of agreement on QWL with the UAW since 1973.

management and labor which will be difficult to negotiate: cost control vs. wage growth, outsourcing vs. employment levels, productivity gains vs. job security. The 1982 contracts create some mechanisms for the two sides to discuss these sensitive issues, although it is much too early to predict confidently what will happen. Nonetheless, the general success of QWL gives reason for hope. Certainly, GM and the UAW remain committed to changing traditional authoritarian practices and adversarial positions. "It is morally right to involve people in the decision-making process," argued Stephen Fuller, "and it would be right even if it didn't lead to improved productivity, profits, and cost."

In sum, while great storms shook the American automobile industry in the 1970s, GM's performance was relatively steady. The reason for this, we suggest, is that GM was better able than its rivals to satisfy the conditions for organizational adaptation and readaptation. It adopted a long-term strategy, supported by the high D and I form and flexible human resource policies from the top to the bottom of the organization. Among the other automakers, Ford appears to be managing more readaptively in the 1980s as well. The success of the Ford-UAW "Employe Involvement Program" (similar to QWL at GM) accounted partly for the company's ability to consummate the 1982 labor contract which increases communication and shares decision-making more evenly between labor and management. This new spirit of cooperation and the mechanisms which now support it portend well for the long term health of the auto industry once it emerges from its immediate crisis.

Conclusion

. . . The first commercial sale of an American automobile took place in the mid-1890s. Early producers entered a rapidly growing market while the capital costs of making cars were very low. Such favorable circumstances quickly attracted many competitors, among them Henry Ford. The industry first grew and then contracted . . . as it grew more difficult to attract start-up capital and as production technology became standardized, favoring larger firms. The oligopoly that eventually emerged by the end of the 1920s resulted from two different kinds of strategies: Ford's emphasis on mass production targeted at a single market, the low-income buyer, and General Motors' strategy of consolidation, tight financial control, and a diversified product line aimed at different segments of the buying public. These two generic strategies, which one writer has dubbed "Fordism" and "Sloanism," were successful until recent events pushed the auto industry into Area 6 conditions.

The Big Three automakers' long dominance of the American market in the four decades after 1930 might suggest that they were well adapted to their environment. Certainly, for most of that time, the business environment was supportive. The spectacular crisis of the late 1970s should give pause for further reflection, however. If the automakers were so well adapted to their environment, then why have their recent performances been so dismal? Admittedly, the business environment of the 1980s is dramatically different from that of the 1960s, yet this change—despite the suddenness of the energy crisis—did not take place overnight. Surely a truly adaptive organization ought to have responded better. What happened?

Our answer is that the internal conditions of the auto companies did not match the external conditions for readaptation during the industry's maturity. None of the companies pursued a readaptive strategy by attempting to balance innovation and efficiency. Only GM went into the 1930s with a high D and I structure. Even there, however, as DeLorean remarked, the integration maintained by the committee structure became stifling while internal differentiation withered. By the 1960s GM's high D and I organization was really low D and high I. Moreover, human resource practices typical of the auto industry did not engage the involvement of most employees in the larger purposes of the business. Bureaucratic and market incentives for managerial personnel were not balanced by clan mechanisms. In terms of power, the finance and, secondarily, the marketing groups overly dominated production and R&D. Finally, the standoff between management and labor did not reflect a working balance of power between the top and the bottom levels of the companies. Autoworkers were well paid, but were given little responsibility or opportunity to improve the production process.

It should not be surprising, then, that the auto industry was unable to sustain satisfactory levels of both efficiency and innovation over time. The great temptation for any organization, of course, is that when it achieves success in one of these areas the other suffers. While the automakers achieved production efficiency, product innovation slowed. Short-term success spawned short-term incentives, which bred more short-term success, a vicious circle until a series of shocks brought the entire automobile industry skidding toward a crash in 1980.

The history of the U.S. auto industry therefore contains an important lesson for managers, workers, and public policymakers: Stability and adaptation are not the same things and are not necessarily mutually supportive. Organizational readaptation involves both striving for resources and learning from experience. It is a difficult goal to achieve and maintain, even in favorable circumstances.

Ideas for Discussion

1. Describe the audience for this article.
2. Why do prescriptive approaches to correcting flaws in the auto industry prove unsatisfactory?
3. What forecast do these authors offer for the future of U.S. auto making?

Topics for Writing

1. As Alfred Sloan, respond to Lawrence and Dyer's chapter in the format of your choice.
2. Write a report on the impact of management decisions on workers' lives.

Andrea Gabor
America Rediscovers W. Edwards Deming

W. Edwards Deming lived in Washington, D.C. He graduated from Yale in 1928 and received a number of honorary degrees for his work in consulting and statistics. He died in December 1993. Deming served organizations around the world, and in 1950, the Union of Japanese Scientists and Engineers established the Deming Prize for outstanding leadership. This portion of the introduction to Andrea Gabor's The Man Who Discovered Quality *(Times Books: Random House, 1990) sets out the sense of Deming's strategic management theories clearly and concisely.*

> I have never known a concern to make a decided success that did not do good, honest work, and even in these days of the fiercest competition, when everything would seem to be a matter of price, there lies still at the root of great business success the very much more important factor of quality. The effect of attention to quality upon every man in the service, from the president of the concern down to the humblest laborer, cannot be overestimated.
>
> —*Andrew Carnegie*

THE NAME AND reputation of W. Edwards Deming first came to the attention of a few Detroit auto executives in 1978, when they began taking fact-finding missions to Japan to figure out how the Japanese automakers were outclassing them in car quality and design. In January 1981 Donald E. Petersen, the new president of Ford, made an urgent appeal to Deming. Would Dr. Deming be willing to come to Ford and help the company lick its quality problems? It appeared to be a highly unusual SOS, to say the least. Half a world away, Deming's name had become almost synonymous with Japanese quality. In the United States, however, he was virtually unknown in executive suites and boardrooms. Moreover, the octogenarian's professional life had been spent as a statistician and academic. He had never held a full-time job in a corporation, never built a company, never even bothered to market his services.

Ford, it turned out, was one of the first major U.S. companies to "discover" Deming's expertise in quality management. The association between

Detroit's second-largest automaker and America's forgotten quality pioneer would change history for both of them. Less than a decade after their first encounter, Ford would be hailed as a model of American management, and Petersen would lay much of the credit at Deming's feet. "We are moving toward building a quality culture at Ford, and the many changes that have been taking place here have their roots directly in Dr. Deming's teachings," says Petersen, who was CEO until 1990. Meanwhile, Ford's success would help turn Deming into the most sought-after quality expert in America, his ideas serving as an inspiration to hundreds of companies.

Deming swooped into Detroit like a tornado that had spent eighty-one years gathering strength. He ripped the lid off prevailing assumptions about the reasons for the United States' competitiveness problems, which had long focused on Japan's manufacturing advantages. First, the theory went, Japan was beating the United States because of low labor rates. But when Japanese wages began to reach parity with U.S. paychecks, nervous Japan watchers pointed to the country's spanking-new factories and its state-of-the-art manufacturing equipment. Then, when General Motors spent close to $70 billion on new technology and acquisitions and *still* lost market share, the nervous Japan watchers seized on the soft yen, certain that the competition's advantage lay in cheap exports. But even before the yen reached an all-time high against the dollar in the late 1980s, cooler heads were beginning to wonder whether they'd been looking at the right symptoms at all.

The day thirty Ford executives gathered, on a chilly February afternoon in 1981, for their first meeting with Deming, they were still convinced their problems lay somewhere between John Doe's paycheck and Cincinnati Milacron's machine tools. They were expecting to hear about cars, about how to transform manufacturing plants that were turning out automobiles with at least 4.5 "things gone wrong" per car, according to one of Ford's traditional quality indices, into operations that could produce trouble-free vehicles. Deming, they knew, had made a name for himself forty years before by popularizing a method of statistical analysis that could help minimize variation and control the quality and consistency of manufacturing output. Although this system, known as statistical process control (SPC), had enjoyed a brief popularity in the United States, American companies had somehow never gotten the hang of it. Deming, the Ford executives thought, would help them apply such techniques properly.

The Ford men couldn't have been further off the mark. While the guru touched on the importance of statistical theory and statistical thinking in that first meeting, he didn't want to talk about cars or the reject rates on the production line. Nor did he deliver conventional bromides about quality, such as that everything would be okay if everyone just worked a little harder. Instead, what Deming really wanted to know about were processes and people and how they were managed at Ford. He wanted to know about the executives sitting in the room, and what they understood their responsibilities to be—to the company, to their employees, and to the customer. Deming has developed a philosophy of quality management that is rooted in an understanding of the power and pervasiveness of variation and how it affects the process, that delicate interaction of people, machines, materials, and the environment. All systems are subject to some amount of variation that leads to inconsistency and, eventually, to an erosion of both process and product quality. Inconsistency makes it difficult for management to predict how its systems and strategies will perform, and the degradation of quality inevitably results in a loss to the organization. Deming's teachings on variation give management the vital knowledge it needs to recognize when a problem is the result of an isolated glitch in an otherwise well-run organization and when it is the result of deep-rooted systemic problems. Thus, an understanding of variation is vital to managing change.

Deming started out looking at processes through the lens of the scientist, studying the effects of variation on a multitude of individual processes. This unique perspective ultimately led him to develop an all-encompassing quality blueprint that helps management hone the focus of the company and, ultimately, improve and optimize the organization as a whole. Deming's system, known as the Fourteen Points, ties together disparate process-oriented management ideas into a single, holistic vision of how companies can anticipate and meet the desires of the customer by fostering a better understanding of "the process" and by enlisting the help of every employee, division, and supplier in the improvement effort.

Deming and his ideas about variation have revived, and redefined, SPC, a way of determining whether a process is producing predictable results and a basic tool for identifying both immediate systemic problems and opportunities for improvement. One of the most powerful characteristics of this methodology is that, if used properly, data derived through SPC can literally be used to predict how a process will function in the future, thereby making it possible to avoid quality problems before they happen. Consider a simple example on a production line: By monitoring the minute fluctuations over time in the dimensions of supposedly identical parts produced by a machine, SPC can in many cases predict *when* the machine's tool bit is likely to wear out, *before* it begins producing faulty products. Ultimately, an understanding of variation can give management a powerful predictive tool that it can use to help manage change in almost every business process.

Today, at companies such as Ford, General Motors, and Florida Power & Light, which have all come under the influence of Deming, traditional SPC has given way to a kind of analysis, variation control, and improvement methodology that embraces far more than just manufacturing. Achieving major improvements in the quality of a part coming off the assembly line can rarely be accomplished by upgrading *only* the production process. It generally requires improvements in design, which in turn call for more detailed market research and close coordination with suppliers—in other words, process optimization throughout the organization.

Tuning in to the Customer

* * *

A company that has a well-defined customer focus and that successfully manages the knowledge gained through incremental improvements is much more likely to trigger innovation than is an organization that is satisfied with the status quo. America's inability to best the Japanese in product innovation, even though we continue to lead in basic science and invention, points directly to the fact that U.S. companies lag far behind the Japanese in mastery of "the process." Just as SPC helps employees analyze, understand, and improve individual processes, so companies that harness such knowledge throughout an organization are more likely to come up with innovations and less likely to squander the seeds of creativity within their ranks. C. K. Prahalad and Gary Hamel refer to this process as recognizing and developing "core competencies"; it was, for example, Honda's ability to exploit its strength in designing lightweight, powerful engines that made it possible for the upstart motorcycle manufacturer to break into the auto business with a unique, winning product. By contrast, the failure of both Xerox and Procter & Gamble to capitalize on potential blockbuster inventions in the 1970s was attributable in large part to their failure to manage institutional knowledge or "core competencies." Although Xerox did some of the earliest work in personal computers, it was unable to translate its inventions into leadership in the computer business because top management didn't understand the significance of the innovations made by its computer scientists. And P&G began to commercialize sucrose polyester (SPE), a low-calorie fat substitute that could be used to make everything from low-cal cookies to ice cream, only some twenty years after researchers first discovered the substance. In what was then a relatively rigid organizational structure at P&G, which thought of itself as a maker of foods, not food ingredients, SPE became an "orphan." As a result, Olestra, as the commercial version of SPE became known, would still be wending its way through the FDA in 1990, when NutraSweet won approval for its fat substitute, Simplesse, and years after its first patent had expired. It's noteworthy that the top management at both companies came from marketing and financial backgrounds, and thus lacked the knowledge to make difficult decisions about R & D. Their organizations failed at least one crucial test of Demingism: making sure *all* the key corporate disciplines are represented in the decision-making process. As the pace of product development quickens, there is less and less margin for sloppiness in innovation. Both companies have since begun to adopt a holistic approach to quality management that is modeled, to a great extent, on Deming's teachings.

Significantly, an understanding of variation and its effects on an organization can serve as a common language for bridging communications barriers and what often seem like the antithetical interests of different departments within the same company. By contrast, accounting, America's primary corporate language today, more often than not excludes nonfinancial employees and is of little use when it comes to understanding manufacturing's concern over operations, the marketing department's analysis of the customer's needs, or R & D's interest in a new avenue of research.

* * *

In Deming's eyes, the ultimate victim of traditional American management isn't the consumer, who can always cross the street to buy a Toyota, but the American employee, whose job is jeopardized by mismanagement and who is often blamed for management's mistakes. That is one reason Deming has become notorious for publicly insulting executives; his attack on Jim McDonald, the former president of GM, at an industry conference in the early 1980s, is among his more infamous tantrums. That is also why Deming is most eloquent (and most patient) when he is addressing a group of hourly workers. Indeed, contrary to what many executives assume, hourly workers are almost always receptive to Deming's message—in part because he understands their powerlessness to change the system without management's support and because he believes in their good faith. But they also believe in Deming because an organization that is managed for continuous quality improvement is one that can make their jobs more interesting and fulfilling, and often easier.

Deming's suspicion that American management hasn't sufficiently grasped the importance of his message has led him to develop a style that sometimes resembles a maddening, mischievous mixture of Huey Long and *enfant terrible*. He demands absolute loyalty from his followers and brooks no dissent. He is willing to take on only a few clients—companies in which top management has demonstrated its willingness to adopt his philosophy. While he remains eager to learn from respected colleagues and experts, he has encapsulated that philosophy in a set of commandments that his followers must accept without question, on pain of banishment from his circle.

For the vast majority of corporate executives, the four-day seminars Deming gives as often as thirty times a year are about the only chance to see the guru in action. Executives from leading companies around the country flock to the seminars like Holy Rollers to a revival meeting. Each of Deming's one-man shows, which boast little in the way of props or visual aids, attracts five hundred to a thousand managers. He introduces his audience to his Fourteen Points, which are the blueprint of his philosophy. And for four straight days Deming stands at a lectern hammering away at the importance of anticipating the desires of customers, winning the trust and involvement of employees, reducing variation, and constantly improving both processes and products.

The Fourteen Points, which range from working closely with suppliers to eliminating numerical quotas, represent the key components of quality management that Deming believes a company must follow in order to achieve continuous improvement and to "get the customer to come back and bring a friend." There are several reasons why Deming and his Fourteen Points deserve a prominent place in the annals of American management.

First, Deming was extremely influential in shaping what the United States has come to think of as the Japanese management method. He initially went to Japan in 1947, at the behest of General Douglas MacArthur's occupying government, to help conduct a census and to assess the needs of the war-torn country. In the following years he returned repeatedly to lecture on quality and statistical theory, and in 1951 the Japanese created the Deming Prize, an award for quality that has been influential in shaping Japanese business practices. Although experts have long assumed that his principal influence was in teaching the Japanese about statistics, far broader themes emerge from his early lecture notes. Deming talked about the importance of market research, working closely with suppliers, and the need to control variation in every process of a business. Moreover, by gaining broad acceptance in Japan, Deming's teaching and the Deming Prize criteria helped institutionalize TQC and an approach to quality management known as "policy deployment" or *kaizen*; the two concepts created the framework for Japan's quality revolution.

There was also something about the way Deming approached the Japanese. When he stood before the recently vanquished enemy of the United States, Deming spoke not as a conquerer but as a man who had grown up poor, the son of Wyoming pioneers, and who understood the hardships involved in building something from nothing. He would, for example, buy out the American PXs and distribute food and sweets to the Japanese he met during his travels, many of whom were homeless and living on the streets. Deming bridged the cultural divide as few other Americans at that time did. Tatsuro Toyoda, executive vice president of Toyota Motor Corporation and scion of the family that founded the company, remembers Deming best for his sensitivity to the unique needs of other countries and cultures.

Deming also has become by far the most influential proponent of quality management in the United States. While both Joseph Juran and Armand V. Feigenbaum have strong reputations and advocate approaches to quality that in many cases overlap with Deming's ideas, neither has achieved the stature of Deming. One reason is that while these experts have often taken a very nuts-and-bolts, practical approach to quality improvement, Deming has played the role of a visionary, distilling disparate management ideas into a compelling new philosophy. While his knowledge and technical expertise are beyond dispute, he believes it is more useful to ask questions than to give answers, an approach that has gained popularity as executives of troubled companies grow wary of easy solutions. Amid the growing din of consultants who have come to profess expertise in everything from "leadership" to "quality" during the 1980s, Deming's voice has emerged as the country's most legitimate and trusted source of advice. And he has been sought out by companies around the world. His four-day seminars attract hundreds of managers from such companies as AT&T, P&G, and Xerox. Executives at a division of Dow Canada credit Deming with helping to stanch the flow of red ink at their company in the early 1980s. The former head of Nashua Corporation, a New Hampshire maker of office paper and equipment, says Deming's ideas saved some of Nashua's most troubled product lines from the ravages of foreign competition. After watching its market share being eroded throughout the 1980s, GM seized on a quality management strategy that it modeled on Deming's philosophy. Several business schools, including those of Fordham and Columbia universities in New York, have revised their curricula or updated existing courses to include a heavier emphasis on quality and process management. Even the Department of Defense has called on Deming to conduct seminars for generals and top Pentagon officials, and is fashioning a new quality management strategy that draws on Deming's philosophy.

Deming's enormous legitimacy also stems from the messianic zeal and total commitment with which he pursues his vision of management. Deming works out of a basement office in his Washington home, where his only employee is Ceil Kilian, his secretary of more than thirty years. Deming has never solicited a single client. Yet the demands on Deming's time and his dedication are so great that through the long months of his wife Lola's illness, before her death in 1987, Deming continued to spend most weekdays on the road, frequently interrupting client meetings to phone home. Well into his eighties, Deming's usual business itinerary includes a visit to at least three cities each week. He spends weekends working at home in Washington and Mondays in New York City, where he frequently teaches a seminar at Columbia University in the mornings, then takes the subway to NYU. Then he spends three or four days either visiting a client company or conducting a four-day seminar. And while he lavishes time on his students, often taking groups of them out for an Indian meal in Greenwich Village, where he maintains a small, sparsely furnished studio apartment, getting on Deming's client roster is almost impossible. Deming, who was knifed in the ribs one evening when he refused to give his wallet to a mugger on his way to mailing a letter in Greenwich Village, has become a virtual legend among students and followers.

Deming works closely with only a handful of companies—Chrysler Corporation is one he was forced to turn away for lack of time. Client companies are ones Deming has agreed to visit every month because the top executives have demonstrated their commitment to his philosophy and because they are willing to meet regularly with him, answer his questions, and take his advice. Client companies must send their executives to Deming's four-day seminars. And they must hire a master statistician of Deming's choosing to help train both workers and management in the proper use of statistical theory and methods and to help guide them through his philosophy.

Deming's Fourteen Points

Deming's modest goal in formulating the Fourteen Points was to create the management equivalent of the Ten Commandments. They do, in fact, crystallize the key quality management practices that have come to be accepted at most high-quality companies in the United States and Japan. None of the commandments stands on its own. Each is part of a holistic guide to building customer awareness, to reducing variation, and to nurturing constant change and improvement throughout a corporation. The Fourteen Points lead to detailed analysis of everything from customers' desires to the decision-making process itself, which is often fraught with erroneous assumptions.

While analysis is unquestionably a staple of most businesses, that which increasingly distinguishes the winners from the losers in our information-laden world is the incisiveness of the analysis. In the competition for global markets, the Japanese outanalyzed the United States, then mustered the corporate willpower to find better and more efficient ways of acting on their conclusions. They were able to do this in part because an understanding of variation and the concept of continuous improvement gave every member of the company a common focus when discussing problems and changes.

The Fourteen Points are, in fact, based on the following six principal ideas:

1. Quality is defined by the customer. Improvement in products and processes must be aimed at anticipating customers' future needs. Quality comes from improving the process, not from "inspecting out" the shoddy results of a poorly run process.
2. Understanding and reducing variation in every process is a must.
3. All significant, long-lasting quality improvements must emanate from top management's commitment to improvement, as well as its understanding of the means by which systematic change is to be achieved. Improvement *cannot* come merely from middle managers' and workers' "trying harder." Neither quality improvement nor long-term profitability can be achieved through wishful thinking and arbitrary goals set without consideration for how they are to be achieved within the context of an organization's process capabilities.
4. Change and improvement must be continuous and all-encompassing. It must involve every member in an organization, including outside suppliers.
5. The ongoing education and training of all the employees in a company are a prerequisite for achieving the sort of analysis that is needed for constant improvement.
6. Performance ratings that seek to measure the contribution of individual employees are usually destructive. Given a chance by management, the vast majority of employees will take pride in their work and strive for

improvement. But performance-ranking schemes can impede natural initiative. For one thing, by their very nature they create more "losers" than "winners" and thus batter morale. And since they don't take into account natural variation, they are inaccurate and unfair, and are perceived as such by employees.

In this rendering of Deming's Fourteen Points, the points themselves have been reordered in the interest of highlighting some of the synergies between them. For example, in Deming's book, his exhortation to "improve constantly . . . the system of production and service" is number five. However, it appears here right after Deming's point number one, constancy of purpose, both because these two ideas are very closely linked and because Deming's definition of constant improvement is central to his philosophy.

Establish Constance of Purpose

Constancy of purpose, on a macro level, entails an unequivocal long-term commitment to invest in, and adapt to, the challenging requirements of the marketplace. It is the antithesis of managing for short-term financial gain. Constancy of purpose, on a micro level, entails the systematic fine-tuning of every function in a corporation around the changes in company strategy and product line that are needed to meet long-term market needs.

Deming's concept of constancy begins and ends with the customer. While U.S. companies initially turned to Deming because they had lost control of their processes and discovered that they were producing far more faulty products than the competition was, they soon discovered that eliminating defects isn't enough to capture markets. Success depends on how well a company evaluates the processes, products, and markets of today to figure out what the customer will want tomorrow, and whether a company has the management conviction to change accordingly. It requires a commitment to long-term strategies and the analytical know-how to accurately gauge where organizational changes need to be made.

U.S. companies may think this obvious. But the evidence of the marketplace shows that many have become so sidetracked by short-term interests that even if they do have a long-term strategy it often lacks commitment from top management and is frequently undermined by contradictory policies and actions. GM's lack of a consistent vision led the company to spend some $70 billion on acquisitions and new technology as a way to help stem its market share erosion and to improve quality, while at the same time it pinched pennies on product development projects like the Fiero. This lack of constancy undermined the Fiero and GM's need to build cars of distinction, while at the same time it diminished the value of the company's technology investments.

Once top management has correctly identified its problems, it must figure out the best way to deploy its improvement strategy in such a way as to make certain that the entire company stays on course and maintains its constancy of purpose. Florida Power & Light learned this lesson, and will show how even the best-laid quality management plans were nearly derailed when FP&L left middle management out of the planning process.

Improve Constantly and Forever Every System of Production and Service

During his early trips to Japan, Deming introduced an approach to process analysis and improvement that the Japanese applied to the development of products, and ultimately to the establishment and execution of strategic plans. The device is known in Japan as the Deming cycle, even though it was originally developed by Walter Shewhart, a Bell Laboratories physicist who was Deming's friend and mentor. The Deming cycle has become both a metaphor for constancy of purpose and a principal method used to achieve continual improvement.

The Deming cycle involves constantly defining and refining the wishes of customers, and is at the same time a vehicle for rallying every function in the business around these desires. Thus, the Deming cycle hinges on the constant cooperation of different departments, including research, design, production, and sales, so that the corporate eye never wavers from either the customer or any part of the process that might affect the integrity of the product being built for the customer. As applied to product development, the Deming cycle, which has also become widely known as the PDCA cycle, for "plan, do, check, act, and analyze," works something like this:

1. Plan the product with the help of consumer research, and design it.
2. Make the product and analyze it.
3. Market the product.
4. Test how the quality, price, and features of the item are received by consumers, both those who buy the product and those who choose not to buy it.

The final step inevitably leads to redesign and improvement. At leading Japanese companies and a few U.S. pioneers in quality management, the process of analysis behind the Deming cycle has permeated every level of decision making from product development to strategic planning.

The Deming cycle is conceptually the mirror opposite of the traditional U.S. approach, which could be described as follows: Design, make, and sell, sell, sell.

Eliminate Numerical Goals and Quotas, Including Management by Objective

The problem with management by objective (MBO), as it is generally practiced, is that an organization can usually achieve almost any objective it wishes to, in the short term, by paying a high enough price, including, in extreme cases, destroying the system itself. By definition, MBO focuses on the end goal rather than the process. For example, almost any company that is losing money can show a profit if it juggles the books and sells off its healthiest operations. Long-term, however, that company has probably made its situation worse. As Deming puts it, "A quota is a fortress against improvement of quality and productivity. I have yet to see a quota that includes any trace of a system by which to help anyone to do a better job."

Similarly, Deming argues that workers should never be subjected to a quota because they can work only as well as the system permits—assuming it is in control. Deming uses one of his students at NYU and her job as a telephone reservations clerk for an airline as an example. "She must take 25 calls per hour. She must be courteous. . . . She is continually plagued by obstacles: (a) the computer is slow in delivery of information that she asks for; (b) it sometimes reports no information, whereupon she is forced to use directories and guides. Christine, what is your job? Is it: To take 25 calls per hour? Or to give callers courteous satisfaction? . . . It cannot be both."

Deming cites a litany of even more absurd examples. Take, for instance, the city of Alexandria, Virginia, where police officers are given outstanding ratings for issuing twenty-five or more traffic tickets and twenty-one or more parking tickets per month. The obvious conclusion is that whether or not a driver gets written up may depend as much on how close an officer is to filling his unofficial quota as it does on the driver's driving and parking habits. Then there is the federal mediator who is rated on the number of meetings he attends annually. Deming likes to cite a U.S. Postal Service buyer who is rated on the basis of the number of contracts she negotiates during the year—the system clearly discourages complex, long-term agreements that might be in the better interest of the Postal Service.

Drive Out Fear so that Everyone May Work Effectively for the Company

"Our managers aren't interested in good news, they only want to know the bad news." No one familiar with U.S. business culture could ever mistake this statement as being typical of the attitude of American management. In fact, it was used by a Japanese quality expert, Ichiro Miyauchi, to describe a crucial difference between management as it is practiced in Japan and in the United States. A Japanese manager, well versed in quality management, isn't interested in "good news" because it is unlikely to reveal opportunities for improvement. He is interested only in "bad news," which offers a gold mine of improvement possibilities.

American management must learn to appreciate the opportunities in bad news. But first it must begin to change a business culture that has grown used to killing the messenger of bad tidings. "No one can put in his best performance unless he feels secure. *Secure* means without fear, not afraid to express ideas, not afraid to ask questions. . . . A common denominator of fear in any form, anywhere, is loss from impaired performance and padded figures," writes Deming in *Out of the Crisis.*

Institute Leadership

Leadership is a natural corollary of managing *without* fear. "The aim of leadership should be to help people, machines and gadgets to do a better job."

Leadership, by Deming's definition, involves transforming the role of both the manager and the production supervisor from that of cop to coach. Thus, management reviews at companies such as Xerox and FP&L have shifted away from their earlier, virtually exclusive focus on the results of financial performance. Today, management reviews at these companies involve a discussion of problems and potential solutions. . . .

End the Practice of Awarding Business Largely on the Basis of Price

No division or company works in a vacuum. Variability can creep into the process through the parts shipped by suppliers, adding inspection and correction costs. Therefore, continual improvement to a system can only be accomplished if the suppliers, whose output constitutes as much as 80 percent of many finished products, are able to deliver at a predictable, and continuously improving, level of quality. Therefore, Deming calls on companies to "move toward a single supplier for any one item on a long-term relationship of loyalty and trust." While Deming acknowledges that it is not always practical to use a single supplier, he insists that the customer company work closely with suppliers in order to convey their needs effectively and to help the vendor improve the quality of its goods while reducing overall cost to the buyer. When quality and consistency are the most important objectives, this often means abandoning the practice of awarding supplier contracts to the lowest bidder. "Price has no meaning without a measure of the quality being purchased," says Deming. In this regard he points out that, for example, the United States, by awarding contracts for the purchase of mass transit equipment to the lowest bidder, has acquired a slew of erratic equipment and "retarded by a generation expansion of mass transit in the U.S."

Moreover, companies that want to move to just-in-time inventory control procedures as a way to hold down costs have no choice but to adopt Deming's supplier directive. Eliminating just-in-case inventories means that a manufacturer must be able to rely on his supplier to deliver top-quality goods precisely on time. Indeed, the point of working closely with suppliers is to get total low cost, rather than just a low purchase price.

Break Down the Barriers Between Departments

Ford chairman Don Petersen refers to this process as "dismantling chimneys." It involves the mobilization of individual corporate fiefdoms to cooperate on common objectives as defined by customer needs and the company's improvement priorities. The need to dismantle chimneys is born of the realization that just as outside suppliers hold enormous sway over a customer company's ability to meet its own quality objectives, various divisions and functions within a corporation—its internal customers and suppliers—affect one another's ability to maintain consistency and control. This logic is behind Xerox's recent reorganization of its sales, service, and administrative functions.

Institute Training on the Job

Controlling a process requires a detailed understanding of the system in question and how variation can affect it. Therefore, it is useful to train as many members of the corporation as possible to recognize when a system is in control or drifting out of control. Workers can do this even more efficiently than quality engineers. In addition, workers and managers need to be trained to identify problems and improvement opportunities. However, learning to understand the effects of variation is just the first step in the training process. Most companies find they must also teach employees who for years have worked in functional fiefdoms how to operate in multidisciplinary teams. "Industry desperately needs to foster teamwork. The only training or education on teamwork our people receive in school is on the athletic field. Teamwork in the classroom is called cheating."

For companies that want to pursue this holistic approach to quality management, the biggest hurdle often involves convincing most of the personnel in nonmanufacturing disciplines that they too create and are affected by variation. Employees in such areas as marketing and product development often believe that training in the use of statistics to spot different kinds of variation applies only to manufacturing. Similarly, subjecting workers to months of training sessions, as many companies do, is useless unless managers go through the same training process. That has certainly been the experience of FP&L and Xerox.

People also tend to overestimate their ability to analyze and pinpoint the information they need to make decisions and to solve problems. Deming talks at great length about the importance of developing operational definitions. Percy Williams Bridgman, a physicist whose book *Reflections of a Physicist* Deming admires, puts it this way: "One of the chief purposes of an operational analysis is to recover the complexities of the primitive situation. . . . Einstein recognized that such apparently simple concepts as length and time have multiple meanings, so that there are different kinds of length for example optical length and tactual length, and that the precise meaning involves the procedure used in obtaining lengths or times in concrete instances. This attitude toward meanings is what I have called 'operational.' . . . A thoroughgoing application of the operational analysis of meanings puts in our hands, I believe, the possibility of eventually eliminating failure of agreement on meanings as a source of friction in human affairs. No more potent instrument of good will can be imagined."

Another reason Deming focuses so heavily on training is that achieving consistency in the output of employees is as important as reducing the variation in the items produced by two different machines or delivered by two different suppliers. As with everything else in a Deming-oriented process, the goal is to bring workers into statistical control—that is, to have their work be as uniform and predictable as possible. Deming argues that once a work group is performing in a stable and predictable manner, defects and problems that occur are the fault not of workers, but rather of the system. Once the performance of the work force is under control, management and workers can begin to search for more efficient ways to perform a job.

Eliminate the annual rating or merit system

Corporate America loves to reward achievement. Deming believes, however, that rating and merit systems are unfair and counterproductive. This is the most controversial and intriguing of his Fourteen Points. It seems to repudiate the American promise that anyone can be a star if only he works hard enough. While it dovetails with the belief in "intrinsic motivation" put forth by other management theorists, it is also a direct outgrowth of the theory of variation.

Deming contends that the time-honored system of performance appraisals, bonuses, and other reward systems that brand a few employees winners and encourage constant competition in the ranks is fundamentally unfair and ultimately harmful to the interests of both companies and employees. He believes that if the system in which people work is predictable—and if management has done its job well in selecting employees—then over time most employees will perform at about the same level, and that only a very few will perform exceptionally well or poorly. Moreover, the influence of variation is such that it is impossible to accurately measure the overall performance of individuals within a variable process. While Deming has been attacked for his views on the subject, he has also begun to attract followers in some surprising places. Both American Cyanamid and GM recently abolished appraisal systems that required managers to rate their employees on a bell curve. In addition, both companies have moved toward systems in which employee evaluations are not ranked numerically at all. Compensation at both companies is increasingly based on the assumption that although there is variation in people, it is impossible to separate the performance of the individual from that of the system, and that employees should therefore be paid on the basis of their experience and responsibilities rather than according to some numerical ranking. Both companies based their new appraisal and compensation schemes on research done within their own companies showing that long-term corporate success often accrues to organizations that foster teamwork and an environment in which an entire group of employees is encouraged to shine, rather than just a select few. GM and American Cyanamid found that ratings, because they are often based on quotas and can almost never be administered fairly, frequently discourage future performance to the detriment of the entire company.

Institute a Vigorous Program of Education and Self-Improvement

Deming's views on training stem from his understanding of variation and his conviction that training is linked directly to a company's ability to maintain and improve processes. While his ideas about education and self-improvement seem to follow directly from the logic of training, they also represent a far more personal view of the nature of work and motivation. A human being deserves to take pride in his work, Deming often says, quoting from Ecclesiastes. A devout Protestant, Deming believes that this pride, or "joy" as he often refers to it, comes from self-improvement and that it is the company's job to offer opportunities for continuous education.

Similarly, Deming believes that a mutual covenant is established between a company and its employees. Just as the employee accepts the responsibility of performing a job to his or her best ability, the company has an obligation to make sure the individual is given meaningful work to do. Deming may have been influenced in his views by the Japanese concept of "lifetime employment," but his sensitivity to the plight of the worker has been evident ever since his days as a student intern at AT&T's Hawthorne plant in the 1920s, which happened to coincide with the beginning of the famous Hawthorne Experiments.

Eliminate Slogans and Exhortations

Since workers alone can do little to change the system, the burden of improvement rests with management. Slogans and exhortations, on the other hand, are at best misleading because they imply that improving quality depends on added effort by individual employees, rather than on a well-functioning system. "Such exhortations only create adversarial relationships, as the bulk of the causes of low quality and low productivity belong to the system and thus lie beyond the power of the work force."

Cease Dependence on Mass Inspection

Deming calls on management to stop depending on inspection. "Routine 100 percent inspection to improve quality is equivalent to planning for defects, acknowledgement that the process has not the capability required for specifications." Deming often quotes Feigenbaum, who estimates that 15 to 40 percent of the manufacturer's cost of U.S.-made products pays for the "waste embedded in it." Of that cost, handling damage alone can equal 5 to 8 percent. By spotting problems in the system early, you can nip them in the bud. Huge savings can be achieved if a system is functioning as it should—if it isn't creating faulty products, if it isn't generating waste, and if the inspection function can be radically reduced. At Nashua Corporation, manufacturing workers trained in statistics helped improve the processes the company uses to make paper products. Deming advocates working toward the virtual elimination of inspection in all but a few critical cases, such as the production of semiconductor chips, in which the production of one or two faulty products is likely, but letting them get into customers' hands is intolerable. Similarly, some inspection is needed to study variation.

However, quality experts point out that as international competition has raised quality standards, many Japanese companies that have honed their processes so that they produce virtually no defects have, nevertheless, reestablished inspection systems in the unlikely event that a faulty product goes down the line.

Adopt the New Philosophy

While many companies have come to understand that they must conduct their businesses differently, few have grasped the enormity of the task that faces them. Some companies adopt SPC in their manufacturing operations only to discover that marketing and sales have been left out of the improvement loop. For example, Ford Motor Company didn't realize a jump in customer satisfaction until it matched advances in manufacturing quality with better product features that were the result of the marketing department's involvement in an overall quality improvement effort. Yet other companies spend millions retraining their workers, but neglect to educate managers about their new role in the process. Quality management, Deming style, is a holistic philosophy that must be adopted in its entirety if it is to work at all.

Deming's philosophy has been widely hailed throughout corporate management, yet in the United States it has rarely been adopted in its totality because of the magnitude of change it requires. "It can be very difficult to make significant changes, especially when you have been in the habit of doing things differently for decades, and especially when the very success that brought you to the positions you now hold is rooted in doing some things, frankly, the wrong way," Deming told a gathering of Ford executives in 1982. "It is going to be hard for you to accept . . . that you were promoted for the wrong reasons a time or two."

Create a Structure in Top Management to Accomplish the Transformation

Every job in an organization is part of a process. And only by understanding the role each job plays in the company's customer-driven strategy can the process be improved. Thus, to achieve transformation, companies must be committed to analyzing every project and every step of a process with a view to constantly bettering it, along the lines of the Deming cycle described earlier.

Since 1981, when he made his Detroit debut, Deming has attracted an enormous management following in the United States. Hardly a single major U.S. company exists that has not been touched by his ideas, either because the companies have themselves learned from them or because their competitors have. Some companies, including Ford, GM, Nashua Corporation, and FP&L, as well as dozens of Japan's Deming Prize winners, have been profoundly affected by Deming's theories concerning the interaction of people, processes, and variation.

Yet nationally, the transformation brought about by Deming's evangelism has been more evolutionary than revolutionary. U.S. companies have been far slower to accept the new management agenda than their Japanese counterparts have. Most U.S. firms have assiduously avoided turning for help to the Japanese themselves. And so, for the most part, the progress of the United States in adopting the quality management principles espoused by Deming was for a long time only as rapid as the stamina and travel schedule of Deming himself would allow.

Ideas for Discussion

1. How would you profile Deming's eventual successful reception as a management theorist?
2. Analyze the structure of Deming's fourteen points of strategic planning.
3. What features of Gabor's style indicate the audience of this book?

Topics for Writing

1. Locate other readings in this book that share features of Deming's ideas for an essay intended to place Deming in context.
2. Select one or more of Deming's points and provide your own analysis of its usefulness to you as a manager.
3. Write your own theory of the role of the customer in manufacturing or business.

Jeffrey Pfeffer
Organizations as Physical Structures

Jeffrey Pfeffer graduated from Stanford University in 1972 and is a professor of business at the University of California at Berkeley. In 1981–82, he was visiting professor at the Harvard Business School. He is the author of five books on organizational management and behavior. His 1982 Organizations and Organizational Theory, *from which the excerpt is taken, won the Terry Book Award in 1984. In 1989, Pfeffer was awarded the Richard D. Irwin Award for Scholarly Contributions to Management.*

ORGANIZATIONS ARE, in many instances, physical entities. They have offices, buildings, factories, furniture, and some degree of physical dispersion or concentration. They come to define spatial as well as social distances between individuals and subunits. They vary not only in terms of their organizational design, the formal network of relationships among roles, tasks, and activities, but also in their physical arrangements. As Collins (1981) has noted, the physical characteristics of organizations are among their more enduring, and activities come to be associated with certain places (for example, meetings in specific conference rooms at prescribed times, informal interactions in other locations, and so forth). In spite of this physical reality of organizations, there has been relatively little systematic work on linking the physical aspects of organizations into organization theory more generally (see Becker, 1981, for an attempt to begin such an analysis). Perhaps it is because the physical effects are almost too obvious to be of interest—after all, people do not walk through partitions or shout through walls, so one might say that the effects of partitions and walls are clear. Perhaps it is also because, as Becker (1981) has argued, the work of designing structures has been placed in the hands of architects and interior designers—and thus outside the purview of organizational analysis. Nevertheless, the effects of physical design are both important and pervasive. What limited literature does exist is strongly suggestive that the physical aspects of organizations are critical in affecting numerous aspects of their functioning. Furthermore, these physical characteristics place constraints on and constitute the context in which social interaction occurs. Thus, the analysis of organizations would seem to begin profitably by considering the physical reality of organizations as social entities.

Measures and Dimensions

To incorporate physical design characteristics into organizational analysis, it is first necessary to have some metrics or dimensions for describing the physical characteristics of organizations. There are some of these currently available in the literature, but it is clear that research is needed to expand and refine this list. The dimensions presented are all those that would seem to have some effects on behavior in organizations, but the list is far from exhaustive.

Physical arrangements can be first of all characterized by their size. Are the buildings or offices in the buildings large or small? Many firms have implicit and, on occasion, explicit standards assigning so many square feet of space to different departments or to employees of different ranks. Interestingly, this space assignment is most often hierarchically based, which means that the actual requirements of the work are not taken into account to any great degree; rather, status in the organizational hierarchy is the most important determinant of space. Size is an important dimension of the environment because of the symbolic and expressive effects of large size. As Becker (1981: 9) has noted, "Our physical surroundings serve symbolic and expressive purposes as well as instrumental ones." The competition for building height becomes a competition for status and prominence among organizations, as the competition for office space becomes a competition for status among individuals and subunits within the organization. Thus, business schools covet their own buildings, and if the buildings are large, that is all the better. A colleague in sociology was persuaded from taking an offer at another university in part by the promise to set up a research institute for him, with its own distinct building. The fact of the building provided symbolic assurance as to the importance of the endeavor, particularly on a campus in which space was the most critical and scarce resource. Similar events are commonplace in the medical school at the same university, in which the rewards offered for remaining are frequently in the form of more laboratory space and even, on occasion, separate buildings and facilities. Thus, size is the first dimension to be considered in assessing the physical nature of organizations.

Related to size, but distinct from it, is the quality of the physical space. Is it spartan or richly decorated? If size is assessed in square feet per employee or square feet per division or office, then quality might be assessed as the dollars spent per square foot on decoration and finish. Again, hierarchy is an important determinant of the quality of the space. At a major San Francisco law firm, when the firm moved into new quarters, each employee was given a budget to be spent in decorating his or her office, with furniture, paintings, and the like. The amount of the budget differed between the lawyers and the paraprofessionals; within the ranks of the lawyers, between the partners and the associates; and within the ranks of the partners, by length of service with the firm. Thus, one can tell a great deal about the individual one is visiting just by the characteristics of the decor of the office.

Size and quality are important symbolic aspects of physical space, but they also have effects on work quantity and quality. Another important symbolic dimension of space, which also has implications for performance, is the flexibility of the space. At the building or system level, this reflects whether walls and partitions are readily movable, so that new arrangements can be designed. At the office or subsystem level, this reflects things such as the availability of electrical outlets and walls and furniture configurations that permit rearrangement. Flexibility also is evident in varying degrees in classroom layouts. Most readers will be familiar with the two extremes—on the one hand, fixed desks and chairs, as found in amphitheater-type classrooms, in which no rearrangement is possible; on the other hand, separate desk-chair assemblies that are not fixed to the floor and that can be rearranged in any configuration without difficulty. Classrooms are also more flexible to the extent they have blackboards on several walls (providing more opportunity for changing what is the "front" of the room) and multiple doors. Flexibility permits the design of space to fit the people and the tasks currently occupying the space. A poignant illustration of lack of flexibility and its consequences comes to mind. I once taught at a university in a classroom in which there were wooden desk-seat assemblies that were linked together by a board running through the legs. One might presume this was done to make it more difficult to steal the chairs and their associated desk tops. In any event, the seats were attached too close for easy comfort if people occupied every chair. One student, finally tiring of having to sit uncomfortably close to his neighbors, brought a saw to class and separated his desk from the others, reintroducing flexibility in seating arrangements in a way not quite anticipated by the designers of the furniture.

A fourth important dimension of physical space is arrangement. Arrangement has several aspects, one of the more important being the distance between people or facilities. Thus, we can ask, How far is it from the personnel department to the controller's office? That question can be answered in terms of the adjacency in vertical space (what is on the same floor) as well as horizontal space, in terms of how many feet it is from one office or one department to another. Within offices, arrangement consists of how the furniture is oriented. Is the desk between the door and the office occupant, for instance, or is the desk against one of the side or back walls, which would produce a different physical orientation between the office occupant and a visitor to the office? In classrooms, one might ask if the furniture was arranged in a circle, or if the desks were all oriented toward the front of the room. Arrangement has both symbolic, evocative effects on persons as well as consequences for the amount and type of social interaction that occurs.

One is, other things being equal, more likely to interact with those who are physically close or adjacent.

A fifth, related dimension, is that of privacy. Privacy is in part a function of the amount of space per person and the arrangement of that space. But, it is also a function of the use of walls, partitions, solid versus glass doors, and so forth, as well as the degree of sound- or noise-proofing designed into the construction. Privacy, as with many of the other dimensions, has symbolic as well as substantive importance. The symbolic importance comes from the fact that privacy often indicates hierarchical rank. A private rather than a shared office may be reserved for professionals rather than clerical workers and for those more senior in the chain of command. The substantive impact can occur because of the effects of the presence of others, or the social facilitation effect (Zajonc, 1965), on task performance.

Location is the sixth dimension that is important in understanding the physical dimension of organizations. By location we mean where, in terms of quality of the neighborhood, type of neighborhood, the organization is set, that is, the placement of the organization and its various facilities. For example, Safeway, a retail grocery store chain with annual sales of well over $10 billion, has its corporate headquarters in an older, three- or four-story building located in the produce market and industrial warehouse section of Oakland, California. There are no other corporate headquarters nearby. By contrast, Fidelity Financial, the parent company of Fidelity Savings and Loan, a $2 billion California savings and loan, built a new headquarters building near Kaiser Center in Oakland, a multistory building near Lake Merritt and numerous other high-rises. This was done even as Fidelity incurred substantial financial difficulties. In further contrast, other companies are willing to pay premium prices to have their headquarters in San Francisco. Thus, Shaklee, the food supplement concern, left a headquarters building in Emeryville, on the bay with a view of San Francisco, to build its own building in San Francisco at substantially increased occupancy expense. Similarly, one can inquire about satellite facilities to see where they are placed, in what kind of setting, and near what other buildings. Location also has symbolic and substantive consequences. The substantive consequences include the fact that location may impact the labor market in which the organization recruits, as well as its ease or difficulty of attracting employees.

These six aspects of physical space—size, quality, flexibility, arrangement, privacy, and location—are only a beginning in terms of the types of dimensions one might use to describe the physical aspects of organizations. Moreover, it should be clear that the measurement and dimensionalization of these variables remains as an important task. Nevertheless, they do represent one set of important aspects that can be studied. They have been selected because all have at least some literature associated with them treating their consequences. But, first, it is necessary to understand the determinants of these dimensions of the physical manifestation of organizations.

Some Causes of Particular Physical Dimensions

Interestingly, although there has been some limited research investigating the consequences of some of these physical dimensions of organizations, there is almost no work exploring how and why organizations come to be physically located and structured the way they are. The two sets of factors that have been considered are the requirements for space necessitated by the technology and the work flow, and the effects of power and influence on design.

The effects of technology on design appear most strongly in the layout of machine shops, assembly lines, and process production facilities. But even here, the technical requirements are scarcely as binding as one might think. After all, automobiles can be assembled as they roll down an assembly line, or, as in some Volvo plants, they can be assembled by groups of workers operating in a more circular, as contrasted with linear, spatial arrangement. Moreover, there are numerous instances of design failures from a technical point of view, as in a hospital in Berkeley, California, in which the X-ray facilities were located quite far away from surgery, and this in a newly designed facility. Thus, technical considerations have some effect on the design of organizations in terms of setting space requirements and arrangement, but they are far from the most important factor in most circumstances. As Braverman (1974) has argued, there are typically a variety of technical arrangements that can be used to accomplish some work. Thus, technical determinism is inadequate as a basis for understanding organizational physical structures.

More important are considerations of power, influence, and social control. As Becker (1981) has argued:

> . . . a major function of the physical setting of organizations can be seen as an attempt to visually and physically reduce the ambiguity of social position and power within the organization by marking distinctions among job classifications with clear signs of spatial privilege (size of office, quality of furnishings) and by minimizing distinctions within job classifications by rendering all environmental support identical (1981: 25–26).

One determinant of design is the job classification system of the organization. Presumably, more elaborate and differentiated job hierarchies are matched with more elaborate and differentiated gradations of physical space. Another determinant of design is the attempt to achieve control through the use of physical constraints on behavior. Becker (1981: 25) argued, "Social control, in terms of the maintenance of established patterns of influence among persons occupying different positions in the organizational hierarchy . . . becomes a critical attribute and major form determinant of the environmental-support system." He goes on to note:

... the goal of design becomes one of removing as many options as possible from the workers so that discretionary activities are difficult or impossible. ... If individual differences cannot be stamped out, and attempts to do so are counterproductive, why do they continue? The answer seems to lie in a view of control as a value in its own right, irrespective of its effect on efficiency. The absence of control is equated with an inherent tendency toward disorder and dissolution in the organization. It is also seen as undermining a necessary social order (1981: 68–69).

One might speculate, then, that the technical deskilling noted by Braverman (1974) and the shift in the form of control hypothesized by R. Edwards (1979) should be manifest not only in terms of changing technical conditions of work and in changing forms of control but also in changing physical arrangements and structures associated with organizations. As a simple example, the shift from personal, hierarchical control to technical and bureaucratic control would clearly permit more geographic dispersion and more physical separation between the supervisors and those being supervised. The shift toward technical control would seem to place greater importance on the design of physical facilities from the point of view of achieving control over behavior. This might be seen in reduced flexibility in aspects of the design of work settings as well as more separation among workers accomplished through noise levels or physical arrangements.

Power and influence are at once symbolically represented by physical arrangements and produced by those very arrangements. Thus, it seems reasonable to argue that "decisions about the nature and use of space and equipment in organizations are political ones (e.g., they concern allocation of scarce resources on the basis of values)" (Becker, 1981: 58). Consequently, it would seem to be reasonable to search for the causes of physical arrangements in those factors of power and influence that have been used to account for the allocation of other resources in organizations as well as to account for the changing patterns of control and the organization of work across organizations.

Some Consequences of Physical Design

There have been more studies of the consequences of physical arrangements, particularly the use of open-space office arrangements, than of the causes of variation in physical designs. However, even in this instance, the research is fairly sparse and much remains to be done. In general, three types of dependent variables have been treated as effects of physical arrangements: the amount of interaction that occurs in a social system, the affective reaction to the job and the organization, and the affective reaction and orientation to those with whom one interacts.

If we consider first the quantity of interaction, the typical supposition underlying the literature has been that more interaction is better than less.

Interaction often leads to interpersonal attraction (Newcomb, 1956; Thibaut and Kelley, 1959; Zajonc, 1968). This may be because interaction has the effect of increasing attitudinal similarity (Newcomb, 1956), an important basis for interpersonal liking (Byrne, 1969). Interpersonal attraction is an important component of effective interpersonal communication, which is necessary for coordination, and also provides rewards for staying in the social system and thereby supplies an important source of social system maintenance.

Interaction, in terms of both quantity and pattern, is profoundly affected by variables such as distance and layout of physical space. Festinger, Schacter, and Back (1950), studying friendship patterns in a student housing complex, found that interaction tended to follow a distance relationship, with those who were in closer physical proximity interacting more and being more likely to be friends. They also found that the design of the buildings in terms of features like having only one stairway (forcing more contact on the single stairway) and having doors that opened to a central hallway, so that people were more likely to run into each other, increased interaction. Studying interaction among people in different seating arrangements around tables, Sommer (1959: 257) noted, "the trend in all the data is that people sitting in neighboring chairs . . . will be more likely to interact than people sitting in distant chairs." Although side-by-side seating in fact produced less interaction than some other (corner, for instance) arrangements, in general, the relationship between distance and the amount of interaction was again found. Vertical distance, as in being on separate floors of a building, is much more disruptive to interaction than horizontal distance. Thus, other things being equal, one would expect to observe more interaction in an organization located on fewer, larger floors than one located in a building having more and smaller floors.

The effect of open-space or open-plan offices on the amount of interaction has also been investigated. In many instances, perceptions or attitudes rather than actual data on interactions have been collected. This leads to reports on what people expect to result from such arrangements rather than what does, in fact, occur. Brookes and Kaplan (1972) observed an increase in reported group sociability after a change to an open-plan office, while Ives and Ferdinands (1974) reported that most employees who moved to an open office believed that communication had increased after the move. However, contrary findings have been more frequently reported, particularly when better data are collected. Oldham and Brass (1979) found that reported interaction did not increase after the change to an open office arrangement. Oldham and Brass argued that for important and meaningful interaction to occur, some degree of privacy was necessary. This privacy is lacking in open office arrangements, and thus these arrangements hinder the interpersonal communication process. Clearwater (1980) found that the landscaped, open office was less adequate on several dimensions of communication behavior, including interoffice communication, communication with different divisions, interaction with supervisors, and the development of friendships with coworkers, than conventional office designs. The rhetoric of enhanced communication may mask the real reasons for moving to open office arrangements. Canty (1977) reported that open office arrangements typically had 50 percent of the square footage per employee of conventional arrangements, which results in a substantial savings of money. And, Becker (1981: 59) suggested that "a major reason for moving to the new form of office environment is to create the impression of increased efficiency through the adoption of the latest in office design." Cost reduction and symbolic affirmation of modern management may outweigh the actual effects on interaction of the open office arrangement.

The effect of physical arrangements on interaction has also been investigated in classroom settings. Sommer (1969), summarizing these studies, reported that interest and involvement in the class, as well as performance, can be predicted from where in the lecture hall the individual happens to be sitting. Persons sitting in more distant locations perform less well and interact less often in class. Ironically, even the Western Electric studies, which are so often cited as evidence of the potency of social factors over the physical environment, provide evidence for the power of physical arrangements to shape interaction patterns and the resulting social consequences.

> The men were working in a room of a certain shape, with fixtures such as benches oriented in a certain way. They were working on materials with certain tools. These things formed the physical and technical environment in which the human relationships more likely to develop in some ways than in others. For instance, the sheer geographical position of the men within the room had something to do with the organization of work and even with the appearance of cliques (Homans, 1950: 80–81).

The effect of physical arrangements on reactions to the job and the organization have also been investigated. Of course, one such effect can be mediated through the extent to which social interaction leads to the development of friendship ties, and we have just seen that this interaction is itself affected by the physical arrangements. There are direct effects as well. Becker (1981) noted:

> As a direct support system for work activities, the physical setting was identified in a national Harris poll of office workers, in 1977, as a major impediment to efficiency. Major problems found were unsuitable office furniture and inadequate office tools, equipment, and materials. Distractions and the lack of privacy were the most negative attributes of offices because these prevented adequate concentration. These kinds of problems influence one's ability to carry out work effectively, as well as acting as "hygienic" factors . . . that influence the total work experience (1981: 55).

Oldham and Brass (1979) reported that the move to the open space office location reduced job satisfaction. Sloan (1972) found that providing individuals with flexibility to arrange and structure their own space led to more positive attitudes toward the job and the work organization. It clearly makes sense that an attractive office or setting that provides both privacy and flexibility for the individual and encourages significant interpersonal interactions will result in more favorable affective responses to work.

Physical arrangements also impact how individuals orient and relate to others. Several studies, for instance, have investigated how physical conditions and the physical setting impact interpersonal perception. Maslow and Mintz (1956) investigated the effects of a beautiful, average, or ugly room on the judgment of energy and well-being in photographs of faces. Subjects rated energy and well-being as higher in the beautiful than in the average or ugly room. In a follow-up study, Mintz (1956) had two examiners test others in a beautiful and an ugly room, alternating rooms between sessions. Mintz found that scores on the rating task were higher in the beautiful room; moreover, in 27 out of 32 instances, the examiner in the ugly room finished before the one working in the attractive room. Sauser, Arauz, and Chambers (1978) had subjects make salary recommendations for simulated candidates in a noisy and a quiet setting. Salaries were significantly higher in the quiet room. Griffitt (1970) found that interpersonal attraction responses were more negative under high and uncomfortable temperature conditions than under conditions of moderate temperature. These studies taken together all indicate that physical arrangements and conditions do impact people's responses to others. Affect is reduced and harsh judgments are increased when the individuals making the judgments are physically uncomfortable because of noise, crowding, temperature, or the quality of the setting.

The effect of office arrangements on affective responses to others has also been investigated. One critical issue in office layout involves the placement of the desk and whether in the interaction one party sits behind the desk—in a position of power—or whether a more collegial, side-by-side arrangement is used. Zweigenhaft (1976) studied faculty office arrangements and the effect of desk location on student interactions and perceptions. He found, in a survey of students, that faculty who placed their desks between themselves and the students were rated less positively on student-faculty interaction than those who used alternative designs (for example, the desk against one wall). Joiner (1971), studying office arrangements in a sample of English organizations, reported that higher-status occupants tended to place the desk between themselves and the door more often and that faculty members as contrasted with businessmen tended to use that kind of arrangement less. D. E. Campbell (1979) found that office furniture arrangement had little effect on student ratings of faculty, although plants and wall posters led to positive ratings and clutter led to negative ratings. Campbell employed an experimental design in which photographs of office arrangements were rated by students. This research avoids confounding variables that might affect the other studies—for instance, that persons who use office arrangements to maintain power relationships are also more distant in other ways as well. Using a similar type of experimental design involving the rating of photographs, Morrow and McElroy (1981) had 100 student subjects rate slides of faculty offices that varied in terms of their tidiness, desk arrangement, and presence of status symbols. Morrow and McElroy (1981: 648) found that desk arrangement did significantly affect subjects' feelings of visitor comfort and visitor welcomeness. They noted that "it is apparent that office occupants do convey nonverbal messages to their visitors through office design" (1981: 650).

Sommer (1969) reported that different seating arrangements are used in situations of casual conversation, competition, or coaction (working together). He has argued that the very setting, because of its association with these different interaction patterns, may help to produce a different type of interaction depending on the seating arrangement. Thus, putting groups across a table from each other may tend to produce more competitive or adversarial interactions, other things being equal, because of the association of this seating arrangement with competitive interactions in the past.

Of course, settings do much to convey authority and leadership. Lecuyer (1976) reported that when groups with leaders were seated at circular tables, the leader was forced to ease group tension arising from the articulation of conflicting points of view. Leaders of groups seated at rectangular tables (at the head) found their ability to direct the group enhanced because of their position. In a second set of studies, Lecuyer found that positions voluntarily chosen at a rectangular table reflected the social relations that had developed previously in discussions around a circular table. Sommer (1969) has also noted the tendency for leaders to assume positions at the head of the table, and, conversely, for those in such physical positions to be treated with more deference and respect, as though they were the leaders. Thus, spatial arrangements affect power and influence perceptions and, as a consequence, achieved influence in group settings.

Steele (1973) has noted how physical settings can assist in organizational development processes. Taking physical space considerations into account can be productive in terms of both affective reaction of employees and productivity. At the Santa Teresa Laboratories of IBM, the particular requirements of computer programmers in terms of working space, furniture design, and conference room and computer terminal access were taken into account in the physical design of the facility. McCue (1978) reported that the results were very positive.

In Figure 1, we summarize the preceding discussion by presenting the determinants of design, the design dimensions, and the outcomes of physical design we have briefly summarized. This is far from a complete model or explication of the effects of physical space. Rather, the intent is to show the importance of physical aspects of organizations and to indicate how research on such aspects might be focused.

It is important to note that the emphasis on organizations as physical, material entities is quite consistent with the perspective on understanding behavior by examining its context, a theme that we have tried to develop throughout this book. Becker (1981) has summarized this argument well:

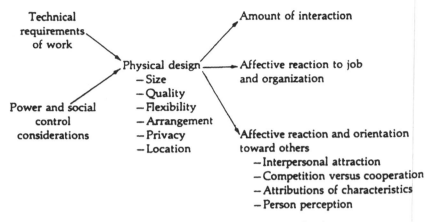

Figure 1 Physical Design and Organizational Behavior

> Ecological psychology was originally developed by Roger Barker (1968) as an alternative to a stimulus-response-oriented model of human behavior. . . . Their most striking conclusion was that the behaviors of children could be predicted more accurately from knowing the situations the children were in than from knowing individual characteristics of the children. This was a major departure from the belief that individuals have stable personality traits that they carry from situation to situation and that, in each situation, guide their behavior in similar ways (1981: 125).

Organizations are, indeed, physical structures that have consequences for interaction. Understanding how such structures emerge and get to look the way they do, as well as understanding their consequences, would seem to be an important place to begin to enrich the analysis of organizations. If settings matter, then we need to pay much more attention to the determinants and consequences of these settings in physical, material terms.

Ideas for Discussion

1. What role does physical dimension play in office relations?
2. What does classroom space have in common with office space?
3. Is Pfeffer's argument legitimate? How so?

Topics for Writing

1. Study your office or classroom space. Prepare a report recommending improvements for better communication.
2. As a manager, share Pfeffer's ideas with your staff in a letter.
3. Research some of the new ergonomic designs for office furniture. As a purchasing agent, write a report recommending that your company buy some of this furniture.

Bowen McCoy
The Parable of the Sadhu

Bowen McCoy is an investment banker and real estate consultant living in Los Angeles. A 1962 Harvard M.B.A., McCoy served with Morgan Stanley until 1985, when he formed his own Buzz McCoy Associates. He received the Ethics Award from the Harvard Business Review *in 1983. McCoy is a frequent contributor to business publications. "The Parable of the Sadhu" appeared in* Ethics in Practice. Managing the Moral Corporation, *edited by Kenneth R. Andrews (1989).*

LAST YEAR, as the first participant in the new six-month sabbatical program that Morgan Stanley has adopted, I enjoyed a rare opportunity to collect my thoughts as well as do some traveling. I spent the first three months in Nepal, walking 600 miles through 200 villages in the Himalayas and climbing some 120,000 vertical feet. On the trip my sole Western companion was an anthropologist who shed light on the cultural patterns of the villages we passed through.

During the Nepal hike, something occurred that has had a powerful impact on my thinking about corporate ethics. Although some might argue that the experience has no relevance to business, it was a situation in which a basic ethical dilemma suddenly intruded into the lives of a group of individuals. How the group responded I think holds a lesson for all organizations no matter how defined.

The Sadhu

The Nepal experience was more rugged and adventuresome than I had anticipated. Most commercial treks last two or three weeks and cover a quarter of the distance we traveled.

My friend Stephen, the anthropologist, and I were halfway through the 60-day Himalayan part of the trip when we reached the high point, an 18,000-foot pass over a crest that we'd have to traverse to reach to the village of Muklinath, an ancient holy place for pilgrims.

Six years earlier I had suffered pulmonary edema, an acute form of altitude sickness, at 16,500 feet in the vicinity of Everest base camp, so we were understandably concerned about what would happen at 18,000 feet. Moreover, the Himalayas were having their wettest spring in 20 years; hip-deep powder and ice had already driven us off one ridge. If we failed to cross the pass, I feared that the last half of our "once in a lifetime" trip would be ruined.

The night before we would try the pass, we camped at a hut at 14,500 feet. In the photos taken at that camp, my face appears wan. The last village we'd passed through was a sturdy two-day walk below us, and I was tired.

During the late afternoon, four backpackers from New Zealand joined us, and we spent most of the night awake, anticipating the climb. Below we could see the fires of two other parties, which turned out to be two Swiss couples and a Japanese hiking club.

To get over the steep part of the climb before the sun melted the steps cut in the ice, we departed at 3:30 A.M. The New Zealanders left first, followed by Stephen and myself, our porters and Sherpas, and then the Swiss. The Japanese lingered in their camp. The sky was clear, and we were confident that no spring storm would erupt that day to close the pass.

At 15,500 feet, it looked to me as if Stephen were shuffling and staggering a bit, which are symptoms of altitude sickness. (The initial stage of altitude sickness brings a headache and nausea. As the condition worsens, a climber may encounter difficult breathing, disorientation, aphasia, and paralysis.) I felt strong, my adrenaline was flowing, but I was very concerned about my ultimate ability to get across. A couple of our porters were also suffering from the height, and Pasang, our Sherpa sirdar (leader), was worried.

Just after daybreak, while we rested at 15,500 feet, one of the New Zealanders, who had gone ahead, came staggering down toward us with a body slung across his shoulders. He dumped the almost naked, barefoot body of an Indian holy man—a sadhu—at my feet. He had found the pilgrim lying on the ice, shivering and suffering from hypothermia. I cradled the sadhu's head and laid him out on the rocks. The New Zealander was angry. He wanted to get across the pass before the bright sun melted the snow. He said, "Look, I've done what I can. You have porters and Sherpa guides. You care for him. We're going on!" He turned and went back up the mountain to join his friends.

I took a carotid pulse and found that the sadhu was still alive. We figured he had probably visited the holy shrines at Muklinath and was on his way home. It was fruitless to question why he had chosen this desperately high route instead of the safe, heavily traveled caravan route through the Kali Gandaki gorge. Or why he was almost naked and with no shoes, or how long he had been lying in the pass. The answers weren't going to solve our problem.

Stephen and the four Swiss began stripping off outer clothing and opening their packs. The sadhu was soon clothed from head to foot. He was not able to walk, but he was very much alive. I looked down the mountain and spotted below the Japanese climbers marching up with a horse.

Without a great deal of thought, I told Stephen and Pasang that I was concerned about withstanding the heights to come and wanted to get over the pass. I took off after several of our porters who had gone ahead.

On the steep part of the ascent where, if the ice steps had given way, I would have slid down about 3,000 feet, I felt vertigo. I stopped for a breather, allowing the Swiss to catch up with me. I inquired about the sadhu and Stephen. They said that the sadhu was fine and that Stephen was just behind. I set off again for the summit.

Stephen arrived at the summit an hour after I did. Still exhilarated by victory, I ran down the snow slope to congratulate him. He was suffering from altitude sickness, walking 15 steps, then stopping, walking 15 steps, then stopping. Pasang accompanied him all the way up. When I reached them, Stephen glared at me and said: "How do you feel about contributing to the death of a fellow man?"

I did not fully comprehend what he meant.

"Is the sadhu dead?" I inquired.

"No," replied Stephen, "but he surely will be!"

After I had gone, and the Swiss had departed not long after, Stephen had remained with the sadhu. When the Japanese had arrived, Stephen had asked to use their horse to transport the sadhu down to the hut. They had refused. He had then asked Pasang to have a group of our porters carry the sadhu. Pasang had resisted the idea, saying that the porters would have to exert all their energy to get themselves over the pass. He had thought they could not carry a man down 1,000 feet to the hut, reclimb the slope, and get across safely before the snow melted. Pasang had pressed Stephen not to delay any longer.

The Sherpas had carried the sadhu down to a rock in the sun at about 15,000 feet and had pointed out the hut another 500 feet below. The Japanese had given him food and drink. When they had last seen him he was listlessly throwing rocks at the Japanese party's dog, which had frightened him.

We do not know if the sadhu lived or died.

For many of the following days and evenings Stephen and I discussed and debated our behavior toward the sadhu. Stephen is a committed Quaker with deep moral vision. He said, "I feel that what happened with the sadhu is a good example of the breakdown between the individual ethic and the corporate ethic. No one person was willing to assume ultimate responsibility for the sadhu. Each was willing to do his bit just so long as it was not too inconvenient. When it got to be a bother, everyone just passed the buck to someone else and took off. Jesus was relevant to a more individualistic stage of society, but how do we interpret his teaching today in a world filled with large, impersonal organizations and groups?"

I defended the larger group, saying, "Look, we all cared. We all stopped and gave aid and comfort. Everyone did his bit. The New Zealander carried him down below the snow line. I took his pulse and suggested we treat him for hypothermia. You and the Swiss gave him clothing and got him warmed up. The Japanese gave him food and water. The Sherpas carried him down to the sun and pointed out the easy trail toward the hut. He was well enough to throw rocks at a dog. What more could we do?"

"You have just described the typical affluent Westerner's response to a problem. Throwing money—in this case food and sweaters—at it, but not solving the fundamentals!" Stephen retorted.

"What would satisfy you?" I said. "Here we are, a group of New Zealanders, Swiss, Americans, and Japanese who have never met before and who are at the apex of one of the most powerful experiences of our lives. Some years the pass is so bad no one gets over it. What right does an almost naked pilgrim who chooses the wrong trail have to disrupt our lives? Even the Sherpas had no interest in risking the trip to help him beyond a certain point."

Stephen calmly rebutted, "I wonder what the Sherpas would have done if the sadhu had been a well-dressed Nepali, or what the Japanese would have done if the sadhu had been a well-dressed Asian, or what you would have done, Buzz, if the sadhu had been a well-dressed Western woman?"

"Where, in your opinion," I asked instead, "is the limit of our responsibility in a situation like this? We had our own well-being to worry about. Our Sherpa guides were unwilling to jeopardize us or the porters for the sadhu. No one else on the mountain was willing to commit himself beyond certain self-imposed limits."

Stephen said, "As individual Christians or people with a Western ethical tradition, we can fulfill our obligations in such a situation only if (1) the sadhu dies in our care, (2) the sadhu demonstrates to us that he could undertake the two-day walk down to the village, or (3) we carry the sadhu for two days down to the village and convince someone there to care for him."

"Leaving the sadhu in the sun with food and clothing, while he demonstrated hand-eye coordination by throwing a rock at a dog, comes close to fulfilling items one and two," I answered. "And it wouldn't have made sense to take him to the village where the people appeared to be far less caring than the Sherpas, so the third condition is impractical. Are you really saying that, no matter what the implications, we should, at the drop of a hat, have changed our entire plan?"

The Individual versus the Group Ethic

Despite my arguments, I felt and continue to feel guilt about the sadhu. I had literally walked through a classic moral dilemma without fully thinking through the consequences. My excuses for my actions include a high adrenaline flow, a super-ordinate goal, and a once-in-a-lifetime opportunity—factors in the usual corporate situation, especially when one is under stress.

Real moral dilemmas are ambiguous, and many of us hike right through them, unaware that they exist. When, usually after the fact, someone makes an issue of them, we tend to resent his or her bringing it up. Often, when the full import of what we have done (or not done) falls on us, we dig into a defensive position from which it is very difficult to emerge. In rare circumstances we may contemplate what we have done from inside a prison.

Had we mountaineers been free of physical and mental stress caused by the effort and the high altitude, we might have treated the sadhu differently. Yet isn't stress the real test of personal and corporate values? The instant decisions executives make under pressure reveal the most about personal and corporate character.

Among the many questions that occur to me when pondering my experience are: What are the practical limits of moral imagination and vision? Is there a collective or institutional ethic beyond the ethics of the individual? At what level of effort or commitment can one discharge one's ethical responsibilities?

Not every ethical dilemma has a right solution. Reasonable people often disagree; otherwise there would be no dilemma. In a business context, however, it is essential that managers agree on a process for dealing with dilemmas.

The sadhu experience offers an interesting parallel to business situations. An immediate response was mandatory. Failure to act was a decision in itself. Up on the mountain we could not resign and submit our résumés to a headhunter. In contrast to philosophy, business involves action and implementation—getting things done. Managers must come up with answers to problems based on what they see and what they allow to influence their decision-making processes. On the mountain, none of us but Stephen realized the true dimensions of the situation we were facing.

One of our problems was that as a group we had no process for developing a consensus. We had no sense of purpose or plan. The difficulties of dealing with the sadhu were so complex that no one person could handle it. Because it did not have a set of preconditions that could guide its action to an acceptable resolution, the group reacted instinctively as individuals. The cross-cultural nature of the group added a further layer of complexity. We had no leader with whom we could all identify and in whose purpose we believed. Only Stephen was willing to take charge, but he could not gain adequate support to care for the sadhu.

Some organizations do have a value system that transcends the personal values of the managers. Such values, which go beyond profitability, are usually revealed when the organization is under stress. People throughout the organization generally accept its values, which, because they are not presented as a rigid list of commandments, may be somewhat ambiguous. The stories people tell, rather than printed materials, transmit these conceptions of what is proper behavior.

For 20 years I have been exposed at senior levels to a variety of corporations and organizations. It is amazing how quickly an outsider can sense the tone and style of an organization and the degree of tolerated openness and freedom to challenge management.

Organizations that do not have a heritage of mutually accepted, shared values tend to become unhinged during stress, with each individual bailing out for himself. In the great takeover battles we have witnessed during past years, companies that had strong cultures drew the wagons around them and fought it out, while other companies saw executives supported by their golden parachutes, bail out of the struggles.

Because corporations and their members are interdependent, for the corporation to be strong the members need to share a preconceived notion of what is correct behavior, a "business ethic," and think of it as a positive force, not a constraint.

As an investment banker I am continually warned by well-meaning lawyers, clients, and associates to be wary of conflicts of interest. Yet if I were to run away from every difficult situation, I wouldn't be an effective investment banker. I have to feel my way through conflicts. An effective manager can't run from risk either; he or she has to confront and deal with risk. To feel "safe" in doing this, managers need the guidelines of an agreed-on process and set of values within the organization.

After my three months in Nepal, I spent three months as an executive-in-residence at both Stanford Business School and the Center for Ethics and Social Policy at the Graduate Theological Union at Berkeley. These six months away from my job gave me time to assimilate 20 years of business experience. My thoughts turned often to the meaning of the leadership role in any large organization. Students at the seminary thought of themselves as antibusiness. But when I questioned them they agreed that they distrusted all large organizations, including the church. They perceived all large organizations as impersonal and opposed to individual values and needs. Yet we all know of organizations where peoples' values and beliefs are respected and their expressions encouraged. What makes the difference? Can we identify the difference and, as a result, manage more effectively?

The word "ethics" turns off many and confuses more. Yet the notions of shared values and an agreed-on process for dealing with adversity and change—what many people mean when they talk about corporate culture—seem to be at the heart of the ethical issue. People who are in touch with their own core beliefs and the beliefs of others and are sustained by them can be more comfortable living on the cutting edge. At times, taking a tough line or a decisive stand in a muddle of ambiguity is the only ethical thing to do. If a manager is indecisive and spends time trying to figure out the "good" thing to do, the enterprise may be lost.

Business ethics, then, has to do with the authenticity and integrity of the enterprise. To be ethical is to follow the business as well as the cultural goals of the corporation, its owners, its employees, and its customers. Those who cannot serve the corporate vision are not authentic business people and, therefore, are not ethical in the business sense.

At this stage of my own business experience I have a strong interest in organizational behavior. Sociologists are keenly studying what they call corporate stories, legends, and heroes as a way organizations have of transmitting the value system. Corporations such as Arco have even hired consultants to perform an audit of their corporate culture. In a company, the leader is the person who understands, interprets, and manages the corporate value system. Effective managers are then action-oriented people who resolve conflict, are tolerant of ambiguity, stress, and change, and have a strong sense of purpose for themselves and their organizations.

If all this is true, I wonder about the role of the professional manager who moves from company to company. How can he or she quickly absorb the values and culture of different organizations? Or is there, indeed, an art of management that is totally transportable? Assuming such fungible managers do exist, is it proper for them to manipulate the values of others?

What would have happened had Stephen and I carried the sadhu for two days back to the village and become involved with the villagers in his care? In four trips to Nepal my most interesting experiences occurred in 1975 when I lived in a Sherpa home in the Khumbu for five days recovering from altitude sickness. The high point of Stephen's trip was an invitation to participate in a family funeral ceremony in Manang. Neither experience had to do with climbing the high passes of the Himalayas. Why were we so reluctant to try the lower path, the ambiguous trail? Perhaps because we did not have a leader who could reveal the greater purpose of the trip to us.

Why didn't Stephen with his moral vision opt to take the sadhu under his personal care? The answer is because, in part, Stephen was hard-stressed physically himself, and because, in part, without some support system that involved our involuntary and episodic community on the mountain, it was beyond his individual capacity to do so.

I see the current interest in corporate culture and corporate value systems as a positive response to Stephen's pessimism about the decline of the role of the individual in large organizations. Individuals who operate from a thoughtful set of personal values provide the foundation for a corporate culture. A corporate tradition that encourages freedom of inquiry, supports personal values, and reinforces a focused sense of direction can fulfill the need for individuality along with the prosperity and success of the group. Without such corporate support, the individual is lost.

That is the lesson of the sadhu. In a complex corporate situation, the individual requires and deserves the support of the group. If people cannot find such support from their organization, they don't know how to act. If such support is forthcoming, a person has a stake in the success of the group, and can add much to the process of establishing and maintaining a corporate culture. It is management's challenge to be sensitive to individual needs, to shape them, and to direct and focus them for the benefit of the group as a whole.

For each of us the sadhu lives. Should we stop what we are doing and comfort him; or should we keep trudging up toward the high pass? Should I pause to help the derelict I pass on the street each night as I walk by the Yale Club en route to Grand Central Station? Am I his brother? What is the nature of our responsibility if we consider ourselves to be ethical persons? Perhaps it is to change the values of the group so that it can, with all its resources, take the other road.

Ideas for Discussion

1. How are the two parts of McCoy's essay integrated?
2. How was McCoy affected by his experience in Nepal? How are we readers to be affected by McCoy's article?
3. What ideas about responsibility does McCoy share with other authors in this book?

Topics for Writing

1. Create or recount an experience of your own in which ethics would influence your course of action.
2. Examine some of the other essays in *Ethics in Practice* for a short review of ideas about ethics in American business.
3. In an article on multiculturalism and the manager, compare McCoy's experiences with those of Lewis and Clark.

Steve C. Wheelwright and W. Earl Sasser, Jr.
The New Product Development Map

Steven C. Wheelwright is a coauthor of Dynamic Manufacturing *(1988) and a professor at the Harvard Business School. W. Earl Sasser, also a Harvard Business School professor, is a student of all aspects of the service industries—including product development, industrial productivity, and globalization. He is coauthor of books on service management with Daryl Wycoff. "The New Product Development Map" (Harvard Business Review May–June 1989) combines Wheelwright's interest in product development in manufacturing with Sasser's studies of service industries to produce a profile of the creation of the Challenger 6000, a Stratovac vacuum cleaner.*

NO BUSINESS activity is more heralded for its promise and approached with more justified optimism than the development and manufacture of new products. Whether in mature businesses like automobiles and electrical appliances, or more dynamic ones like computers, managers correctly view new products as a chance to get a jump on the competition.

Ideally, a successful new product can set industry standards—standards that become another company's barrier to entry—or open up crucial new markets. Think of the Sony Walkman. New products are good for the organization. They tend to exploit as yet untapped R&D discoveries and revitalize the engineering corps. New product campaigns offer top managers opportunities to reorganize and to get more out of a sales force, factory, or field service network, for example. New products capitalize on old investments.

Perhaps the most exciting benefit, though, is the most intangible: corporate renewal and redirection. The excitement, imagination, and growth associated with the introduction of a new product invigorate the company's best people and enhance the company's ability to recruit new forces. New products build confidence and momentum.

Unfortunately, these great promises of new product development are seldom fully realized. Products half make it; people burn out. To understand why, let's look at some of the more obvious pitfalls.

1. *The moving target.* Too often the basic product concept misses a shifting market. Or companies may make assumptions about channels of distribution that just don't hold up. Sometimes the project gets into trouble because of inconsistencies in focus; you start building a stripped-down version and wind up with a load of options. The project time lengthens, and longer projects invariably drift more and more from their initial target. Classic market misses include the Ford Edsel in the mid-1950s and Texas Instruments' home computer in the late 1970s. Even very successful products like Apple's Macintosh line of personal computers can have a rocky beginning.

2. *Lack of product distinctiveness.* This risk is high when designers fail to consider a full range of alternatives to meet customer needs. If the organization gets locked into a concept too quickly, it may not bring differing perspectives to the analysis. The market may dry up, or the critical technologies may be sufficiently widespread that imitators appear out of nowhere. Plus Development introduced Hardcard,® a hard disc that fits into a PC expansion slot, after a year and a half of development work. The company thought it had a unique product with at least a nine-month lead on competitors. But by the fifth day of the industry show where Hardcard® was introduced, a competitor was showing a prototype of a competing version. And within three months, the competitor was shipping its new product.

3. *Unexpected technical problems.* Delays and cost overruns can often be traced to overestimates of the company's technical capabilities or simply to its lack of depth and resources. Projects can suffer delays and stall midcourse if essential inventions are not completed and drawn into the designers' repertoire before the product development project starts. An industrial controls company we know encountered both problems: it changed a part from metal to plastic only to discover that its manufacturing processes could not hold the required tolerances and also that its supplier could not provide raw material of consistent quality.

4. *Mismatches between functions.* Often one part of the organization will have unrealistic or even impossible expectations of another. Engineering may design a product that the company's factories cannot produce, for example, or at least not consistently at low cost and with high quality. Similarly, engineering may design features into products that marketing's established distribution channels or selling approach cannot exploit. In planning its requirements, manufacturing may assume an unchanging mix of new products, while marketing mistakenly assumes that manufacturing can alter its mix dramatically on short notice. One of the most startling mismatches we've encountered was created by an aerospace company whose manufacturing group built an assembly plant too small to accommodate the wingspan of the plane it ultimately had to produce.

Thus new products often fail because companies misunderstand the most promising markets and channels of distribution and because they misapprehend their own technological strengths or the product's technological challenges. Nothing can eliminate all the risks, but clearly the most important thing to do early on when developing a new product is to get all contributors to the process communicating: marketing with manufacturing, R&D with both. Products fail from a lack of planning, planning fails from a lack of information.

Developing a new generation of products is a lot like taking a journey into the wilderness. Who would dream of setting off without a map? Of course, you would try to clarify the purpose of the journey and make sure that needed equipment is available and in order. But once committed to the trip, you need a map of the terrain, something everybody can study—the focus for discussion, the basis for planning alternative courses. Knowing where you've come from and where you are is essential to knowing how to get where you want to go.

Mapping Existing Products

We have often used this analogy of a map with corporate managers involved in product development, and gradually it became clear to us that an actual map is needed, not just an analogy. Managers need a way to see the evolution of a company's product lines—the "where we are"—in order to expose the markets and technologies that have been driving the evolution—the "where we've come from." Such a map presents the evolution of current product lines in a summary yet strikingly clear way so that all functional areas in the organization can respond to a common vision. The map provides a basis for sharing information. And by enabling managers to compare the assumptions underlying current product lines with the ideal assumptions of new research, it points to new market opportunities and technological challenges. Why, for example, should an organization build for department stores when specialty discount outlets are the emerging channels of distribution? Why bend metals when you can mold ceramics?

The first exhibit illustrates a generic map that indicates how the product offerings in one generation may be related to each other. These relations are the building blocks that allow us to track the evolution of product families from one generation to another.

The map categorizes product offerings (and the development efforts they entail as "core" and "leveraged" products, and divides leveraged products into "enhanced," "customized," "cost reduced," and "hybrid" products. (These designations seem to cover most cases, but managers should feel free to add whatever other categories they need.) A core product, first in gray for the engineering prototype, then in black, is the engineering platform, providing the basis for further enhancements. The core product is the initial, standard product introduced. It changes little from year to year and is often the benchmark against which consumers compare the rest of the product line.

Enhanced products, in red, are developed from the core design; distinctive features are added for various, more discriminating markets. Enhanced products are the first products leveraged from the capabilities put in place to produce the core and the first aimed at new or extended market opportunities. Often companies even identify them as enhanced versions, for example, IBM's Display Write 3.1 is an enhanced version of Display Write 3. But a leveraged product isn't necessarily more costly: the idea is simply to get more out of a fixed process—more "bang for the buck." As companies leverage high-end products, they may customize them in smaller lots for specific channels or to give consumers more choice (shown in blue). The cost-reduced model, shown in green, starts with essentially the same technology and design as the core product but is a stripped-down version, often with less expensive materials and lower factory costs, aimed at a price-sensitive market. (Think of the old Chevrolet Biscayne, which was many times the vehicle of choice for taxicabs and business fleets.)

Finally, there is the hybrid product (shown in purple), developed out of two cores. The initial two-stage thermostat products—accommodating a daytime and nighttime temperature setting—were hybrids of a traditional thermostat product and high-end, programmable thermostat lines.

On the generic map, from left to right is calendar time, and from bottom to top designates lower to higher added value or functionality, which usually also means a shift from cheaper to more expensive products.

These distinctions—core, hybrid, and the others—are immediately useful because they give managers a way of thinking about their products more rigorously and less anecdotally. But the various turns on the product map—the various "leverage points"—also serve as crucial indicators of previous management assumptions about the corporate strengths and market forces shaping product evolutions.

A map that shows a proliferation of enhanced products toward the high end, for example, says something important about the market opportunities managers identified after they had introduced the core. A map's configuration raises necessary questions about dominant channels of distribution—then and now. That products could have been leveraged in particular ways, moreover, says something important about in-house technological and manufacturing capabilities—capabilities that may still exist or may need changing. The map generates the right discussions. When managers know how and why they have leveraged products in the past, they know better how to leverage the company in the present.

The First Generation

How can managers plan, develop, and position a set of products—that is, how do they build a dynamic map? With the generic map in mind, let us track offerings from generation to generation, as shown in the second exhibit. Imagine a very simple line of vacuum cleaners, Coolidge Corporation's "Stratovac," introduced, say, in 1952. The core product, the Stratovac, was a canister-type appliance with a 2.5 horsepower motor. Constructed mainly from cut and stamped metals, it was distributed through department stores and hardware chains.

The following year, reaching for the somewhat more affluent suburban household, Coolidge brought out the "Stratovac Plus," an enhanced Stratovac delivered in a choice of three colors, with a 4 horsepower motor and a recoiling cord. In 1959, the company introduced the "Stratovac Deluxe"—a Stratovac Plus with a vacuum resistance sensor which cut off the power when the bag was full and a power head with a rotating brush for deep pile or shag carpeting. By 1959, the basic Stratovac cost $89, the Stratovac Plus, $109, and the Stratovac Deluxe, $159.

To reach the industrial market at $79, Coolidge had decided to offer the "Stratovac Workman," a stripped-down Plus model—one color, no recoiling cord. That was introduced in 1956. And when Deluxe sales rocketed, Coolidge offered Maybel's department store chain a customized version of it, the Stratovac "Maybel's Housekeeper." This came out in 1960, in Maybel's blue gray, with the power head. The price was "only" $129. (Coolidge eventually customized the "Housekeeper Canadian" for the Simpton's chain in Canada, and the "Royal Housekeeper" for the Mid-Lakes chain in England.)

Again, this is a simple product line, but even so, the map raises interesting questions, especially for younger managers who came after this era. Why the Stratovac Plus? Why a proliferation of products toward the high end?

In fact, during the 1950s, most companies marketed home appliances through department stores with product families visibly shaped by the distribution channels. Products stood side by side in the stores, to be demonstrated by a salesperson. The markup was similar for each product on the floor.

What differentiated products in product families at the time was an appliance manufacturer's reach to satisfy more or less obvious customer segments—customers differentiated by factors like income and marital status. (In the 1950s, most vacuum cleaner purchasers were women, with more or less money, time, and patience.)

How Coolidge leveraged its products also points to certain fixed—and not especially unique—manufacturing capabilities. During the 1950s, company engineers designed appliances for manual assembly and traditional notions of economies of scale. By the end of the 1950s, Coolidge acquired new vacuum sensor innovations from the auto industry. It also learned certain flexible manufacturing techniques, making different colors and options possible.

By 1958, Coolidge had solved most of the technical problems of the Stratovac line and had recruited a number of ambitious design engineers to integrate vacuum sensor and power heads into the line. The life cycle of the product—including development time, which stretched back to 1949—was typical for core products of that time: 10 to 15 years. Demand for the Stratovac remained strong throughout the 1950s, and Coolidge sold to department stores in roughly the same proportion as its competition, except for companies organized around the door-to-door trade.

The company's increased (and not fully utilized) technical competence and the steadiness of its key distribution channels are crucial pieces of information to add to the map (see the third exhibit). The map summarizes technical competence in the oval beneath the product lines, and Coolidge's gross sales by distribution channel in the box graph. The fastest growing distribution channel in the industry—in this case, department stores—is shaded for emphasis.

The Second Generation

With so much technical talent in-house, and a society growing increasingly affluent, Coolidge could not be expected to rest on the Stratovac's success indefinitely. Sales were steady, but by the mid-1960s customers assumed there would be some innovations. The age of plastics was dawning; the vanguard of the baby boom was taking apartments; it was the "new and improved" era.

Moreover, marketing people at Coolidge began to detect a new potential market at the low end. People who had relied on their Stratovacs for a decade were looking around for a second, lighter weight appliance for quick cleanups or for the workroom or garage. Lighter weight and cheaper naturally meant more reliance on plastic components.

In the early 1960s, Coolidge managers decided on two product families, each with its own core product (see the fourth exhibit). The design team that had brought out the old core Stratovac would handle the "Stratovac II," and company new hires would design a second line, the all-plastic, mass-produced "Handivac" ("any color, so long as it's beige").

The Stratovac II, introduced in 1968, was heavier and had a 4.3 horsepower motor, resulting in a slightly noisier operation, "jet noise," which the marketing people reasoned would actually increase respect for its power. Half of the case was now plastic for a "streamlined" appearance. The core Stratovac II boasted a new dust-bag system, which virtually eliminated the need for handling dust. A retractable cord was also standard.

The Stratovac II "Sentry," an enhanced version of the core, included electronic controls for variable speed and came in many colors. The Stratovac II "Imperial," like the old Deluxe model, came with the power head. The Stratovac II Workman continued to sell steadily to the light industrial market, as did the Stratovac II Housekeeper line to the department store chains that still sold the vast majority of units.

Most notable about the Stratovac II was how little changed it was, certainly on the manufacturing end. Assembly was still chiefly manual, along the lines of the 1950s—no priority given to modularity, design for manufacturability, or any of the considerations that would drive designers later on. There was some outsourcing of components to Mexico and Taiwan but no real attention to automation. The only significant change in the Stratovac II came in 1973, when inflationary pressures pushed management to develop a fully plastic casing and critical plastic components—in effect, a hybrid developed by merging technologies of the high-end vacuum cleaner with the low-end Handivac.

Handivac, the second core product, introduced in 1969, was something of a disappointment—mostly because of the inexperience of the team managing its development. Reliability was a problem, given Handivac's almost complete dependence on plastic components, components subjected to higher than expected temperatures from an old, slightly updated 2.5 horsepower motor. Weight was also a problem, it was not as light as promised. Mass production lines, which were partially automated, were considered a success when they were finally debugged.

Perhaps the greatest problem with the Handivac, however, was the fact that, like the Stratovac II, it was sold mainly through department stores and hardware chains, where markups were too large to permit it a significant price advantage over the more expensive core product. Handivac sold for $79, while the Stratovac II sold for $99. Handivac managers tried to cut costs by going to an overseas supplier for a lighter weight, somewhat less powerful motor—over the vehement objections of Stratovac II designers, who had depended on Handivac's participation in their motor plant to keep their own costs in line.

Eventually, Handivac introduced a cost-reduced "Handivac 403," which sold for $69, importing a 3.0 horsepower motor and cord subassembly from Japan. The enhanced "405" sold for $83. Handivac engineers began at this time to interact with Japanese manufacturing managers. But there were still no distribution channels where Handivac could enjoy the "price busting" opportunity it needed. The most promising channel, though hardly dominant, was the growing chains of catalog stores, which sold the Handivac 403 for $63, a 10% reduction in the department store price.

The Third Generation

During 1976 and 1977, a number of external and internal pressures led to a redesign of the entire product line. Department stores were still the major source of revenue, but competitors were proliferating and the Stratovac II group felt the need to offer an increasing number of more enhanced and more customized products to maintain demand at the profitable high end. Consumers would pay a premium, marketing people believed, only if the company could produce so many versions that all customers felt they were getting the right color with the right options. Moreover, Coolidge had canvassed Stratovac II customers, who hadn't appreciated the "jet sound," as designers had assumed. Bulk was also a problem, as was the vacuum's unattractive look.

Inside the company, Coolidge's two design teams had become more cooperative, particularly as the advantages of molded plastic became obvious to everyone. The hybrid Stratovac II, which had been redesigned in plastic wherever possible, was something of a victory for the young Handivac designers over the more traditional group. Flexibility and cost were the keys to satisfying many markets, and plastics answered both needs. Eventually the more traditional designers also came to see the advantage of going to Japan for a smaller, lighter, more reliable motor—and for a number of subassemblies critical to the company's goal of offering arrays of options.

Concurrently in the mid-1970s, the Handivac designers were pressing for a complete merging of the design engineering teams and for studying Japanese manufacturing techniques. They argued that if flexibility, cost, and quality were going to be crucial, the manufacturing people would have to become more involved in product design. The young guard also believed that Coolidge could produce motors domestically—at required levels of quality—if it adopted certain innovations in machine tool and winding automation and instituted statistical process control at its existing motor plant.

Where the younger design group still lacked credibility, however, was on the bottom line. Top management was reluctant to give up on a two-track approach when the Handivac group had failed to deliver an appliance that made even as much as the Housekeeper line. The number of catalog stores was growing, and newer discount appliance chains were springing up in big cities, but the Handivac faced intense competition. Could the younger designers hope to come in with enough products, offering enough features, and at low enough costs to meet this competition?

In the end, Coolidge management decided to develop two core product families in its third generation (see the gatefold). The Stratovac II team redesigned the high-end vacuum cleaner in six models, the "Challenger 6000" series. All appliances in this series came with a power head and a new bag system. By steps—6001, 6002, and upward—consumers could buy increasingly sophisticated electronic controls. And they could order the 6004 and 6005 in an array of colors.

The 6000 series was constructed almost entirely of molded plastic. Manufacturing came up with an automated way of applying hot sealant to critical seams, and the Challenger's motor was quieter. Top management agreed with the younger engineers that a more advanced motor factory could be constructed in the United States. The design teams didn't merge, but they found themselves working more closely together and increasingly with manufacturing.

The traditional design group simultaneously came out with the "Pioneer 4000" series. This was a middle-range product, somewhat smaller than the Challenger 6000, and not offering a power head. The marketing people felt that department stores would want a cost-reduced model to compete with the proliferating "economy" products that discount chains were now offering. (The 4001, 4002, and 4003 were distinguished, again, by electronic controls.) The Pioneer 4000 series was leveraged largely from the Challenger 6000 as a cost-reduced version.

Since both series offered stripped-down models, Coolidge did not introduce a specific industrial product and eliminated the Workman. Coolidge executives also believed that it was no longer worthwhile to customize models for particular department stores where margins were shrinking, so they eliminated the Housekeeper line.

A year after they introduced the Challenger 6000, the Handivac team brought out its new series of products, the "Helpmate." With minor modifications, Helpmate was customized as "Helpmate SE," targeted at different low-end market segments—college students, apartment dwellers, do-it-yourselfers, and the industrial market. The cleaner was lightweight. Attachments varied, as did graphic design: the company expected a Spartan gray color and a longer hose to appeal to commercial customers and bright pastels and different size brushes to appeal to women college students.

The key to the Helpmate line, however, was its manufacturing. The motor was no longer outsourced, and designers worked with manufacturing engineers on modular components and subassemblies. Top management agreed to set aside manufacturing space in the assembly plants for cellular construction of the Helpmate so that the company could respond quickly to demand for particular models. And Helpmate came in at two-thirds the price of the Pioneer 4000.

There was still some debate among Helpmate's product development team members about most likely channels. Some saw it designed only for discount chains and catalog stores, which by 1978 had pretty much eclipsed hardware stores. Others saw the Helpmate as a low-end product for department stores too. In the end, Helpmate was a smash in the discount stores and all but disappeared from department stores.

The Next Generation?

Imagine that Coolidge managers are gathered in 1985 to consider the company's future. Their three-generation map has simplified a great deal of information—information the managers might intuitively understand but could not have looked at so clearly before. Where can they go from here?

Looking at their map, it's clear that Coolidge's product offerings are not appropriately matched to the new environment. They have aimed most of their products at department stores, and now discount chains are growing at a tremendous rate. They had devoted too much attention to figuring out how to leverage products at the high end, when the big battle was shaping up at the low end. Now Coolidge's managers wonder how long it will be before power options and accessories show up on cheaper, sturdier import lines distributed to high-volume outlets.

More growth in the company's manufacturing capabilities is obviously very important now. The map indicates the growing reciprocity between design and manufacturing engineers, owing largely to the initiatives of the younger design group. It would not be hard to imagine a merging of all engineering groups and the use of temporary dedicated development teams at this point. Product life cycles have obviously been shrinking; designers have to think fast now and cooperate across functional lines. To bring out a new line of inexpensive products that are both reliable and varied in options, Coolidge will need automated, flexible manufacturing systems. This development means bringing all parts of the company together—designers with marketing, manufacturing with both. It means, interestingly enough, a need for even clearer, more complete new product development maps.

The finished product development map presented here may appear elementary, but managers who have mapped their products' evolution have experienced substantial payoff in several areas. First, the map can be extremely useful to product development efforts. It helps focus development projects and limit their scope, making them more manageable. The map helps set specifications and targets for individual projects, provides a context for relating concurrent projects to one another, and indicates how the sequence of projects capitalizes on the company's previous investments. These benefits do much to minimize the likelihood of encountering two of the pitfalls we identified at the outset of this article, the moving target and the lack of product distinctiveness.

A second important benefit is the motivation the map provides the various functional groups—all with a stake in effective product development—to develop their own complementary strategies. As illustrated in the Coolidge Corporation example, the product development map raises a number of issues regarding distribution channels, product technology, and manufacturing approaches that must be answered in all parts of the company if the map is to represent the organization's agreed-on direction.

This point brings up the need for "submaps" in each functional area. In the Coolidge case, the first couple of product generations may not have shown the need for a more careful distribution channel map, but by the third the need is painfully clear. Capturing other strategic marketing variables in, say, a price map, a competitive product positioning map, and a customer map would enable the marketing function to identify and present important trends in the marketplace, define targets for future product offerings, and provide guidance for developing and committing sales and marketing resources.

Equally apparent by the third generation is the need for supporting maps in design engineering. A set of design engineering submaps can produce a clearer sense of the mix of engineering talent the company requires, how it should be organized and focused, and the rate at which the company should bring new technologies into future product generations. These maps would not only help managers integrate design resources with product development efforts but would also ensure that they hire and train new employees in a timely and effective manner and that they focus new project tools (such as computer-aided engineering) on pressing product development needs. The key is achieving technical agreement in advance of product development.

Toward the end of the third generation at Coolidge, the map reveals the need for more detailed manufacturing functional maps to bring out issues raised in the "critical skills" oval. Such maps would focus on strategic issues relating to manufacturing facilities, vendor relationships, and automation technology.

Again, the development of such functional submaps not only benefits manufacturing but also helps the company maximize the return on new product development resources. The most interesting and useful benefits will come out of debates about what to put in the submaps.

Submaps capture the essence of the functional strategies, and when integrated with the new product development map, serve to tie those functional strategies together and provide both a foundation and a process for achieving a company's business strategy. The whole process facilitates the cross-functional discussion and resolution of strategic issues. How often have well-intentioned functional managers met to discuss their various substrategies only to have those from other functions tune out within the first two minutes, as the discussion becomes too technical, too detailed, or simply too parochial to comprehend?

Mapping provides a process for planning that avoids too much detail (like budgeting) and too much parochialism (like traditional functional strategy sessions). Managers will inevitably develop linkages across the organization by going through the steps of selecting the resources or factors to develop into a map, identifying the key dimensions to capture in the map, reviewing historical data to understand the relationships of those dimensions, and examining what is likely to drive future versions of the map. Functions can share their maps to communicate, refine, and agree on important product strategy choices. It is the sharing of functional capabilities—capabilities applied in a systematic, repetitive fashion to product development opportunities—that will become the company's competitive advantage.

Ideas for Discussion

1. What does constructing an actual development map reveal about new product creation?
2. What observations can you make about the writers' style and intended audience?
3. How would you summarize Coolidge's strategic planning?

Topics for Writing

1. Provide an executive summary for this article.
2. Write an essay on the importance of effective communication in the R and D process.
3. Either individually or in groups, identify a new type of product or service and trace its development in an essay modeled after Wheelwright and Sasser's for oral or written delivery.

Made in the USA
Middletown, DE
01 August 2022